This Civil Text replaces the now deleted

LAND WARFARE PROCEDURES - COMBAT ARMS (Engineers) 5-1-4 GEOLOGY

This text was funded by Land Warfare Development Centre - Doctrine Wing, procured by RAE Doctrine and Development Cell and is authorised for use by the undersigned.

D. M. LUHRS
LTCOL
CO/SI SME
29 JUL 04

GEOLOGY FOR CIVIL ENGINEERS

TITLES OF RELATED INTEREST

Construction Methods and Planning
J. R. Illingworth

Contaminated Land - Problems and Solutions
Edited by T. Cairney

Engineering the Channel Tunnel
Edited by C. Kirkland

Engineering Treatment of Soils
F.G. Bell

Foundations of Engineering Geology
A.C. Waltham

Geology of Construction Materials
J.F. Prentice

Pile Design and Construction Practice
M.J. Tomlinson

Piling Engineering
W.G.K. Fleming, A.J. Weltman, M.F. Randolph and W.K. Elson

Rock Mechanics for Underground Mining
B.H.G. Brady and E.T. Brown

Rock Slope Engineering
E. Hoek and J.W. Bray

Rutley's Elements of Mineralogy
C.D. Gribble

The Stability of Slopes
E.N. Bromhead

Soil Mechanics
R.F. Craig

Underground Excavations in Rock
E. Hoek and E.T. Brown

GEOLOGY FOR CIVIL ENGINEERS

Second Edition

A.C. McLean C. D. Gribble

University of Glasgow

First published 1979 by E & FN Spon, an imprint of Chapman & Hall
Second edition 1985
Reprinted 1990, 1992, 1995, 1997

Reprinted 1999 by E & FN Spon, an imprint of Routledge
11 New Fetter Lane, London EC4P 4EE
29 West 35th Street, New York, NY 10001

Printed and bound in Great Britain by T. J. International Ltd, Padstow, Cornwall

British Library Cataloguing in Publication Data
A catalogue record for this book is available from the British Library

ISBN 0-419-16000-0 (pbk)

Printed on acid-free text paper, manufactured in accordance with ANSI/NISO Z39.48-1992 and ANSI/NISO Z39.48-1984 (Permanence of Paper)

This book is
dedicated to the memory of
Dr Adam McLean

Preface to the second edition

Adam McLean and I were asked by Roger Jones of Allen & Unwin to consider producing a second edition of our book after the first edition had been published for a few years. Critical appraisals of the first edition were sought, and I am most grateful to Professor Van Dine and Dr Drummond for their many detailed and helpful comments. I should also particularly like to thank Dr Bill French, who pointed out where corrections were required and also where additions (and subtractions) to the text could gainfully be made without changing the original flavour of our book. I have incorporated most of these helpful suggestions and hope that the text has been improved, but any mistakes and inaccuracies are mine.

At the beginning of the revision Adam McLean became ill, and the illness got progressively worse until, in March 1983, he died. In memory of all the enjoyment we had with the first edition, I should like to dedicate the second edition to Adam with my respect.

Colin Gribble
Glasgow, September 1983

Preface to the first edition

The impulse to write this book stemmed from a course of geology given by us to engineering undergraduates at the University of Glasgow. The course has changed, and we hope improved, during the twenty years since one of us was first involved with it. It was essentially a scaled-down version of an introductory course to science undergraduates; it is now radically different both in content and in the mode of teaching it. Our main thought, as we gradually reshaped it, was to meet the special interests and professional needs of budding civil engineers. It is a matter for serious debate as to whether time should be found within an engineering course for classes of a broad cultural nature. Our experience in teaching indicates that the relevance of subject matter to the vocation of those taught usually increases their interest and enthusiasm. Furthermore, in engineering curricula which are being crowded by new and professionally useful topics, we doubt whether a place would have been found for a general course on geology which discussed, for example, the evolution of the vertebrates or the genetic relationship of the various basic plutonic rocks. On the other side of the scale, we have firm beliefs that educated men and women should be aware of the Theory of Natural Selection and its support from the fossil record, and should be aware of other major scientific concepts such as plate tectonics. We have found some space for both of these in our book. Other apparent digressions from what is obviously relevant may serve a professional purpose. For example, civil engineers must have an insight into how geologists reach conclusions in making a geological map, in order to evaluate the finished map. Similarly, they should appreciate how and why geologists differentiate between (say) gabbro and diorite, not because these differences are important for most engineering purposes but so that they can read a geological report sensibly and with the ability to sift the relevant from the irrelevant information.

Our course and this book are essentially an introduction to geology for civil engineers, which is adequate for the needs of their later careers, and on which further courses of engineering geology, soil mechanics or rock mechanics can be based. They are not conceived as a course and text on engineering geology. We have, however, extended the scope of the book beyond what is geology in the strict sense to include engineering applications of geology. This is partly to demonstrate the relevance of geology to engineering, and partly in the expectation that the book, with its appendices, will also serve as a useful handbook of facts and methods for qualified engineers and other professionals who use geology. The reactions of the majority of those who reviewed our first draft reassured us that our ideas were not peculiar to ourselves, and that we were not the only teachers of geology who felt the need for a textbook tailored to them. Other views ranged from a preference for altering the book to make it a comprehensive account of the whole of geology largely devoid of material on engineering, to a preference for a more radical change along the lines we were following, which would have produced an introductory text in engineering geology rather than geology. The balance of opinion seemed reasonably close to our own prescription, though we are grateful for the many constructive suggestions that have led to major changes of content and arrangement as well as minor amendments. If we have not ended at the centre of the many opinions that colleagues and friends have kindly given us, it is because at the end of the day we have special interests and views ourselves, and it is our book. We hope that you will find it useful and readable.

ADAM McLEAN
COLIN GRIBBLE
Glasgow, August 1978

Acknowledgements

We wish to thank the friends and colleagues who assisted us generously and patiently by their advice, by their critical reading of our text and by their encouragement. We considered carefully all the points that they made, and many significant improvements from our original draft are witness to this, just as any persistent failings, and any errors, are our own responsibility. A special thank you is due to Professor W. Dearman of the University of Newcastle, Professor P. McL. D. Duff of the University of Strathclyde, Dr I. Hamilton of Paisley College of Technology, Dr D. Wilson of the University of Liverpool, and Professor Boyd of the University of Adelaide, for reading critically the entire text and making a host of useful comments. We were fortunate in being able to discuss particular sections of the book with friends, whose specialised knowledge was a source of expert opinion and information, and we thank all of them sincerely. They include Mr R. Eden, Assistant Director BGS; Mr N. Dron of Ritchies Equipment Limited, Stirling; Mr C. I. Wilson, Dunblane; and Dr G. Maxwell of the University of Strathclyde. We are grateful to Professor B. E. Leake of our own department at the University of Glasgow for help and encouragement; to other colleagues there, particularly Dr J. Hall, Dr B. J. Bluck and Dr W. D. I. Rolfe; to the two typists, Mrs D. Rae and Mrs D. MacCormick, who prepared the draft copy; and to the wife of one of us, Mrs Beatrice McLean, who did most of the preparation of the Index–Glossary as well as offering help at all stages. Last, but not least, we acknowledge the courteous shepherding of Mr Roger Jones of George Allen & Unwin from the start of it all, to this point.

The second edition could not have been produced without the very great help and guidance I received from Roger Jones and Geoffrey Palmer of George Allen & Unwin. I also wish to thank Mary Sayers, whose careful editing of the revised text unquestionably improved the final product, and Beatrice McLean who helped with the Index–Glossary for this edition. Finally I should like to thank Professor Bernard Leake of my own department for his help and encouragement at a particularly difficult time, Dr Brian Bluck for his guidance on sedimentary rocks and processes, the secretaries of Glasgow University Geology Department – Irene Wells, Dorothy Rae, Irene Elder and Mary Fortune – who typed the entire book a second time, and my sister, Elizabeth, who proof read the entire book.

C.D.G.

Contents

List of tables

1 *Introduction*

1.1 Role of the engineer in the systematic exploration of a site

The investigation of the suitability and characteristics of sites as they affect the design and construction of civil engineering works and the security of neighbouring structures is laid out in British Standard Code of Practice for site investigations (BS 5930 : 1981, formerly CP 2001). The sections on geology and site exploration define the minimum that a professional engineer should know.

The systematic exploration and investigation of a new site *may* involve five stages of procedure. These stages are:

(1) *preliminary investigation* using published information and other existing data;
(2) *a detailed geological survey* of the site, possibly with a photogeology study;
(3) *applied geophysical surveys* to provide information about the subsurface geology;
(4) *boring, drilling and excavation* to provide confirmation of the previous results, and quantitative detail, at critical points on the site; and
(5) *testing of soils and rocks* to assess their suitability, particularly their mechanical properties (**soil mechanics** and **rock mechanics**), either *in situ* or from samples.

In a major engineering project, each of these stages might be carried out and reported on by a consultant specialising in geology, geophysics or engineering (with a detailed knowledge of soil or rock mechanics). However, even where the services of a specialist consultant are employed, an engineer will have overall supervision and responsibility for the project. The engineer must therefore have enough understanding of geology to know how and when to use the expert knowledge of consultants, and to be able to read their reports intelligently, judge their reliability, and appreciate how the conditions described might affect the project. In some cases the engineer can recognise common rock types and simple geological structures, and knows where he can obtain geological information for his preliminary investigation. When reading reports, or studying geological maps, he must have a complete understanding of the meaning of geological terms and be able to grasp geological concepts and arguments. For example, a site described in a geological report as being underlain by clastic sedimentary rocks might be considered by a civil engineer to consist entirely of sandstones. However, clastic sedimentary rocks include a variety of different rock types, such as conglomerates, sandstones and shales or mudstones. Indeed it would not be unusual to find that the site under development contained sequences of some of these different rock types – say, intercalated beds of sandstone and shale, or sandstone with conglomerate layers. Each of these rock types has different engineering properties, which could affect many aspects of the development work such as core drilling into, and excavation of, the rock mass, and deep piling into the underlying strata.

The systematic testing of the engineering properties of soils and rocks lies between classical geology and the older disciplines of engineering, such as structures. It has attracted the interest of, and contributions from, people with a first training in either geology or engineering, but has developed largely within departments of civil and mining engineering and is usually taught by staff there. These tests, and the advice about design or remedial treatment arising from them, are more naturally the province of the engineer, and fall largely outside the scope of this book. The reasons for this lie in the traditional habits and practices of both fields. The engineer's training gives him a firm grounding in expressing his conclusions and decisions in figures, and in conforming to a code of practice. He also has an understanding of the constructional stage of engineering projects, and can better assess the relevance of his results to the actual problem.

These reasons for the traditional divisions of practice between geology and engineering must be qualified, however, by mentioning important developments during the last decade. An upsurge of undergraduate and postgraduate courses, specialist publications and services in engineering geology, initiated or sponsored by departments of geology or by bodies such as the Geological Society of London, has reflected an awakened interest in meeting fully the geological needs of engineers and in closing the gaps that exist between the two disciplines.

1.2 Relevance of geology to civil engineering

Most civil engineering projects involve some excavation of soils and rocks, or involve loading the Earth by building on it. In some cases, the excavated rocks may be used as constructional material, and in others, rocks may form a major part of the finished product, such as a motorway cutting or the site for a reservoir. The feasibility, the planning and design, the construction and costing, and the safety of a project may depend critically on the geological conditions where the construction will take place. This is especially the case in extended 'greenfield' sites, where the area affected by the project stretches for kilometres, across comparatively undeveloped ground. Examples include the Channel Tunnel project and the construction of motorways. In a section of the M9 motorway linking Edinburgh and Stirling that crosses abandoned oil-shale workings, realignment of the road, on the advice of government geologists, led to a substantial saving. In modest projects, or in those involving the re-development of a limited site, the demands on the geological knowledge of the engineer or the need for geological advice will be less, but are never negligible. Site investigation by boring and by testing samples may be an adequate preliminary to construction in such cases.

1.3 The science of geology

Geology is the study of the solid Earth. It includes the investigation of the rocks forming the Earth (**petrology**) and of how they are distributed (their **structure**), and their constituents (**mineralogy** and **crystallography**). **Geochemistry** is a study of the chemistry of rocks and the distribution of major and trace elements in rocks, rock suites, and minerals. This can lead to an understanding of how a particular rock has originated (**petrogenesis**), and also, in the broadest sense, to a knowledge of the chemistry of the upper layers of the Earth.

The distribution of rocks at the Earth's surface is found by making a **geological survey** (that is, by **geological mapping**) and is recorded on **geological maps**. This information about rocks is superimposed on a topographic base map. Knowledge of the nature and physical conditions of the deeper levels of the planet can be gained only by the special methods of **geophysics**, the twin science of geology; the term 'Earth sciences' embraces both. From the theory and methods of geophysics, a set of techniques (**applied geophysics**) has been evolved for exploring the distribution of rocks of shallower levels where the interests of geologists and geophysicists are most intertwined.

Knowledge of the Earth at the present time raises questions about the processes that have formed it in the past: that is, about its history. The interpretation of rock layers as Earth history is called **stratigraphy**, and a study of the processes leading to the formation of sedimentary rocks is called **sedimentology**. The study of fossils (**palaeontology**) is closely linked to Earth history, and from both has come the understanding of the development of life on our planet. The insight thus gained, into expanses of time stretching back over thousands of millions of years, into the origins of life and into the evolution of man, is geology's main contribution to scientific philosophy and to the ideas of educated men and women.

1.4 The aims and organisation of this book

This book defines essential terms, explains concepts, phenomena and methods of argument, and shows how to reach conclusions about the geology of a site and to appreciate its relevance to an engineering project. It is envisaged as a text to accompany an introductory course for engineering undergraduates. It also contains additional information that will be of use to students who intend carrying their study of applied geology beyond a basic course. At the same time, the book is intended to be more than a narrow professional manual, and it is hoped that it will advance the general scientific education of students by presenting, for example, the nature and use of inductive reasoning in science.

The book is arranged so that first the rocks and soils that form the Earth are described, followed by the factors that control their distribution within it. Next it shows how their distribution at one place may be determined, and finally it discusses the relative importance of geological factors in some types of engineering project.

The wording is as succinct as possible. Academic geologists have manufactured words in abundance to describe their science, and applied geologists have not only added to the vocabulary but have also acquired a jargon – sometimes only local in use – from their contacts with miners. Since the development of an ability to read geological reports is an aim of the book, it would be contradictory to omit ruthlessly every geological term that seems inessential to the concept or general argument under discussion, however tempting such drastic editing may be. To guide the student in acquiring a basic geological vocabulary, the important terms are printed in bold type, usually at their first occurrence. In addition they are listed in the index, which therefore also serves as a glossary. Again, since the book is meant to serve the double purpose of reading and later reference, there are appendices of some factual details that might otherwise have clogged the text. With much of this information, it is enough that the engineer should understand, for example, how and why properties vary among the common rock types, with only a sense of the order of magnitude of numerical values.

References and selected reading

British Standards Institution 1981. Code of Practice for Site Investigations. BS 5930:1981 (formerly CP 2001).
Edwards, R. J. G. 1971. The engineering geologist. *Q. J. Engng Geol.* **4**, 283–316.
Glossop, R. 1969. Engineering geology and soil mechanics. *Q. J. Engng Geol.* **2**, 1–5.
Military Engineering 1976. *Applied geology for engineers.* Vol. 15. London: HMSO.
Rawlings, G. E. 1971. The role of the engineering geologist during construction. *Q. J. Engng Geol.* **4**, 209–20.
Taylor, R. K. 1971. The functions of the engineering geologist in urban development. *Q. J. Engng Geol.* **4**, 221–40.

2 *Minerals and rocks*

2.1 The common rock-forming minerals

2.1.1 The properties of minerals

A mineral is a naturally occurring inorganic substance which has a definite chemical composition, normally uniform throughout its volume. In contrast, rocks are collections of one or more minerals. In order to understand how rocks vary in composition and properties, it is necessary to know the variety of minerals that commonly occur in them, and to identify a rock it is necessary to know which minerals are present in it. Two techniques are employed to identify minerals:

(a) *the study of a hand specimen* of the mineral, or the rock in which it occurs, using a hand lens (×8 or ×10) and observing diagnostic features; and
(b) *the examination of a thin slice* of the mineral, ground down to a thickness of 0.03 mm, using a microscope, the rock slice being mounted in transparent resin on a glass slide.

The former method is by far the most useful to an engineer, since proficiency in the use of a microscope requires an amount of study out of proportion to its future benefit, except for the specialist engineering geologist. However, examination of rocks in thin section will provide excellent details of rock textures, some of which are difficult to see in the hand specimen. In hand-specimen identification, some features are purely visual (for example, the colour of the mineral) but others, such as hardness, have to be assessed by simple tests. (If the mineral grains are large enough, and an accurate value is needed, they may be removed from the rock and measured in a laboratory.)

A mineral specimen can be an object of beauty in those occasional circumstances where it forms a single crystal or cluster of crystals. The requirements are that the mineral has been free to grow outwards into the solution or melt from which it formed, not obstructed by other solid matter, nor hindered anywhere around it by a shortage of the constituents needed for growth. In such an environment, it develops a regular pattern of faces and angles between the faces, which is characteristic of a particular mineral. The study of this regularity of form, and of the internal structure of the mineral to which it is related, is called **crystallography**. In most mineral specimens, the local conditions have hindered or prevented some of the faces from developing, or the surface of the mineral is formed simply from the fractures along which it was broken off when collected. Even in these specimens, there is the same regular internal arrangement of atoms as in a perfect crystal of the same mineral. The specimen is **crystalline** even though it is not a crystal. Furthermore, in an imperfect crystal, where some faces have developed more than others to produce a distorted external form, the angles between the faces are still the same as in a perfect crystal.

A study of the regularity of crystal forms, including the values of interfacial angles, shows that all crystals possess certain **elements of symmetry**. These elements include:

(a) **a centre of symmetry**, which a crystal possesses when all its faces occur in parallel pairs on opposite sides of the crystal. A cube, for example, possesses a centre of symmetry but a tetrahedron does not.
(b) an **axis of symmetry**, which is a line through a crystal such that a complete rotation of 360° about it produces more than one identical view. There are four types of axis of symmetry: a **diad** axis, when the same view is seen twice (every 180°); a **triad** axis, when the same view is seen three times (every 120°); a **tetrad** axis (four times – every 90°), and finally a **hexad** axis (six times – every 60°).
(c) a **plane of symmetry**, which divides the crystal into halves, each of which is a mirror image of the other without rotation.

On the basis of the number and type of symmetry elements present in naturally formed crystals, seven **crystal systems** have been proposed, to which *all* minerals can be assigned.

Twinning in crystals occurs where one part of a crystal has grown or has been deformed such that its atomic structure is rotated or reversed compared with the other part. Multiple twinning occurs and is a diagnostic property in the plagioclase feldspars (see Section 2.1.3).

As well as crystallography (form) and twinning, other important properties are used to identify minerals in hand specimens, as follows:

COLOUR AND STREAK

The **colour** of a mineral is that seen on its surface by the naked eye. It may depend on the impurities present in light-coloured minerals, and one mineral specimen may even show gradation of colour or different colours. For these reasons, colour is usually a general rather than specific guide to which mineral is present. Iridescence is a play of colours characteristic of certain minerals. The **streak** is the colour of the powdered mineral. This is most readily seen by scraping the mineral across a plate of unglazed hard porcelain and observing the colour of any mark left. It is a diagnostic property of many ore minerals. For example, the lead ore, galena, has a metallic grey colour but a black streak.

Figure 2.1 A crystal of calcite showing cleavage.

CLEAVAGE

Most minerals can be cleaved along certain specific crystallographic directions which are related to planes of weakness in the atomic structure of the mineral (see Fig. 2.1). These **cleavage directions** are usually, but not always, parallel to one of the crystal faces. Some minerals, such as quartz and garnet, possess no cleavages, whereas others may have one (micas), two (pyroxenes and amphiboles), three (galena) or four (fluorite). When a cleavage is poorly developed it is called a **parting**.

A surface formed by breaking the mineral along a direction which is not a cleavage is called a **fracture** and is usually more irregular than a cleavage plane. A fracture may also occur, for example, in a specimen which is either an aggregate of tiny crystals or glassy (that is, non-crystalline). A curved, rippled fracture is termed **conchoidal** (shell-like).

HARDNESS

The relative hardness (H) of two minerals is defined by scratching each with the other and seeing which one is gouged. It is defined by an arbitrary scale of ten standard minerals, arranged in **Mohs' scale of hardness**, and numbered in degrees of increasing hardness from 1 to 10 (Table 2.1). The hardnesses of items commonly available are also shown, and these may be used to assess hardness within the lower part of the range. The only common mineral that has a hardness greater than 7 is garnet. Most others are semi-precious or precious stones.

Table 2.1 Mohs' scale of hardness.

10	diamond	carbon				
9	corundum	alumina				
8	topaz	aluminium silicate				
7	quartz	silica	scratches glass			
6	feldspar	alkali silicate	scratched by a file			
5	apatite	calcium phosphate				↑
4	fluorspar	calcium fluoride			↑	penknife
3	calcite	calcium carbonate		↑	penny	|
2	gypsum	hydrated calcium sulphate		fingernail	|	|
1	talc	hydrated magnesium silicate		|	|	|

LUSTRE

Light is reflected from the surface of a mineral, the amount of light depending on physical qualities of the surface (such as its smoothness and transparency). This property is called the **lustre** of the mineral, and is described according to the degree of brightness from 'splendent' to 'dull'. The terms to describe lustre are given in Table 2.2.

Table 2.2 Descriptive terms for the lustre of minerals.

metallic	like polished metal	mainly used in describing mineral ores and opaque minerals
submetallic	less brilliant	
dull		
vitreous	like broken glass	mainly for silicate minerals
resinous	oily sheen	
silky	like strands of fibre parallel to surface	
dull		

CRYSTAL HABIT

The development of an individual crystal, or an aggregate of crystals, to produce a particular external shape depends on the temperature and pressure during their formation. One such environment may give long

Table 2.3 Descriptive terms for crystal habit.

Individual crystals	
platy	broad, flat crystal
tabular	elongate crystal which is also flat
prismatic	crystal is elongated in one direction
acicular	crystal is very long and needle-like
fibrous	long crystals – like fibres
Crystal aggregates (amorphous minerals often assume this form)	
dendritic	crystals diverge from each other like branches
reniform	kidney-shaped
botryoidal	like a bunch of grapes
amygdaloidal	infilling of steam vesicles or holes in lavas by salts carried in solution
drusy	crystals found lining a cavity

needle-like crystals and another may give short platy crystals, both with the same symmetry. Since the mode of formation of a mineral is sometimes a clue to what it is, this shape or **crystal habit** is of use in the identification of some minerals. The terms used to describe crystal habit are given in Table 2.3.

Aggregations of minerals may also show some internal structure formed by the relationship of the crystals to each other. For example, in columnar structure, the crystals lie in columns parallel to each other. In granular structure, the minerals are interlocking grains similar in appearance to the crystals in sugar lumps. In massive structure, the crystal grains cannot be seen by the naked eye.

SPECIFIC GRAVITY

The **specific gravity** or **density** of a mineral can be measured easily in a laboratory, provided the crystal is not too small. The specific gravity (sp. gr.) is given by the relation:

$$\text{sp. gr.} = W_1/(W_1 - W_2)$$

where W_1 is the weight of the mineral grain in air, and W_2 is the weight in water. A steelyard apparatus such as the Walker Balance is commonly used. In the field such a means of precision is not available, and the specific gravity of a mineral is estimated as low, medium or high by the examiner. It is important to know which minerals have comparable specific gravities:

(a) low specific gravity minerals include silicates, carbonates, sulphates and halides, with specific gravities ranging between 2.2 and 4.0;
(b) medium specific gravity minerals include metallic ores such as sulphides and oxides, with specific gravities between 4.5 and 7.5;
(c) high specific gravity minerals include native metallic elements such as pure copper, gold and silver; but these are rare minerals and are very unlikely to be encountered.

TRANSPARENCY

Transparency is a measure of how clearly an object can be seen through a crystal. The different degrees of transparency are given in Table 2.4.

Table 2.4 Degrees of transparency.

transparent	an object is seen clearly through the crystal
subtransparent	an object is seen with difficulty
translucent	an object cannot be seen, but light is transmitted through the crystal
subtranslucent	light is transmitted only by the edges of a crystal
opaque	no light is transmitted; this includes all metallic minerals

REACTION WITH ACID

When a drop of cold 10% dilute hydrochloric acid is put on certain minerals, a reaction takes place. In calcite ($CaCO_3$), bubbles of carbon dioxide make the acid froth, and in some sulphide ores, hydrogen sulphide is produced.

TENACITY

Tenacity is a measure of how the mineral deforms when it is crushed or bent. The terms used to describe it are given in Table 2.5.

Table 2.5 Descriptive terms for the tenacity of minerals.

brittle	shatters easily
flexible	can be bent, but will not return to original position after pressure is released
elastic	can be bent, and returns to original position after pressure is released
malleable	can be hammered into thin sheets
sectile	can be cut by a knife
ductile	can be drawn into thin wires

OTHER PROPERTIES

Taste and **magnetic properties** are diagnostic of a few minerals. Mineral associations are also of use. Some minerals often occur together whereas others are never found together because they are unstable as a chemical mixture and would react to produce another mineral.

Nearly all identification of minerals in hand specimens in the field is made with the proviso that the specimen being examined is not a rare mineral but is one of a dozen or so common, rock-forming minerals, or one of a couple of dozen minerals commonly found in the sheet-like veins that cut rocks. The difference between common quartz and one particular rare mineral in a hand specimen is insignificant and easily missed, but mistakes of identification are presumably as rare as the mineral. The same limits of resolution using such simple techniques mean also that only in favourable circumstances is it possible to identify, for example, which variety of feldspar is present in a fine-grained rock as distinct from identifying feldspar.

Three or four properties are usually sufficient for a positive identification of a particular mineral and there is little point in determining the others. For example, a mineral with a metallic lustre, three cleavages all at right angles, a grey colour and a black streak is almost certainly the common lead ore, galena.

2.1.2 *Silicate minerals*

Of the hundred or so elements known, only eight are abundant at the Earth's surface. These, in decreasing order of abundance, are oxygen (O),

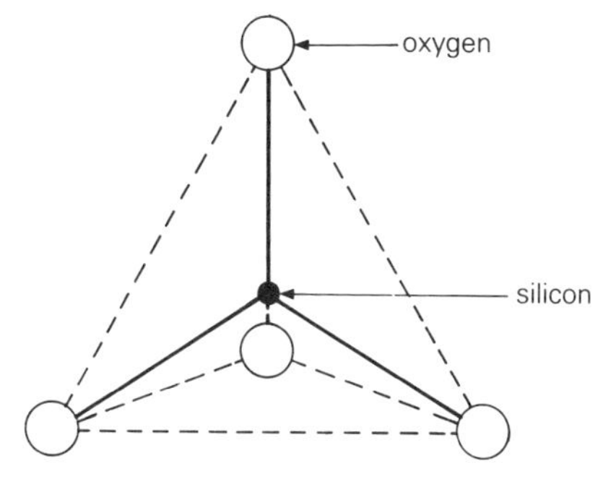

Figure 2.2 The silicon–oxygen tetrahedron. The broken lines represent the edges of the tetrahedron that can be drawn around an $[SiO_4]^{4-1}$ unit.

silicon (Si), aluminium (Al), iron (Fe), calcium (Ca), sodium (Na), potassium (K) and magnesium (Mg). The common **rock-forming minerals** are formed mainly of combinations of these important elements, and most of them are silicates.

All silicate minerals possess the silicate oxyanion, SiO_4^{4-}. This oxyanion resembles a tetrahedron in outline (Fig. 2.2), with a silicon atom at the centre and four oxygen atoms at the corners. Modern classification of silicate minerals is based on the degree of polymerisation of the tetrahedral SiO_4^{4-} groups. In some silicate minerals all four oxygens in a tetrahedron are shared with other tetrahedra. These are called **framework silicates** and include feldspars and quartz. Sharing three oxygens leads to the formation of **sheet silicates**, such as the micas; and when two oxygens are shared, **chain silicates** form, as in the pyroxenes and amphiboles. In other silicate minerals the tetrahedra remain as discrete units sharing no common oxygens. These are called **island silicates** and include olivine and garnet.

The most obvious difference visible in hand specimens between two minerals in a rock is often that one is light coloured and the other is dark coloured. Generally, the dark-coloured silicates contain iron and magnesium as essential elemental constituents, whereas the light-coloured silicates contain aluminium and alkalis. A simple division into these two principal groups of rock-forming silicates is therefore employed in this chapter.

DARK-COLOURED SILICATE MINERALS

Dark-coloured silicate minerals range from vitreous to dull in lustre. Their other properties, observable in hand specimens, are listed in Table 2.6, and each of the important minerals in this group is now discussed.

Olivine Olivine ($[MgFe]_2SiO_4$) is a mineral formed at high temperature which crystallises early from a basic **magma** to form well shaped, rather squat prisms in most of the rocks in which it is present. Magma is hot

Table 2.6 Physical properties of some dark-coloured silicate minerals.

Mineral	Colour	Specific gravity	Hardness	Cleavages
olivine	green or dark green	3.5+	$6\frac{1}{2}$	none (one poor fracture)
pyroxene (augite)	black or brown	3.3	$5\frac{1}{2}$	two
hornblende	black	3.3	$5\frac{1}{2}$	two
biotite	brown	3	$2\frac{1}{2}$	one (perfect)
garnet	red (variable)	3.5+	7	none

liquid rock which, when consolidated, is known as **igneous rock**. Some crystals in igneous rocks may, however, show corroded crystal faces because of a reaction with the surrounding magma before solidification was complete.

Pyroxene Pyroxene ($X_2Y_2O_6$), where X may be calcium, iron or magnesium, and Y is silicon or aluminium, exists in many varieties, but the

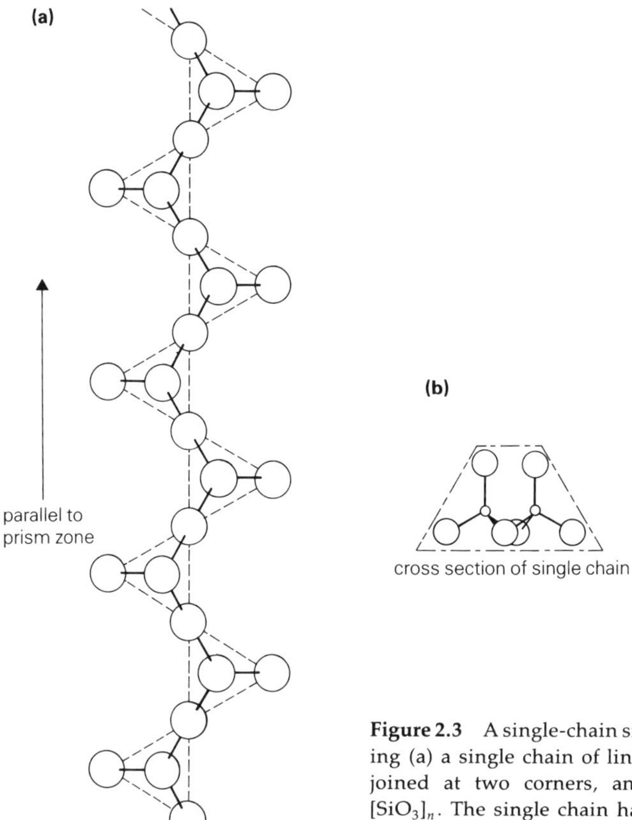

Figure 2.3 A single-chain silicate structure showing (a) a single chain of linked [SiO_4] tetrahedra joined at two corners, and with composition $[SiO_3]_n$. The single chain has a trapezoidal cross section (b) when viewed along its length.

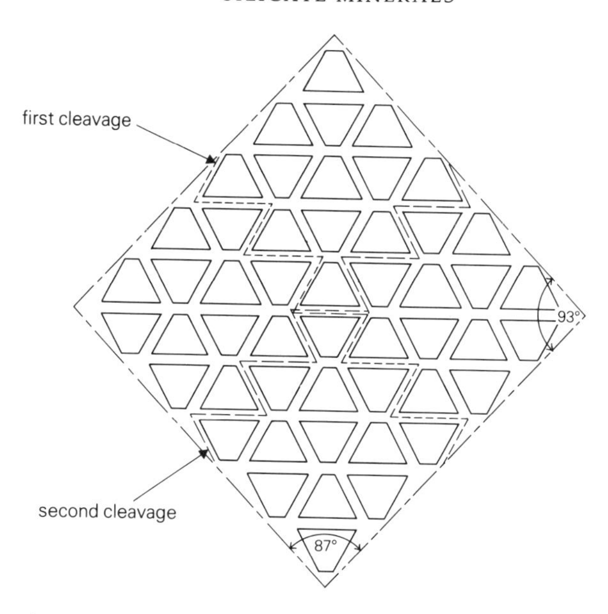

Figure 2.4 Section of atomic structure of augite crystal viewed at right angles to prism zone.

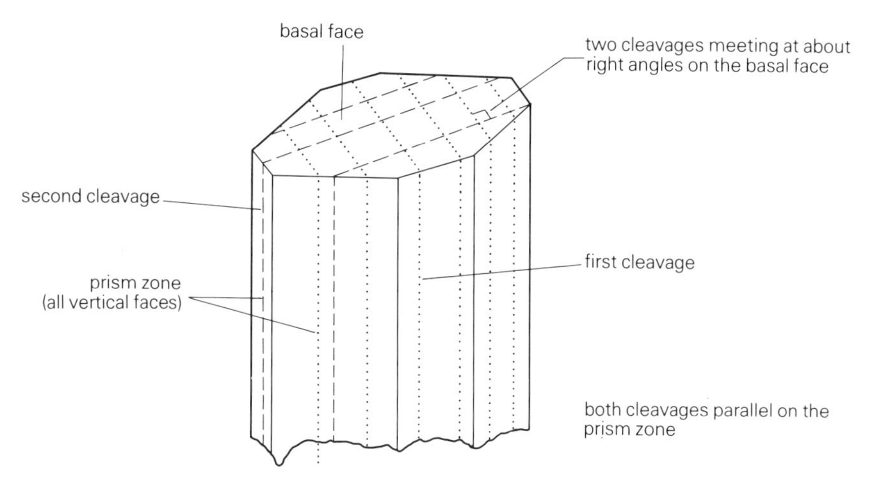

Figure 2.5 Crystal of augite showing two cleavages meeting at approximately right angles on the basal face (compare with Fig. 2.7).

most important one in igneous rocks is the mineral **augite**, the properties of which are listed in Table 2.6. The atomic structure of augite consists of single chains of tetrahedra $[SiO_3]_n$ linked laterally by calcium (Ca), magnesium (Mg) and iron (Fe) cations. The bonds between individual chains are relatively weak and the cleavage directions are parallel to the chains. Augite has two cleavages parallel to the length of the mineral, which are seen to intersect at about 90° on the basal face of its crystal (Figs 2.3, 4 & 5). Augite is common in igneous rocks which have a relatively low percentage of silica, and frequently occurs with olivine. No hydroxyl group (OH) is present in either augite or olivine and both can be described as 'dry' minerals. These minerals are rarely found in sediments since they alter easily when exposed to water and air (Section 2.1.5). They may be present in some metamorphic rocks.

Hornblende Hornblende ($X_{2-3}Y_5Z_8O_{22}(OH)_2$), where X may be calcium or sodium, Y may be magnesium or iron, and Z may be silicon or aluminium, belongs to the amphibole group of minerals, and its main physical properties are shown in Table 2.6. Its structure resembles that of augite and it has two cleavages parallel to its length, which are seen to intersect at about 124° on the basal face of its crystal (Figs 2.6 & 7). This angle of cleavage intersection is the most certain way of distinguishing hornblende from augite in a hand specimen.

Hornblende is a common mineral in igneous rocks (see Section 2.2.2) with average amounts of silica. It crystallises from magma containing appreciable amounts of water and can be said to be a 'wet' mineral, since it contains hydroxyl groups in its structure. It is not very stable when weathered at the Earth's surface and is rarely found in sediments. It is, however, a common constituent of metamorphic rocks (Section 2.2.5).

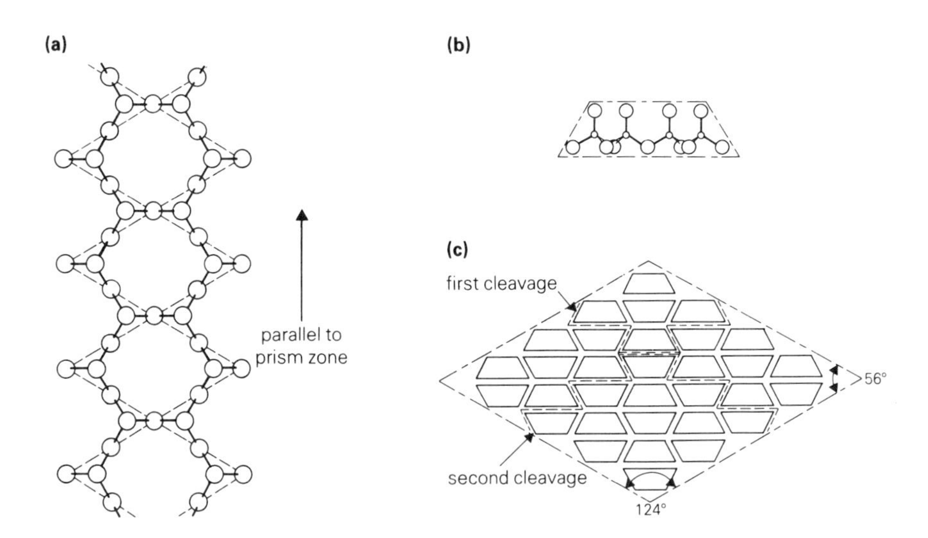

Figure 2.6 Structure of hornblende. (a) Double chain of linked $[SiO_4]$ tetrahedra, with composition $[Si_4O_{11}]_n$. (b) The double chain has trapezoidal cross section (as shown) when viewed along its length. (c) The composite structure of double chains stacked together and linked laterally by Na, Ca, Mg, Fe, and OH ions. Splitting occurs along two preferred directions (i.e. there are two directions of cleavage).

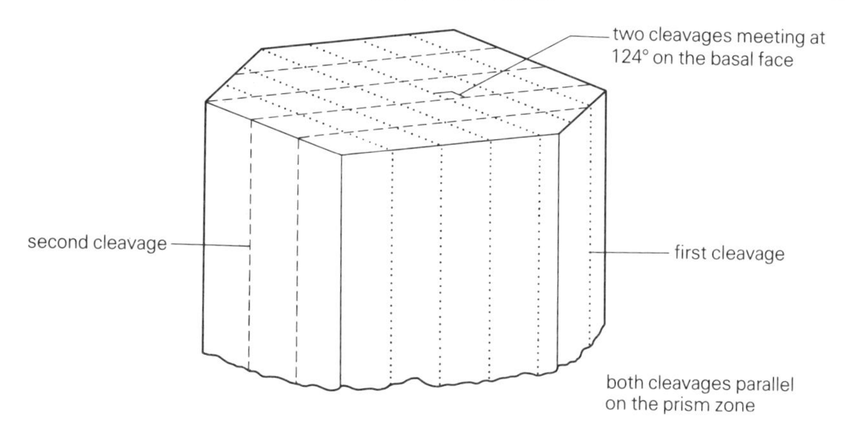

Figure 2.7 Crystal of hornblende showing two cleavages meeting at 124° on the basal face (see also Fig. 2.5).

Biotite Biotite ($K_2[MgFe]_6Si_6Al_2O_{20}[OH]_4$) is a dark-coloured member of the **mica group** of minerals. Its main properties are listed in Table 2.6. Like all micas, the atomic structure of biotite consists of a 'sandwich' with two sheets of linked $[SiO_4]$ tetrahedra forming the outer layers. These are linked together by another layer of metallic cations (Mg^{2+}, Fe^{2+}, Fe^{3+}, etc.) and hydroxyl groups. Each of these 'sandwich' units is linked to another identical unit by a layer of potassium cations. The links are weak bonds and the mineral cleaves easily into flakes along the planes separating the 'sandwich' units (Fig. 2.8). Biotite usually crystallises from a

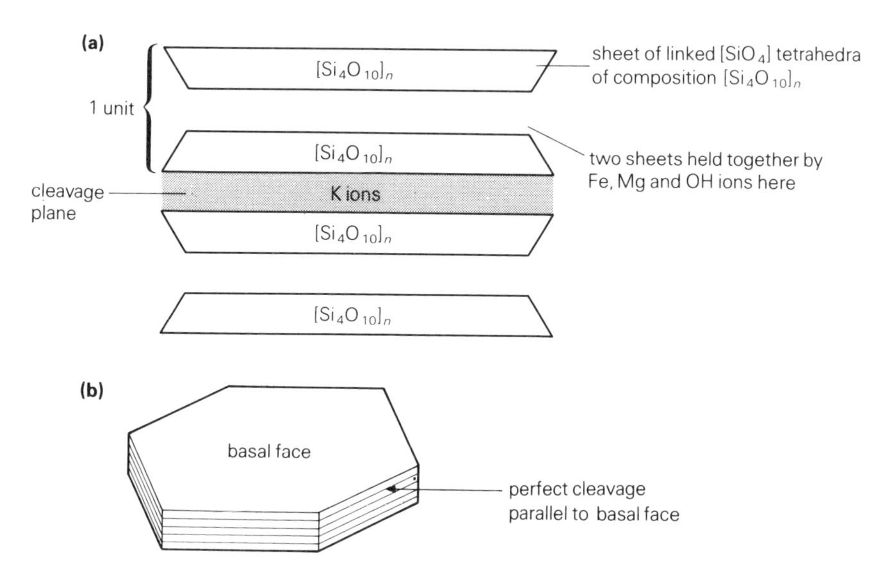

Figure 2.8 (a) The atomic structure of biotite (and other mica-group minerals), and (b) a crystal of biotite showing one perfect cleavage parallel to the basal face.

Table 2.7 Physical properties of light-coloured silicate minerals.

Mineral	Colour	Specific gravity	Hardness	Cleavages
feldspars	white, pink, variable	2.7±	6	two
quartz	colourless, white, red, variable	2.65	7	none
muscovite	colourless	2.7	$2\frac{1}{2}$	one perfect (cf. biotite)

magma containing water at a late stage in solidification. It is common in igneous rocks which are relatively rich in silica, and also in sediments and metamorphic rocks. Biotite is mined for use as an insulating material in certain electrical appliances, from deposits where the crystals are about 1 m across.

Garnet Garnet ($R_3^{2+}R_2^{3+}Si_3O_{12}$), where R^{2+} may be ferrous iron, magnesium, calcium and manganese, and R^{3+} may be ferric iron, aluminium or chromium, has a distribution restricted largely to metamorphic rocks (Section 2.2.5). Its principal properties are given in Table 2.6. The most useful criteria to identify it in a hand specimen are its lack of cleavage and its hardness, which exceeds that of quartz. This makes it a useful high-quality abrasive in such applications as garnet (sand) paper.

LIGHT-COLOURED SILICATE MINERALS

The more important properties of light-coloured silicate minerals are listed in Table 2.7.

Feldspars The chief members of the feldspar group of rock-forming silicates are **K-feldspar** or **potassium feldspar** ($KAlSi_3O_8$), and the **plagioclase feldspars** (composition varies from $NaAlSi_3O_8$ to $CaAl_2Si_2O_8$). The shapes of crystals that occur commonly are shown in Figure 2.9. Feldspars have two cleavages, which can be seen to meet at right angles in certain faces or sections of the crystal (Fig. 2.9). Twinning (see Section 2.1.1) is a diagnostic property. In its simplest form there are two parts of the crystal mutually reversed with respect to each other. In its complex form (multiple or repeated twinning), successive slabs within a crystal are twinned such that every alternate slab has the same orientation of its atomic structure. It is visible on the surface of plagioclase feldspar crystals and is diagnostic of this variety of the mineral group (Fig. 2.10).

Another feature possessed by all feldspar crystals is **zoning**. As they grow by crystallisation of the magma, and as the composition of the remaining liquid is slowly changed, shells of new material (which are different in composition from that of the previous ones) are added to the crystal to give concentric zones, ranging from calcium rich near the core to sodium rich at the periphery (Fig. 2.11).

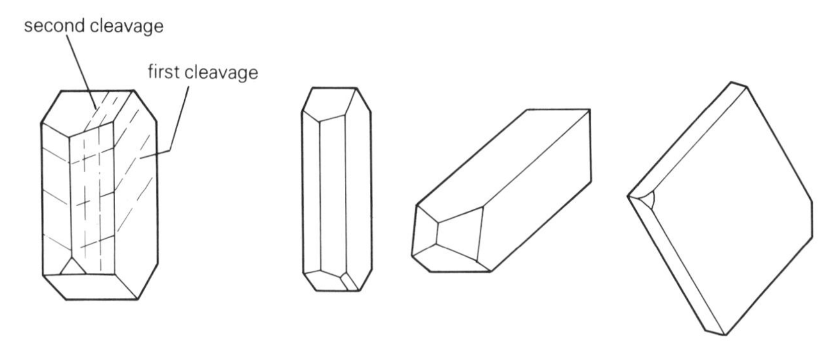

Figure 2.9 Various types of feldspar crystal, showing the two feldspar cleavages.

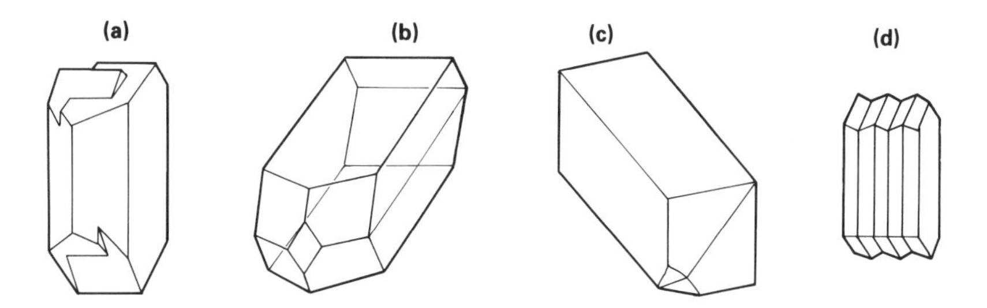

Figure 2.10 Various types of feldspar twin. Three twin forms of orthoclase feldspar are illustrated: (a) Carlsbad twin, (b) Manebach twin, and (c) Baveno twin. Also shown in (d) is an albite twin, a multiple or repeated twinning common to all plagioclase feldspars.

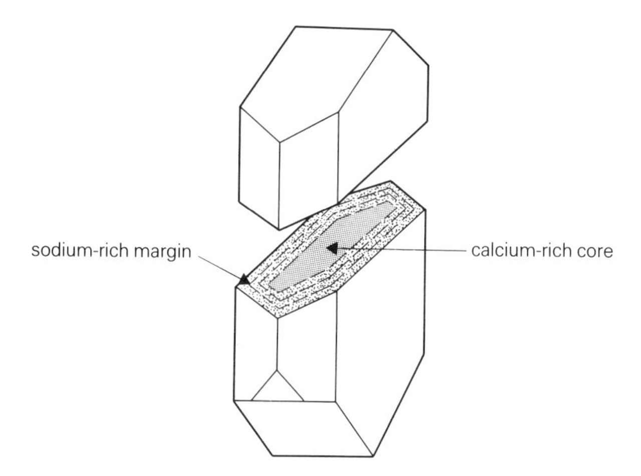

Figure 2.11 Diagrammatic representation of a feldspar crystal, 'split open' to show the nature of zoning.

K-Feldspar (or **orthoclase feldspar**) occurs in igneous rocks which are relatively rich in silica. Plagioclase feldspars are the most abundant and important silicate minerals in igneous rocks and are used for their classification. In silica-rich igneous rocks over 80% of the volume may be feldspars, whereas in silica-poor igneous rocks about half the volume may be plagioclase feldspar.

Quartz Quartz (SiO_2) has an atomic structure built of interlocked tetrahedra. It is colourless when pure, but small amounts of impurities may produce one of a range of coloured varieties. Manganese is present in **rose quartz**, and iron may give purple **amethyst** or red–brown **jasper**, depending on the amount of oxygen combined with it. Other varieties of silica are aggregates of very fine crystals, known in general as **chalcedonic silica** and, specifically, according to their lustre, colour and colour banding, as **chert, flint, opal** or **agate**. One of the differences between them lies in the water content.

A crystal of quartz is six-sided, with terminal pyramids (Fig. 2.12). Quartz (like garnet) never shows any alteration, except in rare cases where a thin skin of the crystal becomes milky white as a result of absorbing water. Quartz is found in rocks such as granite which are so rich in silica that not all of it combines to form silicates. It also occurs with other minerals, including ores, in sheet-like veins associated with granite masses. If its crystals are large enough, say several centimetres in length, then it may itself be mined for use in electrical parts or to make quartz windows which can withstand high pressure. Quartz is now grown artificially to produce large synthetic crystals that can be used commercially. It is also present in most sedimentary rocks (Section 2.2.4) because of its resistance to abrasion when rocks are broken down, and it is an essential component of sandstones. It occurs widely in metamorphic rocks.

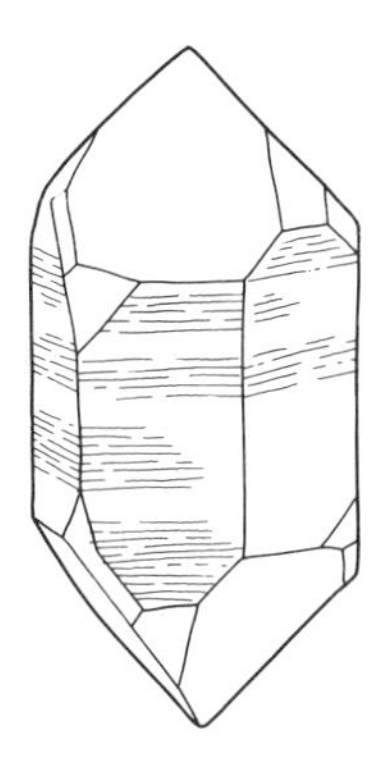

Figure 2.12 A double-ended crystal of quartz. Many quartz crystals in veins are single ended, but all show striations (parallel lines and gouges) across the prism faces.

Chert and flint are varieties of cryptocrystalline silica which can be used as aggregate in concrete if they are weathered. If chert or flint is fresh, it may be alkali reactive and therefore unsuitable to use with Portland cement. Cherts occur as bands or nodules within limestone sequences.

Muscovite Muscovite ($K_2Al_4Si_6Al_2O_{20}[OH]_4$) is a light-coloured member of the mica group, which has a similar atomic structure and crystal form to biotite. It occurs in silica-rich igneous rocks as a 'wet' mineral which crystallises at a late stage, together with quartz. It is particularly common in veins of coarse granite-like rock (pegmatite) and may be mined from them to be used as sheets having good thermal or electrical insulation. It is also present in many sedimentary and metamorphic rocks. Like most other micas, muscovite alters to clay minerals, particularly illite and montmorillonite. *Sericite* is a term used to describe fine-grained white micas. Such white micas are chemically similar to muscovite.

ALTERATION MINERALS

Many silicate minerals alter in the presence of air and water to form new, stable products. The most commonly altered minerals include the ferromagnesian silicates and the feldspars.

Serpentine Serpentine ($Mg_3Si_2O_5[OH]_4$) is an alteration product which forms mainly from olivine. Serpentine is green in colour with low specific gravity (2.6) and hardness ($3\frac{1}{2}$). The fibrous variety of serpentine (**chrysotile**) is a type of **asbestos**. Serpentine forms from olivine in the presence of water and free silica as follows:

$$\text{olivine (Mg-rich)} + \text{water} + \text{silica} = 2\ \text{serpentine}$$

The process of alteration of olivine in the rock is depicted in Figure 2.13. An examination of the structure of serpentine shows it to be mesh-like, with fibrous serpentine crystals in a precise arrangement, as shown in Figure 2.14. The formation of serpentine within a rock involves an increase in volume from that of the original olivine. In some basic igneous rocks this increase in volume gives rise to cracks that radiate out from the mineral and weaken the rock structurally.

The presence of serpentine changes considerably the physical properties of olivine-bearing igneous rocks. Many of the best rocks for road-metal aggregate (including dolerite, basalt and gabbro) contain olivine, and its degree of alteration should always be checked. A small amount of alteration may be beneficial, since the aggregate then bonds better with bitumen, but aggregates in which there is extensive alteration should be avoided.

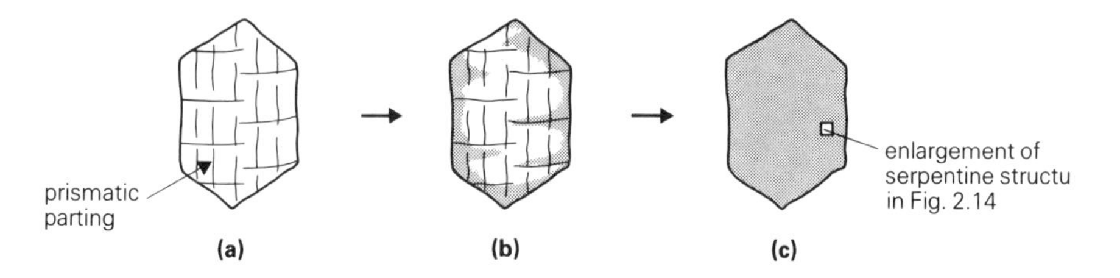

Figure 2.13 Olivine alteration. (a) Olivine showing basal fractures and poor prismatic parting; (b) serpentinisation begins with alteration of olivine along cleavages; (c) serpentinisation complete: olivine crystal is now entirely composed of serpentine.

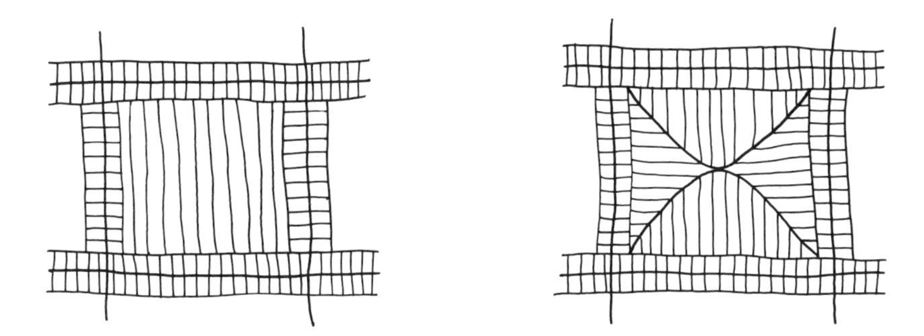

Figure 2.14 Two possible mesh types of serpentine structure.

Recent studies on serpentines have shown that aggregate from serpentinite bodies (rocks composed entirely of serpentine) can be used in concrete. In the Middle East, serpentinite concretes have been employed in buildings with no resulting ill effects.

Chlorite Chlorite ($[MgFeAl]_{12}[SiAl]_8O_{20}[OH]_{16}$) is a sheet silicate closely related to the micas, and is dark green in colour. It has a variable specific gravity of between 2.6 and 3.3, depending upon its composition, and a hardness of $2\frac{1}{2}$. Chlorite forms as a **secondary mineral** from the hydrothermal alteration of ferromagnesian (dark-coloured) silicate minerals, in particular augite, hornblende and biotite. In metamorphic rocks chlorite occurs as a **primary mineral**; that is, it forms from pre-existing (clay) minerals present in the rock, as the rock is slowly subjected to increasing temperature (and pressure).

Many ferromagnesian silicate minerals in sediments will alter to chlorite during the weathering process. Shales and mudstones, particularly, tend to be rich in chlorite. Many joint or fault surfaces in basic igneous rocks may be 'coated' with chlorite, causing these joints to be very weak with very low interparticle angles of friction. Normal silica minerals have interparticle angles of friction in the range 25–35°, whereas chlorite's angle of friction is 13°. Thus the angle of friction between ferromagnesian minerals can be greatly reduced by weathering and chlorite formation.

Clay minerals Clays form mainly by the alteration of other minerals, by the action of weathering. The specific type of clay formed depends upon the composition of the original mineral undergoing alteration and the surface conditions where weathering is taking place. The change is not usually a direct or simple one. Other alteration products which are not strictly clays may be formed as intermediate stages of the weathering process, and one clay mineral may be transformed into another more stable one as conditions change. For example, secondary chlorite, formed by hydrothermal alteration of primary ferromagnesian minerals, will itself alter readily to clay during the weathering process (see Fig. 2.15). Secondary minerals above the dashed line in the figure are formed by late stage magmatic processes, such as hydrothermal alteration, whereas those below the line are usually formed by weathering, although initial clay formation within feldspars may be hydrothermal in origin.

Clay minerals are sheet silicates with densities betwen 2.5 and 3.0, depending on the type of clay and its composition, and with low hardness (2 to $2\frac{1}{2}$ (kaolin) and 1 to 2 for all other clays).

In structure, all clays consist of two fundamental units: sheets of silicon–oxygen tetrahedra, and sheets of aluminium or magnesium octahedra in which each Al^{3+} or Mg^{2+} ion is linked to six hydroxyl (OH^-) anions. The sheet of silica–oxygen tetrahedra (silica sheet) was shown previously in Figure 2.8, and can be further simplified to  when the tetrahedra have their apices pointing upwards, or 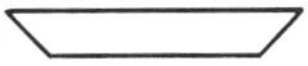 when the apices of the tetrahedra are pointing downwards.

The octahedral layers may consist either of Al and (OH) ions, called a **gibbsite** layer, since gibbsite is a mineral with the composition $Al(OH)_3$, or of Mg and (OH) ions, called a **brucite** layer, since brucite is a mineral with the composition $Mg(OH)_2$. These layers can be further simplified to 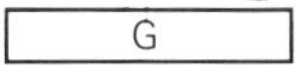 G for a gibbsite layer or 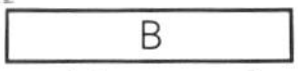 B for a brucite one. Some clays have a structure that consists of layered units from different clays joined together, and these are called **mixed-layer clays**.

Symbolic structures for the clay minerals can now be depicted.

(a) **Kaolin**, $Al_2Si_2O_5(OH)_4$. Two units are held together by attraction (van der Waals' forces). Note that the structure of kaolin is worth comparing with that of serpentine, in which a brucite layer replaces the gibbsite (G) layer of kaolin. Kaolin (or kaolinite) is often termed a 1 : 1 sheet silicate, since one silica layer is coupled with one gibbsite layer.

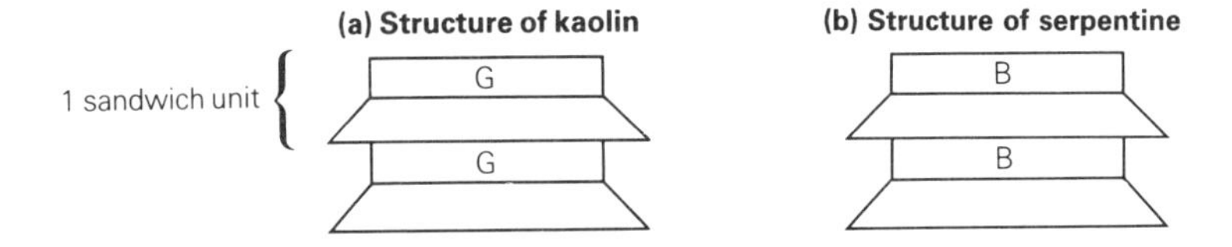

(b) **Illite**, $K_{0.5-1}Al_2(AlSi_3)O_{10}(OH)_2$. Illite, like the micas, is termed a 2:1 sheet silicate since one sandwich unit consists of two silica layers with one gibbsite layer between. The units are joined by K^+ ions.

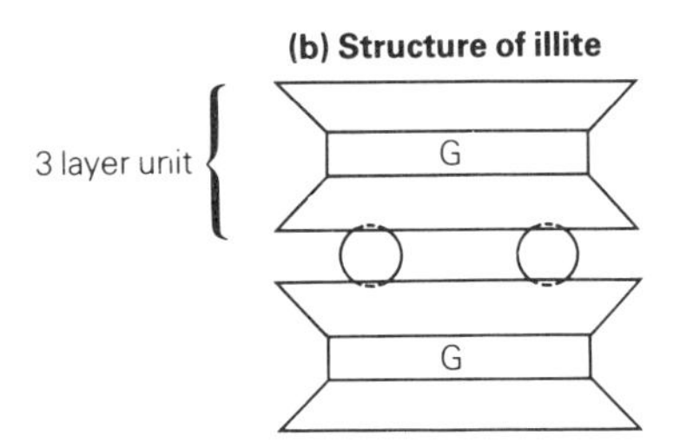

(c) **Montmorillonite**, $[2Al_2(AlSi_3)O_{10}(OH)_2]^{2-}$. In montmorillonite the units are held together by H^+ ions and occasional Na^+ ions. In the gibbsite layer Al can be replaced by Mg. Montmorillonite is a member of the **smectite clays** and is also a 2:1 sheet silicate similar to illite. The structure of montmorillonite is similar to that of **vermiculite**. In the latter the main octahedral layers are brucite (with some Al replacing Mg). Chlorite is also a sheet silicate, but is termed a 2:2 sheet silicate since two silica layers are joined to two brucite or gibbsite ones.

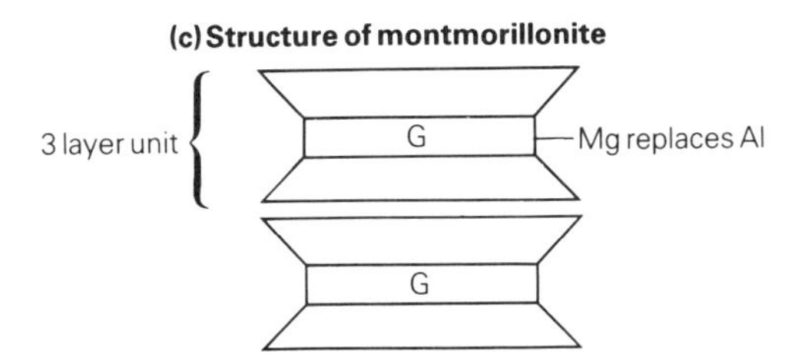

Formation of alteration minerals The common dark-coloured silicate minerals, with the exception of garnet, are relatively susceptible to chemical weathering. Olivine is very susceptible to hydrothermal alteration, and alters in the presence of water to serpentine. It is also susceptible to weathering, other products being formed. Augite, hornblende and other chain silicates are easily penetrated by water, augite being especially susceptible to reaction since it contains no hydroxyl group in its structure. Biotite is the most easily weathered of this group and alters to vermiculite. Garnet is extremely resistant to both chemical attack and abrasion, but is rare in soils because it is rare in most rocks.

Among the light-coloured silicate minerals, the feldspars are susceptible to chemical weathering, and quartz and muscovite are resistant. Feldspars weather to produce clays, the type depending on the composition of the feldspar and the conditions in which weathering has taken place. Figure 2.15 shows the development of clays by the alteration of other minerals. The most important primary minerals from which clays form are the feldspars, and their alteration is now discussed in detail.

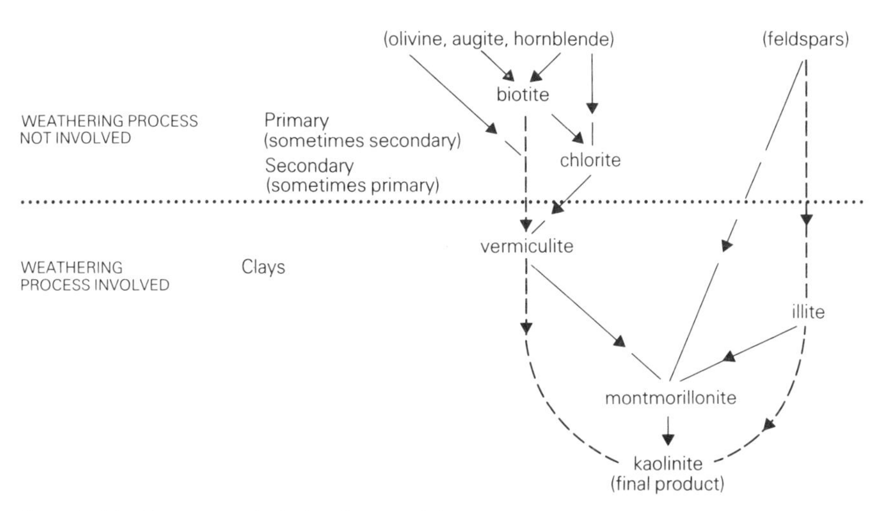

Figure 2.15 Clay mineral relationships and their production from silicate minerals.

Orthoclase changes to illite when the supply of water is not sufficient to remove all the potassium, thus:

3 orthoclase + 2 water = illite + 6 silica + 2 potash

If there is an excess amount of water present, then orthoclase alters to kaolinite as follows:

2 orthoclase + 3 water = kaolinite + 4 silica + 2 potash

Illite may also alter to kaolinite when unlimited water is available. Plagioclase feldspar alters to sericite, a white mica similar to muscovite, during hydrothermal activity, but weathering of plagioclase feldspar produces montmorillonite as follows:

3 plagioclase feldspar + 6 water
= 2 montmorillonite + 3 calcium hydroxide

Montmorillonite will also change to kaolinite if excess water is present. In all cases kaolinite is the final product (see Fig. 2.15). In some tropical and sub-tropical centres, basalt weathers to produce montmorillonite initially, but as weathering continues, kaolinite is formed as the stable product. Leaching of the kaolinite deposit leads to a removal of silica and the final product is **bauxite**.

Vermiculite is a complex hydrous biotite which may form from either chlorite or biotite, and vermiculite also may break down to give kaolinite as the final product.

The effect of exchangeable ion composition and water content on clay mineral properties can be considered in terms of their **plasticity**, defined in terms of the **Atterberg plastic state limits**. The **plastic limit** (PL) is the minimum moisture content at which clays exhibit obvious plasticity by being capable of forming threads (3 mm in diameter) without breaking when rolled by the palm of the hand on a glass plate. The **liquid limit** (LL) is the minimum moisture content at which the clay will begin to flow when subjected to small impacts in a liquid limit device (defined in BS 1377). The Atterberg limits of the clay minerals are shown in Table 2.8. **Plasticity index** (PI) is the difference between the liquid limit and the plastic limit (i.e. LL − PL = PI), and it indicates the moisture content range over which the soil remains plastic. The range of values depends on which exchangeable ions are present; usually higher limits occur in the presence of higher valency exchangeable ions, except in the case of montmorillonite, when the reverse is true. The limit values are indicative of the relative amounts of water absorbed on to the particle surfaces during the duration of the test. The data in Table 2.8 indicate that, in the presence of water, montmorillonites would swell more than illites, which swell more than kaolinites. This has important implications in tunnelling or any civil engineering work that involves excavation of clays in the presence of water.

Clay deposits are quarried in many parts of the world. The kaolin formed from highly weathered granites in Cornwall is extracted for the ceramic industry and is Britain's greatest export, by bulk if not by value. Other important deposits are worked in China, the USA, France,

Table 2.8 Atterberg limits for common clay minerals.

	Plastic limit (%)	Liquid limit (%)
kaolinite	26–37	29–73
illite	34–60	63–120
montmorillonite	51–98	108–710

Malaysia and Czechoslovakia. Montmorillonite is mined in the western USA and elsewhere as a major constituent of the clay assemblage **bentonite**, the industrial uses of which include water softening and making muds for well drilling. It is thixotropic (which means that the clay particles form a weak structure in the mud while it is at rest and give it a jelly-like consistency), but this collapses as soon as the drill starts and disturbs the mud, so that its viscosity drops suddenly. This has the advantage that the sides of the hole are supported and the clay particles stay in suspension during periods when drilling has stopped. Montmorillonite also has important absorbent properties and in another of its impure forms, **Fuller's Earth**, it is used to cleanse wool of natural fats.

Some clay minerals, such as attapulgite and palygorskite, have 'chain-type' structures, and resemble chrysotile and asbestos. These natural minerals can be dangerous when breathed into the lungs, and breathing equipment is needed to protect against respiratory diseases.

2.1.3 *Non-silicate minerals*

For ease of description and identification, the non-silicate minerals can most simply be grouped mainly on the basis of lustre into metallic ore minerals and non-metallic minerals. In general, the former are dark and the latter are light coloured.

METALLIC ORE MINERALS

Nearly all ore minerals are chemically simple substances such as native elements, sulphides (where a metallic element is combined with sulphur), or oxides (where a metal is combined with oxygen). Native elements, such as copper, silver or gold, occur only rarely. Sulphides, except for pyrite, are not quite as rare, but are of importance only when they are relatively concentrated into veins in quantities where they can be worked as sources of metals. The common ore minerals, present in most rocks in small quantities, are oxides, although argillaceous sedimentary rocks and low-grade fissile metamorphic rocks usually contain sulphides.

If iron ores are present in large quantities ($>5\%$) in aggregates from crushed igneous rock, they may weather and discolour any concrete made from them. Magnetite is often present in large quantities in basic igneous rocks (basalts, dolerites) and its presence may affect electrical equipment and compasses.

Sulphides in shales weather easily and may cause expansion and disintegration of such mechanically weak rocks. Thus care should be taken in excavations involving shales – whether stabilising slopes of road cuts or, for example, preventing disintegration while sinking caissons for foundation works – since all such operations expose new surfaces of shale

Table 2.9 Physical properties of some ore minerals.

Mineral	Formula	Colour	Streak	Specific gravity	Hardness	Cleavages
Sulphides						
galena	PbS	grey	black	7.5	$2\frac{1}{2}$	three (parallel to cube faces)
blende (sphalerite)	ZnS	dark reddish brown		4	4	three
pyrite	FeS_2	yellow	black	5	6	none
Oxides						
magnetite	Fe_3O_4	black	black	5+	6±	poor fractures
haematite	Fe_2O_3	brown	brown	5+	5–6	none
limonite	$Fe_2O_3.3H_2O$	earthy	brown	4	5	poor fractures
goethite	FeO.OH	yellow	yellow	4	5+	two (prismatic)

to the air. The properties of the more common ores are shown in Table 2.9. Only the common ore minerals containing iron merit further brief description.

Pyrite (iron sulphide) occurs as cubes in slate and in some silica-rich igneous rocks. It has been worked for its sulphur content (not its iron) in the past, but the large amount of sulphur obtained as a by-product of gas or oil refining has changed this. Another sulphide mineral, **pyrrhotite**, is found in slates and shales and can weather easily.

Magnetite occurs in most igneous rocks, especially basic varieties such as gabbro and basalt, where it may constitute up to 15% of the rock's volume. In an acid rock such as granite, magnetite constitutes only about 1% by volume. It is present in metamorphic rocks but is rarely an important constituent. Magnetite grains are present in sandstones and other sedimentary rocks and may be concentrated by natural panning (it has a higher specific gravity than most mineral grains in sand) to a degree to which it may, in rare cases, form the bulk of the deposit and be worth exploiting as a source of iron.

Haematite occurs widely in all sorts of rocks and is also concentrated into veins, where it may be exploited as an economic iron ore. It is the cement that binds grains together in red sandstones. **Limonite** and **goethite** are formed by the combination of iron oxide with water.

NON-METALLIC, NON-SILICATE MINERALS

The most common minerals in this group are calcite (a carbonate) and gypsum (a sulphate). All have white or colourless streaks. They are often found in veins, together with metallic ores. In veins such minerals, together with fluorite and barite, are called **gangue minerals**, in contrast to the metallic ores. The properties of the more important minerals in this group are listed in Table 2.10.

Table 2.10 Physical properties of some non-metallic, non-silicate minerals.

Mineral	Formula	Colour	Specific gravity	Hardness	Cleavages
barite	$BaSO_4$	white	4.5	3½	two
fluorite	CaF_2	variable	3	4	four (parallel to octahedron faces)
Chemical precipitates					
gypsum	$CaSO_4.2H_2O$	colourless or white	2	2	one perfect
halite (common salt)	NaCl	colourless or variable	2	2	three perfect (parallel to cube faces)
Carbonates					
calcite	$CaCO_3$	white or colourless	2.7	3	three (parallel to rhombohedron faces)
dolomite	$CaMg(CO_3)_2$	off-white	3	4	three (parallel to rhombohedron faces)

Calcite is an essential mineral in limestones and may be present in many sedimentary rocks as the cement matrix. It is the principal constituent of marble. Limestone deposits may be crushed to a fine powder and used as agricultural lime. The neutralising value of any calcite limestone deposit must be known for this purpose, and in many regions agricultural lime with a higher MgO content may be preferred; that is, the limestone will be composed of both calcite and dolomite. In many countries of the Middle East, limestones are widespread and may be the only material available to use as a crushed rock aggregate for concrete or roads. Limestones are also the most important constituent in the manufacture of cement.

Gypsum and its associated mineral, **anhydrite**, are called **evaporites**, being formed by precipitation of salts from sea water that has evaporated. These evaporite deposits are common in rock formations of Permo-Trias age. In desert regions, gypsum may be found on the surface of the desert as clusters of flat crystals called **desert roses**. Salt-flats (**sabkhas** or **sebchas**) are widespread in desert regions (see Section 2.2.4), particularly on the coast, and the engineer must be wary of using any such material in building works. Gypsum is mined throughout the world and has many uses; for example 'panelling' made of crushed gypsum (gyproc) is widely used in buildings for inner wall claddings because of its excellent fire-resistant and soundproofing qualities. Calcium sulphate is also used as plaster of Paris.

Halite or **rock salt** is an evaporite deposit which is commercially mined as salt for use on icy roads in the winter, and can be refined as common table salt.

Barite is usually found as a gangue mineral in veins (often with fluorite, galena and blende), but it can also occur as an evaporite deposit. Large deposits are commercially valuable because barite is extensively used in the oil industry as an additive to drilling muds.

2.2 Rocks

2.2.1 *The nature of rocks*

Rocks are aggregates of one or more mineral. The nature and properties of a rock are determined by the minerals in it (particularly those **essential minerals** which individually make up more than 95% of its volume) and by the manner in which the minerals are arranged relative to each other (that is, the texture of the rock). Weathering, of course, will affect the engineering properties of a rock, and this is dealt with in detail in Chapter 3. An individual rock type or specimen is always described in terms of its mineral composition and its texture, and both are used in the classification of rocks.

According to their manner of formation, rocks are of three main types:

Igneous rocks are formed from magma, which has originated well below the surface, has ascended towards the surface, and has crystallised as solid rock either on the surface or deep within the Earth's crust as its temperature fell.

Sedimentary rocks are formed by the accumulation and compaction of (a) fragments from pre-existing rocks which have been disintegrated by erosion (see Section 3.2.2); (b) organic debris such as shell fragments or dead plants; or (c) material dissolved in surface waters (rivers, oceans, etc.) or ground water (Section 5.2.2), which is precipitated in conditions of oversaturation.

Metamorphic rocks are formed from pre-existing rocks of any type, which have been subjected to increases of temperature (T) or pressure (P) or both, such that the rocks undergo change. This change results in the metamorphic rock being different from the original parental material in appearance, texture and mineral composition.

2.2.2 *Igneous rocks*

MINERALOGY OF IGNEOUS ROCKS

Many different types of mineral occur in igneous rocks, but only about eight (see Table 2.11) are normally present as essential constituents of a rock. Which of the eight are present is controlled primarily by the

Table 2.11 Mineral crystallisation from a magma.

Mineral	Ultrabasic	Basic	Intermediate	Acid
quartz			— — —	————
orthoclase			— — —	————
plagioclase	Ca-rich ————	————	————	———— Na-rich
muscovite				— ————
biotite		— —	————	————
hornblende		— —	————	———— — —
augite	— — ————	————	———— — —	
olivine	————	— —		
	early (high temperature)			late (low temperature)

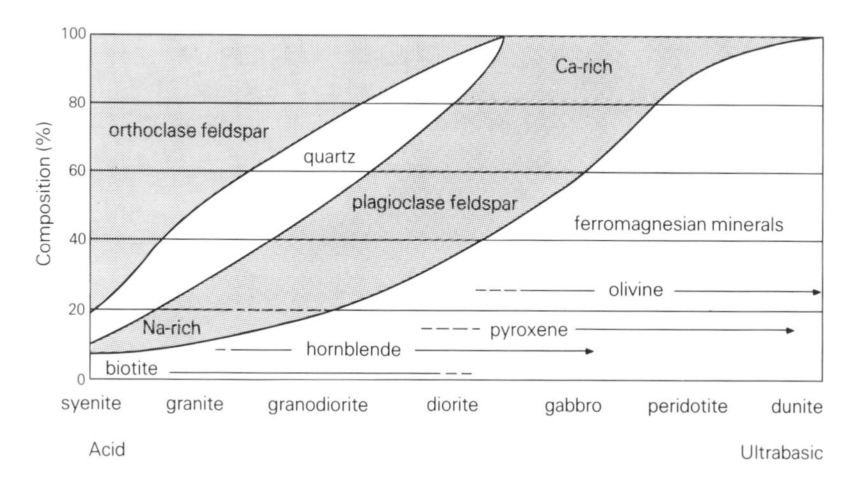

Figure 2.16 Igneous rock composition, based on the proportion of each mineral present.

Table 2.12 Minerals present in the four main groups of igneous rock.

Rock composition	Amount of SiO_2 (%)	Minerals
acid	65	quartz, orthoclase, Na-plagioclase, muscovite, biotite (±hornblende)
intermediate	55–65	plagioclase, biotite, hornblende, quartz, orthoclase (±augite)
basic	45–55	Ca-plagioclase, augite (±olivine, ±hornblende)
ultrabasic	45	Ca-plagioclase, olivine (±augite)

composition of the magma. Each mineral starts to crystallise at a particular temperature and continues to form throughout a limited temperature range as the magma cools. More than one mineral is usually forming at any one time. Since the crystals formed early have a higher specific gravity than the remaining liquid of the magma, they settle downwards. Alternatively, the two fractions, crystals and liquid, may be separated by some other process. As time progresses, different minerals crystallise from the magma. Eventually this process gives rise to a sequence of layers of minerals of different composition, from high-temperature, high-specific gravity minerals at the bottom, to low-temperature, low-specific gravity minerals at the top. Such a magma is said to be **differentiated**. When consolidated, the highest temperature layer of minerals at the bottom, consisting of olivine, calcium-rich plagioclase and often augite, will form an **ultrabasic igneous rock**. At relatively low temperatures, the last of the magmatic liquid solidifies into rock that contains quartz, orthoclase, sodium-rich plagioclase and micas. Because of its relative richness in silica, it is called an **acid rock**. These relationships are given in Table 2.11. A partial classification by grouping igneous rocks according to their mineral compositions, and hence percentage of silica, is shown in Figure 2.16. The boundaries between the four main groups of igneous rock are arbitrary divisions of what are really gradational changes in composition, but they are drawn in such a way that the presence or absence of certain minerals such as quartz and olivine allows a rock to be put at the acidic or basic ends of the scale (Table 2.12).

TEXTURES OF IGNEOUS ROCKS

A variety of textures may occur in igneous rocks. Each reflects the physical conditions under which the rock formed. With few exceptions, igneous rocks are composed of interlocking crystals (only a few of which display a perfect crystal form), and are said to have a **crystalline texture**. The next most important textural feature is the size of the individual crystals, and this is used as a criterion, together with mineral composition, in the most common and simplest classification of igneous rocks. Generally speaking, crystal size is usually related to how long it has taken the magma to solidify completely, and thus how much time individual crystals have had to grow. In **fine-grained rocks**, crystals are on average less than 1 mm across, in **medium-grained rocks** they are between 1 and 3 mm across, and in **coarse-grained rocks** they are over 3 mm across. Most textbooks specify crystal size as 1 to 5 mm for hypabyssal, and over 5 mm for plutonic rocks, but since, in our experience, a better dividing line is 3 mm, this has been used here.

Extrusive igneous rocks have formed (as lavas) by rapid cooling of magma at the surface, and are fine grained. Conversely, most, but not all, fine-grained igneous rocks are extrusive. **Hypabyssal igneous rocks**

form minor intrusions (dykes and sills, see Section 2.2.3) which have solidified below the surface, and have cooled more slowly than extrusive rocks because of the thermal insulation of the surrounding country rocks. They are typically, but not invariably, medium grained. **Plutonic igneous rocks** have formed by the slow cooling of great volumes of magma, typically at depths of a few kilometres, within the Earth. It is erosion throughout geological time that has removed the overlying rocks and revealed these plutonic intrusions at the surface. The heat from their enormous bulk can dissipate only very slowly and they are normally coarse grained.

Sometimes the magma has chilled so quickly that crystals have failed to form. The rock is then a natural glass, and is described as having a **glassy texture**. This texture occurs most commonly in acid extrusive rocks.

Porphyritic texture (Fig. 2.17) exists where larger and smaller crystals are both present in the same rock (see also Figs 2.18 & 19). The smaller crystals may be of fine-, medium- or coarse-grained size and are spoken of collectively as the matrix in which the larger **phenocrysts** are set. The texture is found most commonly in extrusive rocks, but is also sufficiently common in some hypabyssal rocks to merit a special name (**porphyries**) for them. For example, in quartz porphyry the most common phenocryst is quartz. The texture is frequently produced when a rock has cooled in two or more stages, and crystals from the first stage gain a head start in growth over the later-stage crystals of the matrix.

Vesicular and **amygdaloidal textures** (Fig. 2.20) occur most commonly in extrusive rocks (see Fig. 2.21). Gases dissolved in magma under the high pressures found at depth come out of solution and start to expand as the magma rises towards the surface. The gas forces the magma apart initially as a bubble within the rock, and eventually it may force its way upwards through the magma and escape. In a lava flow the rapid congealing of the molten rock hinders or prevents this, and the gas is trapped within the rock in cavities which are called **vesicles**. The process

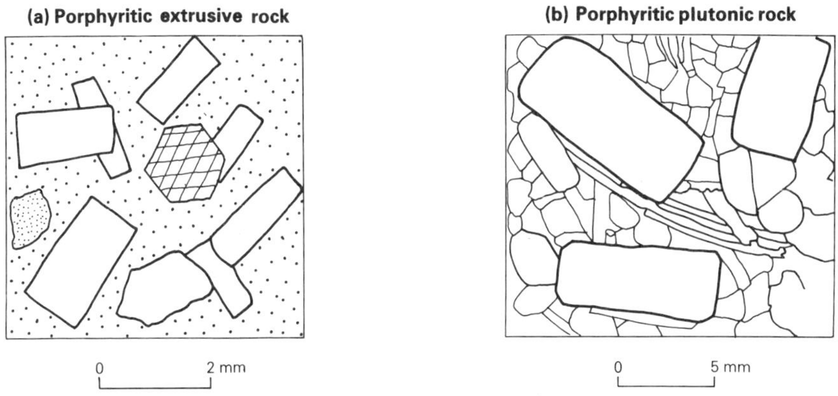

Figure 2.17 Porphyritic rocks.

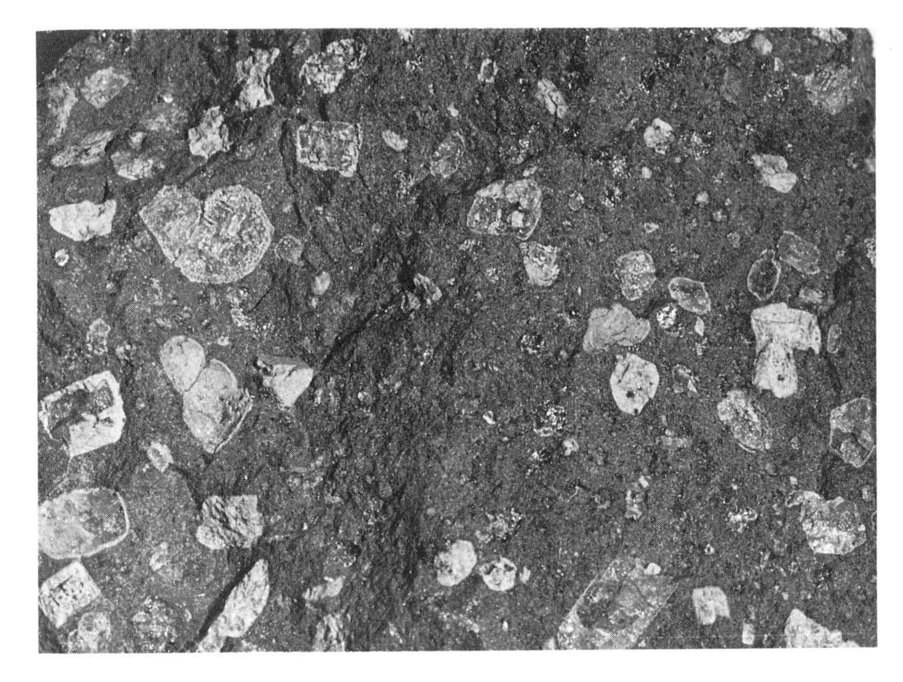

Figure 2.18 Porphyritic texture in fine-grained basalt.

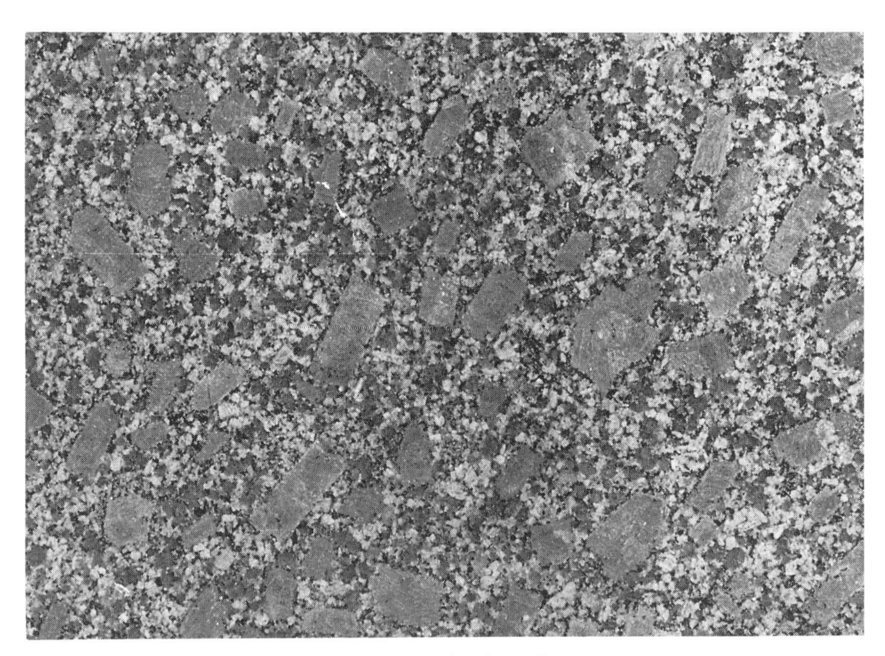

Figure 2.19 Porphyritic texture in coarse-grained granite.

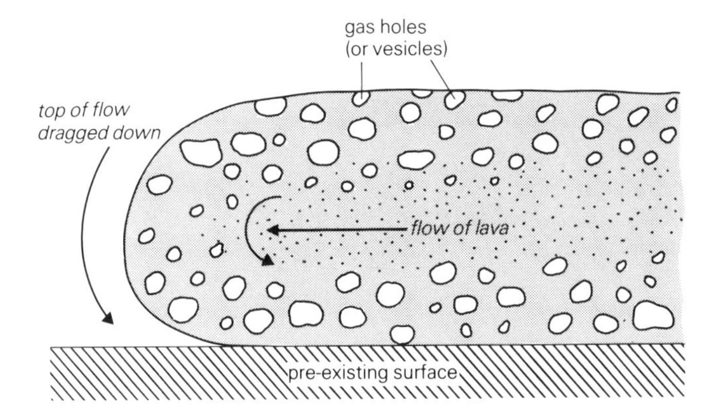

Figure 2.20 Formation of vesicular texture in a lava flow.

results initially in a concentration of vesicles near the upper, chilled, surface of the lava flow. However, the rotation of the solid skin of the flow as it moves forward like the tread of a caterpillar tractor carries the concentration of vesicles to the bottom of the flow, so that eventually both top and bottom are vesicular. These voids in the rock are often filled with later minerals which may have been precipitated from water circulating in the rock just before, or at some time after, consolidation. A vesicle filled with mineral is called an **amygdale**, and the rock has an amygdaloidal texture. Amygdales can be distinguished from phenocrysts because of their rounded shape and variation in amount within the rock.

Figure 2.21 Vesicular and amygdaloidal texture in basalt.

Figure 2.22 Ophitic texture.

Ophitic texture (Fig. 2.22) is produced when plagioclase feldspar and augite crystallise at the same time to produce a felted arrangement of interlocking crystals (see Fig. 2.23). It is normally present in the basic hypabyssal rock **dolerite**, and is the reason for its high crushing strength, and hence for its suitability as roadstone.

Pegmatitic texture is produced where the concentration of water and other fluxes in the late-stage residue of a magma lowers the temperature of crystallisation of the minerals forming in it, allowing an individual crystal to achieve a size well beyond what is meant by 'coarse grained'. In extreme cases, crystals weighing a few tonnes (1 tonne = 1000 kg) have been found. Pegmatitic texture characteristically occurs in veins of very acid igneous rock (**pegmatite**), which are sources of economically important minerals.

Classification of igneous rocks uses silica content and texture to assign each rock to a slot in a framework that demonstrates how one type of rock

Figure 2.23 Ophitic texture in dolerite.

Table 2.13 Classification of normal (calc-alkaline) igneous rocks.

Rock type	Amount of SiO_2 (%)	Grain size: Extrusive	Hypabyssal	Plutonic
acid	>65	rhyolite, dacite	quartz and orthoclase porphyries	granite, granodiorite
intermediate	55–65	pitchstone andesite	plagioclase porphyries	diorite
basic	45–55	basalt	dolerite	gabbro
ultrabasic	<45	various basic olivine-basalts	various basic dolerites	picrite, peridotite, serpentinite, dunite

is related to another. Two rocks that are adjacent in the framework are not totally different from one another, and are likely to have been formed under conditions sufficiently similar for one variety even to grade into the other within a body of igneous rock. The commonly used classification (Table 2.13) is for 'normal', relatively abundant (calc-alkaline) types of igneous rocks. However, in its scheme of progressive chemical change and mineral composition, it cannot accommodate those less common igneous rocks that have crystallised from an alkaline magma where sodium and potassium were relatively abundant. A different series of rocks is formed, each containing large amounts of orthoclase feldspar. They usually have less than 65% silica in their composition. The extrusive variety is called **trachyte**, the hypabyssal variety **felsite**, and the plutonic variety **syenite**, although it should be noted that the name 'felsite' may be used for many rocks, including some rhyolites.

A helpful rule of thumb in identifying rock type is that coarse- and medium-grained acid rocks are usually lighter coloured than their basic equivalents. Acid lavas may, however, be glassy and as dark coloured as any fine-grained basic rock.

Typical engineering properties of igneous rocks are given in Table 2.14, but it must be realised that investigation of any of the rocks given in the table could reveal values outside the limits shown, especially as regards the capacity for water absorption.

2.2.3 *Igneous structures and forms*

EXTRUSIVE ROCKS

Extrusive rocks are formed when molten rock (magma) reaches the surface, along either wide vertical fissures or pipe-like openings in the Earth's crust. **Fissure openings** may vary from a fraction of a kilometre to several kilometres in length. Huge outpourings of magma can be emitted

Table 2.14 Engineering properties of some unweathered igneous rocks.

		Specific gravity	Water absorption (%)	Compressive strength ($MN\,m^{-2}$)
plutonic	granite	2.6–2.7	0.3–0.8	100–240
	diorite	2.7–2.85	0.3–0.8	100–250
	gabbro	2.75–3.0	0.1–0.3	100–250
	ultrabasic varieties	2.8–3.3	0.1–1.5	75–300
hypabyssal	porphyry	2.6–2.7	0.2–1.0	100–350
	dolerite	2.7–2.9	0.1–0.7	120–300
extrusive	rhyolite	2.5–2.7	0.3–2.5	<100–280
	andesite	2.6–2.8	0.2–2.0	<100–350
	basalt	2.7–2.9	0.1–1.0	150–300

from such fissures; for example, the Deccan Plateau in India is composed of extrusive rocks, estimated to have a volume of about 1 million km^3, which originated as magma issuing from surface fissures.

Pipe-like openings on the surface give rise to a typical conical **volcano**, such as Vesuvius or Fujiyama, of the **central-vent** type. An arbitrary scale of intensity of volcanic eruptions from central-vent volcanoes has been suggested, and this shows a variation from a quiet type, such as the Hawaiian volcanoes, to violent types, such as Vesuvius, Stromboli and Mont Pelée. All central-vent volcanoes consist of **volcanic cones**, which mark the magma duct or vent. These are composed of layers of ash and lava, and they vary in the steepness of their sides according to the composition of the lava and the amount of ash and dust ejected. In general, viscous acidic lavas do not flow far from their place of eruption and are highly explosive or violent. They form steep cones. Basic lavas or basalts flow easily to cover large areas, and they may produce no more than a broad swell at the fissures or vent from which they came.

The ash and dust ejected from central-vent volcanoes consist of small blobs of lava blown apart by expanding volcanic gases as the magma neared the surface, and which chilled as they flew through the air. The collective term for this ejected material is **pyroclastic rock**. The dust and fine ash may settle on land or in water to form **tuffs**, which may or may not show bedding. Tuffs are fine-grained, porous, crumbly (friable) rocks. Sometimes the volcanic ash has been welded together by heat from the eruption, and **welded tuffs** are formed. The coarser ash travels a short distance from the volcanic vent, and may fall back into it. After consolidation it is called **agglomerate**. Most pyroclastic rocks are easily and quickly eroded, and volcanic cones are usually ephemeral features on a geological scale of time. The 'ancient volcanic hills' of most areas are not volcanic cones, but the eroded lava-infilled pipes of the volcanoes.

Depending on their composition, **lavas** may have a rough broken surface (scoriaceous lava) or a smooth wrinkled surface (ropy lava) when extruded. Gas bubbles (vesicles) are concentrated by buoyancy at the upper surface of a lava flow while it is molten, and at the lower surface by its flow (Fig. 2.20). In older flows, such vesicles are usually filled with secondary minerals. If flows are poured out on land, they are subject to sub-aerial weathering until the next flow is extruded, and a soil rich in ferric iron (**red bole**) is formed. If the time between eruptions is long enough, this layer may be as much as 1 m thick and may contain fossil plants. Red bole is a weak rock compared with the fresh igneous rock of the flow. A section through a typical lava flow is shown in Figure 2.24.

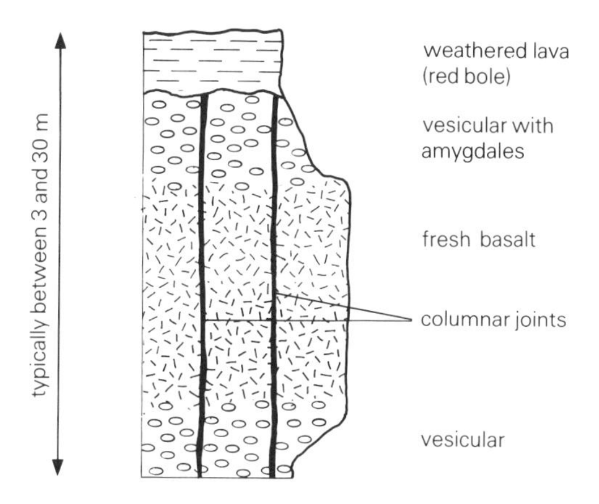

Figure 2.24 A lava flow showing columnar joints, the change in rock type from top to bottom of the flow, and the corresponding relative resistance to erosion of the exposed face on the right.

INTRUSIVE ROCKS

Minor (hypabyssal) intrusions are relatively small igneous structures formed from magma which has penetrated to the uppermost few kilometres of the Earth's crust, but has not reached the surface to be poured out as lava. They are often found in association with volcanic rocks. For example, when volcanic activity ceases, the volcanic vent by which magma has travelled to the surface is choked with once-molten rock which has solidified to form a medium-grained igneous rock, together with any agglomerate that collapsed into the vent. The body of mixed igneous rock, roughly circular in plan and anything from about 100 m to a few kilometres across, is called a **volcanic plug** (Fig. 2.25). It has near-vertical margins and extends down to the magma reservoir from which it came. This might lie below the crust, but more commonly the plug extends only to a major intrusion, originally at a depth of 5–10 km below the surface.

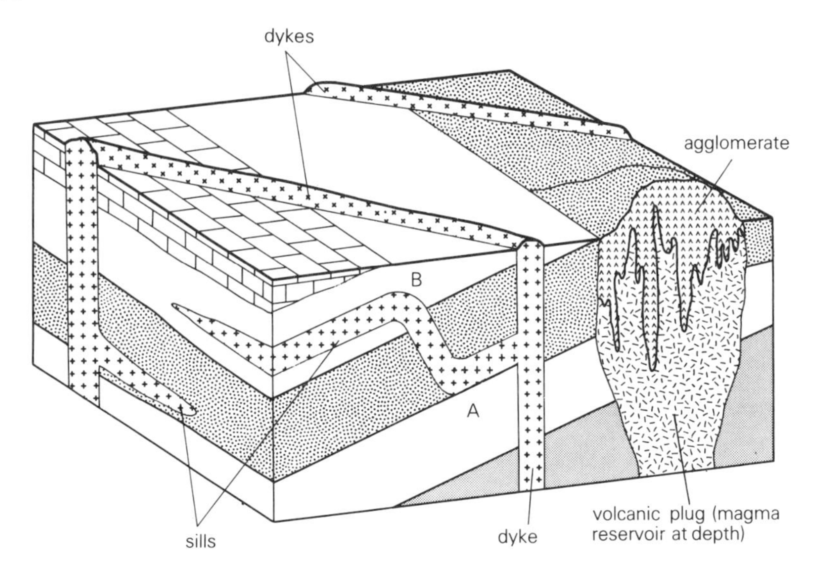

Figure 2.25 Three common types of hypabyssal intrusion: a volcanic plug, a sill and a dyke are shown in sections and on the surface. These igneous rocks are harder than the sediments into which they are intruded, and form ridges and a low hill. The higher levels in the volcanic plug are formed mainly of agglomerate (volcanic breccia) represented by triangles, and the lower levels of basalt (stippled). The sill is transgressive, changing level in the strata between A and B, where it cuts across the bedding discordantly like a dyke. Note the displacement of strata by the sill at AB.

The common hypabyssal intrusions (Fig. 2.25) are sheet-like in form, with widths usually between 1 and 70 m. They are labelled according to whether or not they conform to the structure of the strata in which they are emplaced. A **concordant hypabyssal intrusion** injected along the layering in the country rocks is called a **sill**. A **discordant hypabyssal intrusion** cutting steeply across the layering is called a **dyke** (Fig. 2.25). Most sills are subhorizontal and most dykes are near-vertical, so the terms are often used loosely with this relative orientation in mind. An intrusion (Fig. 2.25) consisting of several segments, mostly concordant but at different levels in the column of strata and linked by discordant segments, is called a **transgressive sill**.

Rapid cooling takes place at the margins of intrusions, where the magma has come into contact with the cold country rocks into which they are injected. As a result, the medium-grained rock (for example, dolerite), which forms the main body of the dyke or sill, grades into fine-grained rock (for example, basalt), or even into a thin glassy selvage. Any small vesicles are most likely to be found at the margins. Columnar jointing (see Section 4.4.1) is often present. It is usually vertical in sills and horizontal in dykes.

In the field or at an engineering site, certain differences between sills and lava flows may enable an engineer to determine which igneous form he is dealing with. These differences are listed in Table 2.15.

A dyke rarely occurs in complete isolation from other igneous intrusions. Characteristically, tens or hundreds of dykes occur as a **dyke swarm** within a linear zone, which is usually tens of kilometres across. Each dyke fills a tension crack (Section 4.4.4) formed at the time of its intrusion. The dykes of a swarm are parallel or subparallel, except where they are deflected locally by secondary stresses at geological boundaries, such as a major fault. A dyke swarm is sometimes related to a major centre of igneous activity, either by intersecting it, or by spreading out from it. The centre may be the site of one or more major granite **plutons**. Dykes 'key in' to the centre by swinging into trends which are radial from it. The frequency of dykes and the total amount of tensional stretching are greater near the centre. Whether or not the magma in a dyke fissure, forcing its way towards the surface, continues to rise beyond any

Table 2.15 The main field differences between lava flows and sills.

Item	Lava flow	Sill
grain size	Matrix fine grained but variations may occur.	Matrix medium–fine grained, depending on thickness of sill, with variation in grain size from margin (finer) to centre (coarser).
thickness	Variable from few metres to tens of metres. Invariably the lava flow will form part of a lava pile (many flows).	Variable, usually tens of metres in depth. Sills usually form distinct and quite separate intrusions.
thermal effects	Lavas tend to heat the underlying rocks.	Both under- and overlying rocks thermally affected.
weathering	Original upper surfaces of lava flows may be weathered (if extruded sub-aerially) leading to the formation of an oxidised skin or bole on top. This is frequently preserved by baking effects from subsequent overlying lavas.	A sill is not originally exposed and red boles do not occur.
textures	Vesicles are common both at the top and at the bottom of a lava flow.	Vesicles rare but occasionally present – never prolific as in lava flows.

The cooling joints present in lava flows and sills are similar in orientation.

particular level or is injected laterally along the layering to form a sill depends on the relative resistance offered to injection by either path. This, in turn, depends on the local geology. Where favourable geological circumstances exist (for example, a gentle downbuckle in a varied sequence of layers), sills branch off from their feeder dykes.

Other types of minor intrusion occur within certain of the major centres of volcanic activity, and they are usually concentric or radial to the centre to which they belong. There are distinctive terms to name them, and the mechanics of their formation are understood, but their occurrence is so limited that it is not appropriate here to do more than mention their existence.

The rocks forming major intrusions have crystallised relatively deeply within the Earth's crust (that is, at depths of at least a few kilometres) and are said to be plutonic. They are coarse grained, except where they have been rapidly cooled near a **chilled margin**. The smallest intrusions of this class are often mushroom-shaped, with a flat base and an upwards bulging roof, but other forms exist. The rocks are commonly acid or intermediate in composition, but basic and ultrabasic bodies of this form also occur.

Plutonic sheet intrusions are of basic composition in bulk, with a gradation from ultrabasic near the base to a small amount of acid rock at the top. Sheets of this type are a few kilometres thick, often downwarping the underlying original rocks because of the weight of magma involved.

The second type of major plutonic intrusion is a great body of granodiorite and granite called a **batholith** (Fig. 2.26). Batholiths are always formed from acid magma, and are characteristic of late igneous activity in mobile belts (see Section 4.6.1), where mountain building is taking place. They extend for tens or hundreds of kilometres along the mountain belt, are tens of kilometres wide, and extend downwards for about 10 km. For example, the Coast Range batholith of the American Rockies, which consists of many intrusions emplaced at the same time, is 2000 km long, 100 km wide and can be seen vertically through a range of 5 km. At no place, however, is the base of a batholith visible. The roof of a batholith may lie below the ground surface, and in that case the only granite outcrops are those of smaller intrusions (commonly between 5 and 10 km across), which rise above the main batholith as **stocks** (Fig. 2.26). They appear to be separate bodies of granite, but are joined at depth. The margins of a batholith usually dip outwards at a steep angle. Close to them the granite often contains blocks of country rock (**xenoliths**), which have been broken off during the rise of the acid magma. Xenoliths vary from a few centimetres to tens of metres across. In this marginal zone, the minerals may be arranged parallel to the edge of the intrusion to give foliation to the granite. This is caused by flow while crystallisation proceeds, before the magma solidifies completely.

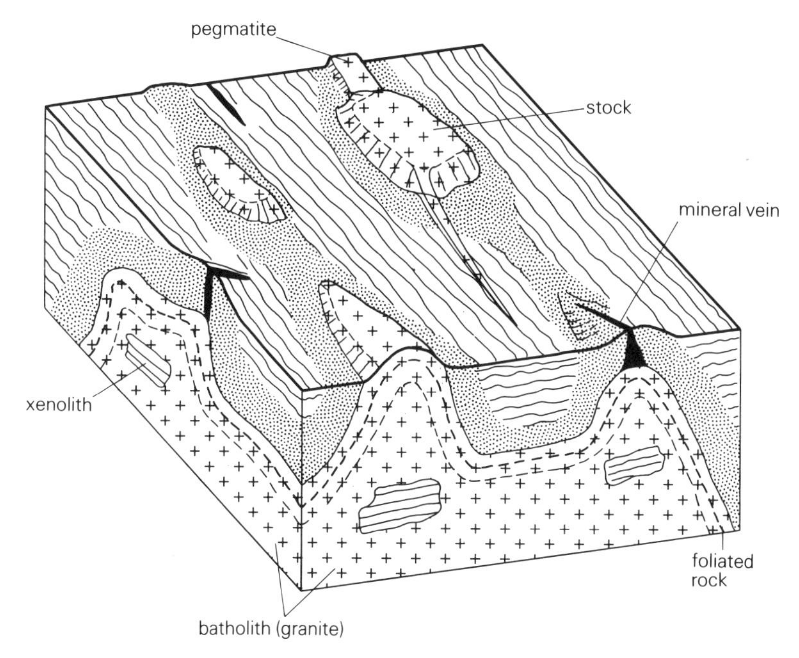

Figure 2.26 The granite of the batholith has been emplaced in country rocks which have suffered contact metamorphism. The thermal aureole is shown by dotting. The granite close to the contact is foliated. Blocks of country rock occur as xenoliths within the granite. Stocks rise from the main mass of granite, and pegmatite and mineral veins are present in the roof.

Heat from the crystallising magma affects a zone of **country rocks** surrounding the intrusion. This zone is called a thermal aureole, and it, and the effects of heat on country rocks (thermal metamorphism), are described fully in Section 2.2.5. The size of a thermal aureole will vary, depending upon the size of the igneous intrusion producing it and the amount of water available from the intrusion and the country rocks. Thus a dolerite dyke 30 m in width is likely to produce thermal effects in the surrounding rock for a distance of less than 1 m beyond its margins, whereas a large plutonic intrusion may produce a thermal aureole from 1 to 2 km in width (Fig. 2.26); however, some granites possess quite narrow aureoles.

The country rocks of the roof are also likely to be affected by the final residue of volatiles, both liquid and gaseous, in the solidifying granite. These collect in the upper parts of the batholith and, at the final stage of crystallisation, they are injected into fissures in the roof to form minor intrusions. They are often referred to, especially if they are narrow (less than 1 m wide), irregular or contain economic minerals, as **veins** rather than dykes or sills. Sometimes the granite intrusions themselves may possess these veins, caused by late-stage igneous liquids being injected into the granite rocks which have just solidified. The composition of these residual fluids approximates to granite modified by a relative enrichment in rare elements. In these veins the individual crystals are able to grow to a relatively large size, and the rock is very coarse grained. The rock (pegmatite) in the veins may contain large crystals of unusual minerals such as topaz, tourmaline, beryl and fluorite. In some regions, hot water with lead, zinc, silver, tin and other economic metals in solution forms **hydrothermal veins** containing **ore minerals** around the granite. Those such as **cassiterite** (tin oxide), which come out of solution at high temperature, are deposited close to, or within, the granite. Those precipitated at lower temperatures, such as galena (lead sulphide), are deposited further away. The temperatures of formation of hydrothermal ore deposits are in the following ranges: 300–500°C for **hypothermal deposits** (cassiterite, arsenic compounds, pyrite and other iron sulphides); 200–300°C for **mesothermal deposits** (lead, zinc, copper, silver and gold); and 50–200°C for **epithermal deposits** (antimony and mercury deposits, and silver and gold compounds). These temperatures should be compared with the magmatic and metamorphic temperatures given in Figure 2.39.

2.2.4 *Sedimentary rocks*

Sedimentary rocks are formed from the solid debris and the dissolved mineral matter produced by the mechanical and chemical breakdown of pre-existing rocks, or in some cases from the skeletal material of dead plants and animals. The processes involved in the disintegration of rocks by weathering and erosion, and the transport of these products to the place where they are deposited, are all discussed in Chapter 3. They merit this special attention in a book for engineers since the deposits ('soils' to an engineer) which have recently formed, or are forming, blanket most of the solid rocks of the Earth, and are the natural material encountered and dealt with in nearly every shallow excavation. These modern deposits are also relevant in discussing the solid sedimentary rocks, which have been produced from similar accumulations in the geological past. The sediment has been transformed into solid rock by compaction as it was buried and compressed by subsequent deposits. In many cases, minerals in solution in the ground water have been precipitated to act as a natural cement, and bond the fragments together. The nature of a sedimentary rock, and its position in a scheme of rock classification, are partly dependent on these original conditions of transport and deposition. Thus it may be described according to the type of environment in which it accumulated: as a **continental deposit** if it were laid down on land or in a lake by rivers, ice or wind; as an **intermediate deposit** if it were laid down in an estuary or delta; or as a **marine deposit** if it were laid down on a seashore, in the shallow waters of the continental shelves, or in the abyssal areas of the deep oceans.

The four major groups of sedimentary rocks are:

(a) **terrigenous sedimentary rocks** (sometimes referred to as **clastic** or **detrital rocks**), which are formed from minerals or rock fragments derived from the breakdown of pre-existing rocks;
(b) **chemical sedimentary rocks,** which are formed from the precipitation of salts dissolved in water;
(c) **organic sedimentary rocks,** which are formed from the skeletal remains of plants and animals and include coal and oil;
(d) **limestones and dolomites,** which are sedimentary rocks consisting of more than 50% carbonate, and can include chemical, clastic and biological material.

MINERALOGY OF SEDIMENTARY ROCKS

The composition of a given rock will depend on the source from which the waste material came, on the resistance, chemical and mechanical, of each component during transport, and on the distance travelled. The main constituents of terrigenous sedimentary rocks are fragments from pre-existing rocks and minerals. These may be fresh and unaltered, or may be alteration products of weathering (see Section 2.1.3), such as clay minerals. Quartz is the most common mineral. It is chemically stable and hard enough to resist abrasion as it is transported. Some rarer minerals, such as garnet, tourmaline and rutile, have similar properties. Feldspar is less stable but may survive for long enough, under favourable circumstances, to be present in deposits that have not been transported far from their source, or which are protected in other ways. The other common rock-forming minerals of igneous rocks generally have an even lower survival rate when exposed to air and water (p. 23).

Some minerals in sedimentary rocks may be formed in the area of deposition. The most common examples are those minerals precipitated from solution to form the chemical sedimentary rocks. They include carbonates (calcite, dolomite and siderite), sulphates (gypsum and anhydrite), chlorides, and silica (chalcedonic minerals). Each may not only be an essential mineral in a particular type of chemical sedimentary rock, but may also be the sole mineral present.

Limestones may possess mineral constituents that include clastic particles, chemically precipitated material and organic remains (shells, corals, etc.).

Diagenesis is the name given to processes that alter the character of a sedimentary rock after it has been deposited, either by interreactions between constituent minerals or by reactions between constituent minerals and pore fluids or other liquids circulating through the sedimentary material. Diagenesis is a low-temperature process, which can occur when the sediment is still in contact with the sea or lake water after its deposition, and also later when direct contact with the original water has been removed. The process continues until the constituents and pore fluids are in chemical equilibrium.

In the sedimentary pile during diagenesis, silica is first deposited at low temperatures between grains. As the sedimentary pile thickens, the deeper sediments then have calcite deposited at the higher temperatures; these deep sediments have been compacted and their porosity has been reduced. If the sedimentary pile is later uplifted, the calcite cement redissolves and the sedimentary rock will possess gaps or voids in its structure. This explains why some sedimentary rocks are much weaker than their appearance would suggest.

Sedimentary rock grains may be bonded together by a cement, which consists of any naturally deposited mineral matter as described earlier. Commonly found cements in sedimentary rocks include calcite and silica, as well as haematite, which gives the rock a red colour. The nature of the cement may be of importance to the engineer since it affects a rock's properties. Thus, for example, old (pre-1880) buildings in Glasgow and Edinburgh are built from block sandstones of Lower Carboniferous age. The Glasgow sandstone is from a local horizon and fairly porous, with some carbonate cement and a low compressive strength (<50 MN m^{-2}), whereas the Edinburgh sandstone, which is also obtained from a local horizon, has a low porosity, is silica cemented and possesses a higher strength (>60 MN m^{-2}). Thus the pre-1800 Edinburgh buildings (the so-called Edinburgh New Town) have survived in better condition than the much younger, Victorian houses in Glasgow, because of the higher resistance of the Edinburgh sandstone to weathering by the industrial atmosphere of these parts of Scotland.

TEXTURES OF SEDIMENTARY ROCKS

Chemical sedimentary rocks generally have a crystalline texture. Some, however, are formed of fragments, and their textures are dependent on the sizes, shapes and arrangement of these fragments. If the rock has been formed from organic debris, then the fragments may consist of particles of shell or wood, but the texture can be described in the same terms as are used for other fragmented rocks.

The **size of grains** is an important textural feature of a terrigenous rock, as an indication of distance between its source and depositional areas, as well as an easily observed property which may be used to distinguish and classify the rock. The coarsest particles are deposited nearest to the source area, and most of the finest particles are carried in suspension to greater distances before they settle. These clay particles, or rock flour (which represents minute particles of the common rock-forming minerals), are deposited only when the current slackens to nearly zero, and in some cases the salinity of the water of an estuary or the open sea makes the minute specks of clay clot (flocculate) into larger particles.

The **degree of roundness** of grains is related to the amount of abrasion suffered during transport, and hence to distance travelled from their source before deposition. Roundness is related to the sharpness or curvature of edges and corners of grains. It is also dependent on the size and hardness of the grains and the violence of impact of one against another. No matter how far they travel, sand grains lack the necessary momentum when they collide in water to produce perfect rounding, and it is only when dry grains are blown by the wind that perfectly rounded grains are produced. Larger fragments, however, may be well rounded after transportation by water for distances of less than 150 km. The degree of roundness is an important property in sand being used to make concrete, or for other engineering purposes. Sand grains deposited from ice are normally more angular than those in river deposits, and the most rounded grains usually occur in sand dunes.

A property associated with roundness of grains is **sphericity**, which defines the degree to which a particle or grain approaches the shape of a sphere. Equidimensional particles have a greater prospect of becoming spherical during transportation than other shapes of particle. Sphericity is controlled by directions of weakness such as bedding planes or fractures. It is also related to size, in that the larger the grain above 8 mm, the lower the sphericity. The relationship of rounding to particles less than 2 mm of high and low sphericity is given in Figure 2.27.

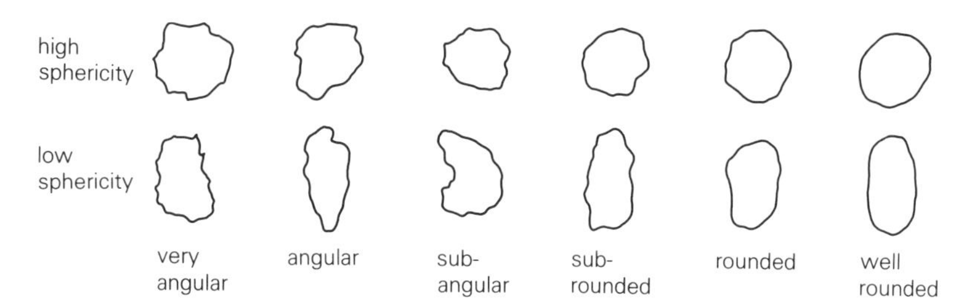

Figure 2.27 Sand-sized particles showing variation in roundness and sphericity.

A positive correlation exists between sphericity and roundness, but other factors, notably particle size, planes of weakness, and particle composition, may have a marked influence on the final shape of the particle when it is finally deposited. Settling velocity will also influence the shape of the deposit.

Most terrigenous rocks contain grains of different sizes (see Fig. 2.28). The relative homogeneity of a rock is expressed as its **degree of sorting**, a well sorted rock consisting of similarly sized particles. In contrast, a poorly sorted rock has a wide range of particle sizes (that is, grades). It

Figure 2.28 Volcanic breccia formed of angular fragments of rock glass in a matrix of volcanic fragments, from Maryvale, Utah, USA. (Reprinted by kind permission of Professor Donald Bowes.)

should be noted that, in engineering practice, a soil such as gravel, containing a wide range of sizes (grades), is said to be **well graded**, and that 'well graded' is opposite in meaning to 'well sorted'.

In the **Unified Classification System (UCS)** used by engineers, soils are classified by the predominant particle size and organic content into gravels (G), sands (S), silts (M), clays (C), organic soils (O) and peats (Pt). Coarse-grained soils (G + S) are subdivided by grading, and fine-grained –clay soils by liquid limit, LL, and plasticity index, (see Section 2.1.3). Table 2.16 gives a more detailed account of mechanical composition scales; coarse and fine aggregates are included for comparison. This table should be compared with Table 2.17, which classifies clastic sedimentary rocks on their grain size. In soil classification most emphasis is put on the <2 mm fraction, and soil subdivisions (called **soil textures**) can be represented in the form of a triangular diagram (Fig. 2.29), where 'loams' include soils with a wide range of particle sizes. This diagram should be compared with Figure 2.33 in which arenaceous sedimentary rocks are classified according to their composition, percentage mud fraction, and so on. Particle sizes in unconsolidated sediments or soils can be plotted as summation curves (Fig. 2.30), the forms of which give more easily digested information than can be obtained from a table of percentages.

Table 2.16 Mechanical composition scales for sands and gravels.

International Society of Soil Science scale		BS 1377: 1974. Methods of testing soils for civil engineering purposes*		BS 812 Part I: 1975. Methods of sampling and testing of mineral aggregates, sands and fillers†	
Fraction (*a*)	Size range (mm) (*b*)	Fraction (*c*)	Size range (mm) (*d*)	Fraction (*e*)	Sieve sizes (mm) (*f*)
stones or gravel	>2	cobbles	200–60		75
					63
					50
				coarse aggregate	37.5
		gravel: coarse	60–20		28
		medium	20–6		20
		fine	6–2		14
					10
					6.3‡
coarse sand	2–0.2	sand: coarse	2–0.6		5
		medium	0.6–0.2		3.35
fine sand	0.2–0.02	fine	0.2–0.06*		2.36
					1.7
					1.18
silt	0.02–0.002	silt: coarse	0.06–0.02	fine aggregate	850 μm
		medium	0.02–0.006		600
		fine	0.006–0.002		425
					300
					212
					150
clay	<0.002	clay	<0.002		75

Notes: *Pedologists sometimes use BS 1377: 1974 and also sometimes divide sand from silt at 0.05 mm following the practice of the US Department of Agriculture.
†*See* also BS 882, 1201: Part 2, 1973. Aggregate from natural sources for concrete.
‡Aggregates are specified by percentages passing various sieves, and there is a small percentage overlap between coarse and fine.

Table 2.17 Clastic sedimentary rock classification based on grain size.

Grain size	Class	Rock types
>2 mm*	rudaceous	conglomerate, breccia
2–0.06	arenaceous	sandstone, arkose, greywacke
0.06–0.002	arenaceous	siltstone
<0.002 mm	argillaceous	shale (clay-rock), mudstone

*More than 30% grains must be greater than 2 mm in size.

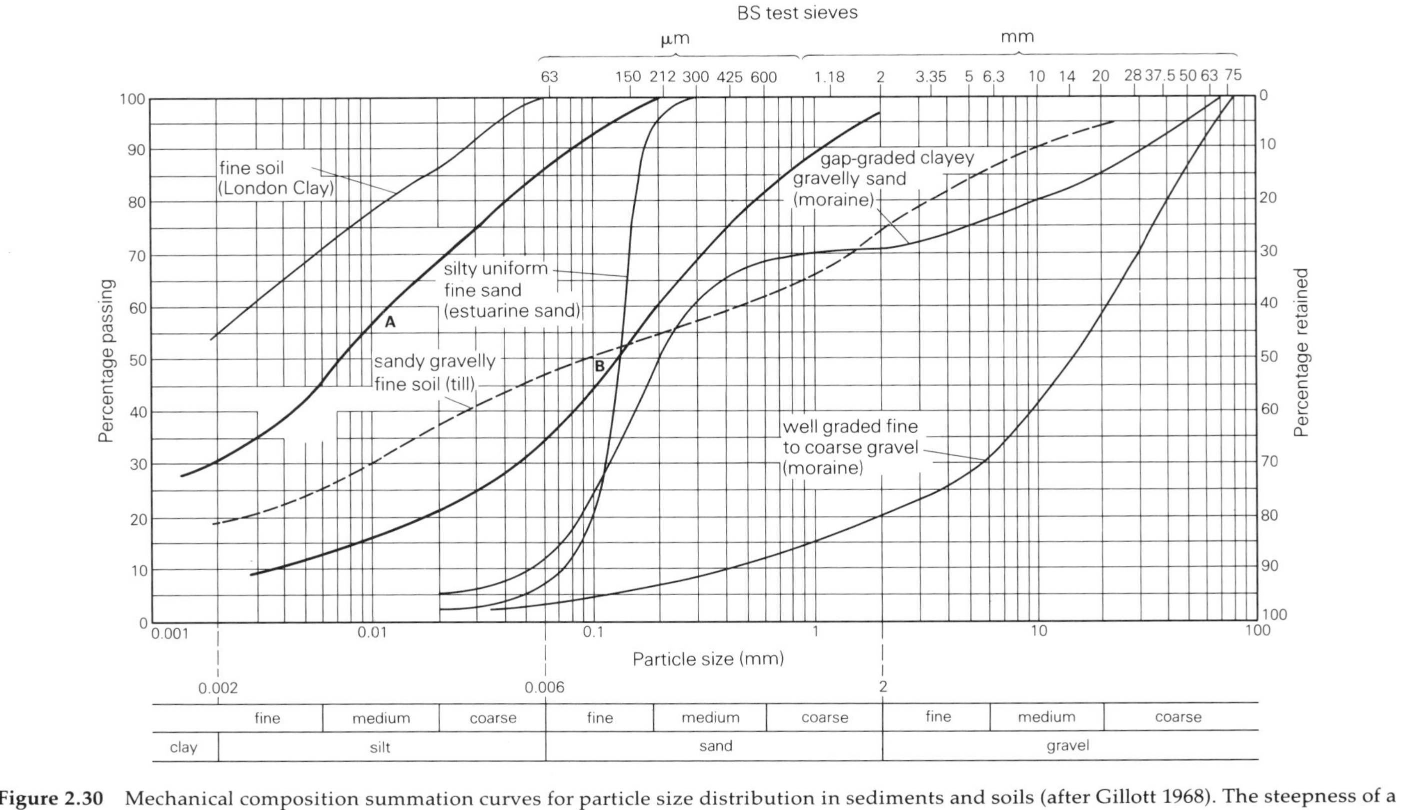

Figure 2.30 Mechanical composition summation curves for particle size distribution in sediments and soils (after Gillott 1968). The steepness of a curve is an indication of the degree of sorting in the material. The curves of frost-susceptible soils usually fall between those of A and B. Grading curves of selected soil types are also included on the chart (from BS 5930: 1981).

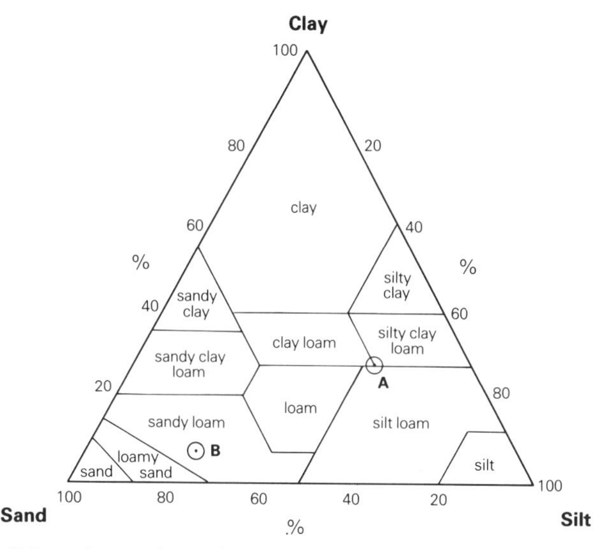

Figure 2.29 Triangular mechanical composition diagram for soils (based on Military Engineering 1976). If the mechanical composition of a material is expressed in terms of sand (2.0–0.06 mm), silt (0.06–0.002 mm) and clay (<0.002 mm), it may be plotted on the diagram as a single point. The method is more convenient than summation curves when large numbers of soils are to be compared. The points A and B represent the corresponding soils in Figure 2.30.

The occurrence of sedimentary structures indicates some variation in composition or texture of sedimentary rock layers in response to changes in the environmental conditions in which the particular sediment was laid down. Two of the most common sedimentary structures, graded bedding and cross bedding, are associated with sorting.

In **graded bedding**, a sediment containing a wide range of grain sizes is sorted vertically such that there is a continuous gradation from coarse particles at the bottom of the sedimentary layer to fine grains at the top (Fig. 2.31). Certain thick sedimentary sequences are characterised by a rhythmic alternation of thin sandstones and shales. The sandstones (or **greywackes**) show graded bedding. These are believed to have been deposited by **turbidity currents**, probably flowing off ocean shelf areas into deep water carrying a slurry of sand-laden muddy water, which forms **turbidites**.

Cross bedding is most commonly found in sandstones and feldspathic sandstones (or **arkoses**), which have been laid down in shallow water or deposited as dunes by the action of wind. Successive minor layers are formed as sand grains settle in the very slow moving, deeper water at the downstream end of a sandbank or delta, and the sandbank grows in that direction. Each layer slopes down stream and is initially S-shaped; however, erosion of the top of the sandbank by the stream leaves the minor layering still curving tangentially towards the major bedding

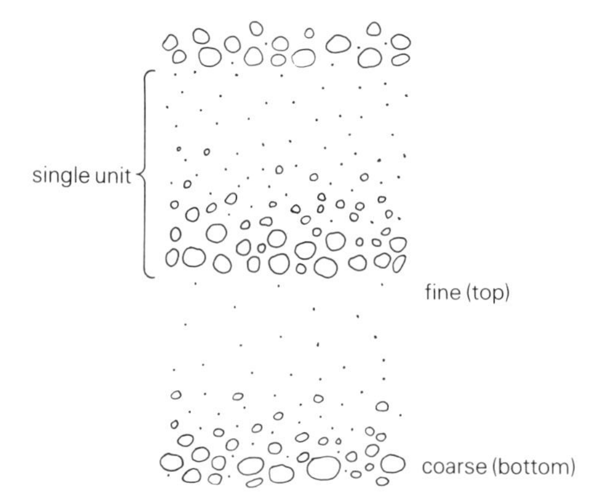

Figure 2.31 Sequence of graded beds.

plane at its base, but truncated sharply at its junction with the upper bedding plane (Fig. 2.32). Because of its mode of origin, it is sometimes referred to as **current bedding**. A typical occurrence is shown in Chapter 4 (Fig. 4.2). In both graded and cross bedding the original top of the bed may be recognised from the asymmetry of the structure within the bed, and either may be used as a **way-up criterion** (see Section 4.3.1).

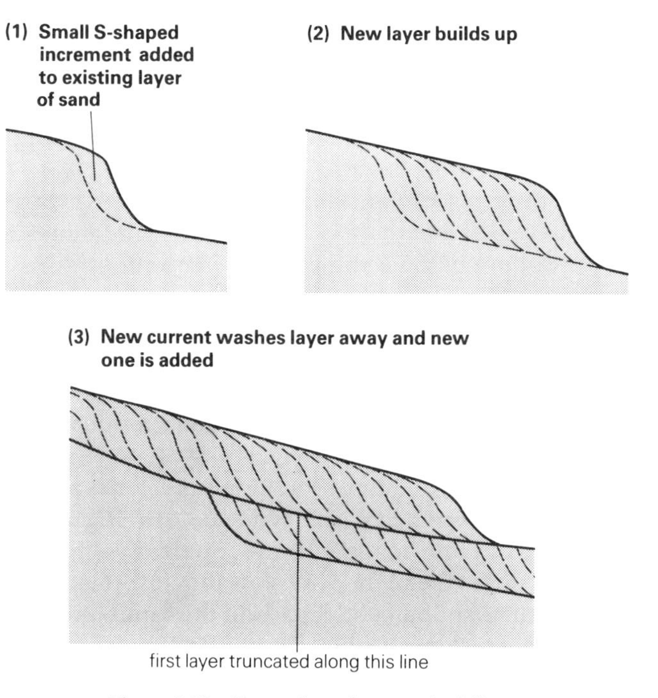

Figure 2.32 Formation of current bedding.

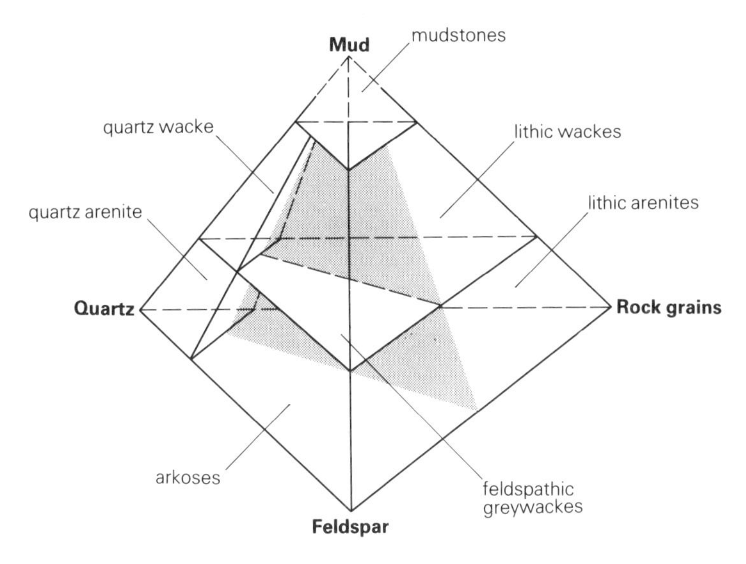

Figure 2.33 Classification of arenaceous sedimentary rocks based on their composition. Arenites and arkoses contain 0–15% mud, and wackes between 15 and 75% mud.

Relationships between the various arenaceous rocks and their four main components (quartz, feldspars, rock grains and mud) are given in Figure 2.33.

For the manufacturer of lime, limestone should contain at least 80% of calcium or magnesium carbonate. However, the geologist uses the term 'limestone' for any rock in which the carbonate fraction exceeds the non-carbonate constituents. If a carbonate rock contains more than 50% of sand-sized quartz grains, the rock is termed a **calcareous sandstone**; if it contains more than 50% clay, the rock would be termed a **calcareous shale**. Limestone is the term properly used for rocks in which the carbonate fraction, which must be greater than 50%, is composed primarily of calcite (or aragonite, which has the same composition as calcite but has a different structure), whereas rocks which are composed mainly of dolomite are termed dolomites. Limestones are formed in many ways. Some are fragmental or clastic and have been mechanically transported and deposited, whereas others are chemical or biochemical precipitates and formed *in situ*.

Classification of some limestones can be represented by means of a simple triangular diagram (Fig. 2.34; cf. Fig. 2.33) in which the four main mineral components (ooliths, shell and other animal remains, 'rock' or 'lithic' fragments including calcareous granules and fragments of calcitic mudstones, and carbonate muds) represent the apices. Calc-wackes are richer in carbonate muds than the other types. Bioclastic limestones include **shelly limestones**, and **oolites** are composed of ooliths, which are minute, layered spherules of calcite deposited around a nucleus of shell

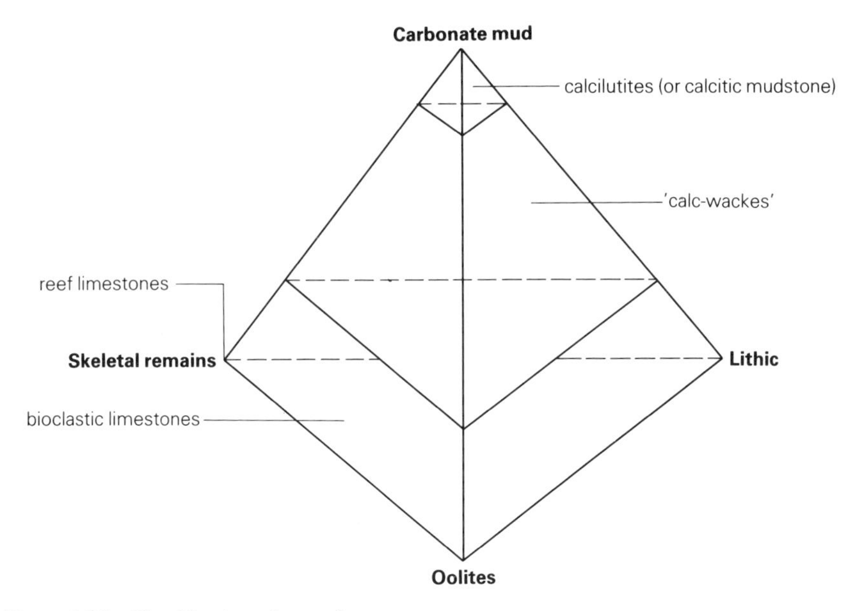

Figure 2.34 Classification of some limestone types.

or mineral grain. **Reef limestones**, which are framework limestones composed only of organic remains (corals, algae, etc.), are plotted at the 'skeletal remains' apex in Figure 2.34. **Marl** is a term used to describe some friable carbonate earths deposited in freshwater lakes. Figure 2.35 gives the classification of marls and their commercial uses. **Tufa** and

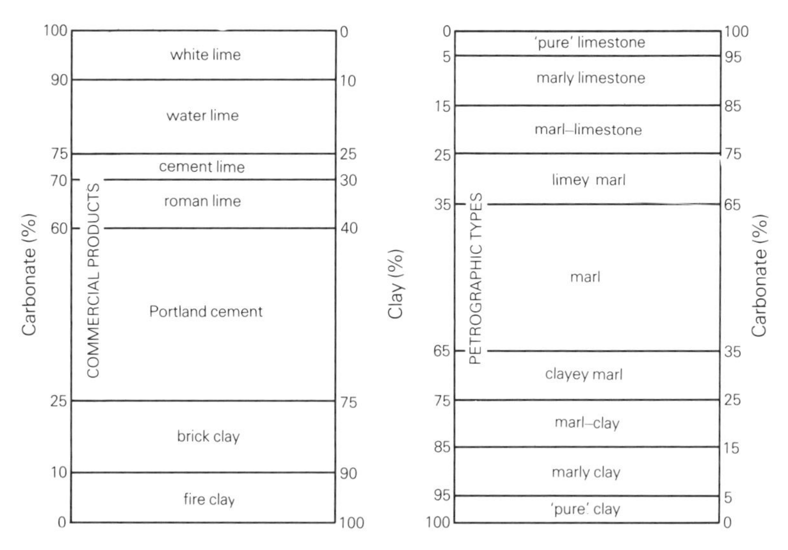

Figure 2.35 Geological and commercial classification of marls, i.e. clay–limestone mixtures.

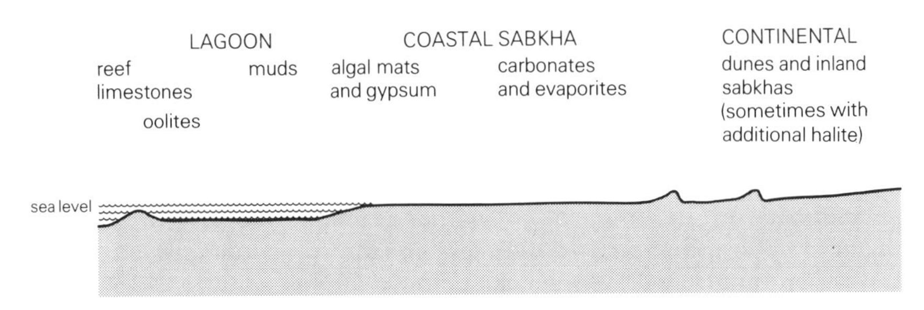

Figure 2.36 Schematic cross section across coastal desert region (about 50 km across).

travertine are limestones formed by evaporation of spring or river waters. Travertine is widely used as indoor cladding panels. **Chalk** is a friable porous calcium carbonate rock found in thick deposits which contain chert or flint nodules.

In coastal desert regions, particularly in the Middle East, broad, coastal salt-flats called sabkhas develop. Carbonate and sulphate minerals are deposited within the sedimentary layers of the sabkha. These chemical precipitates, or evaporites, are laid down in a cycle, with gypsum and anhydrite forming first, and then dolomite forming by reaction between brines and underlying carbonate sediments. A typical coastal cross section is shown in Figure 2.36. Such carbonate–sulphate (carbonate–anhydrite) cycles have been recognised in ancient rocks, e.g. the Middle Devonian Stetter Formation of western Canada, and the Upper Permian Bellerophon Formation of northern Italy.

The engineering properties of some sedimentary rocks are given in Table 2.18. (Note, however, that the data in the table should serve only as a rough guide, and the natural rock involved in any engineering scheme should always be tested.) The porosity, water absorption and uniaxial compressive strength of many sedimentary rocks are highly variable, depending upon how well the rock has been compacted or the grains cemented together.

Table 2.18 Engineering properties of some unweathered sedimentary rocks.

	Density (sp. gr.)	Porosity (%)	Water absorption (%)	Unconfined compressive strength ($MN\,m^{-2}$)
conglomerate	2.5–2.8	1–20	—	—
sandstone	1.9–2.6	5–25	<14.0	20–179
siltstone	2.2–2.5	2–24	<6.0	—
shale	2.0–2.4	10–35	<6.0	5–100
hard limestone	2.5–2.7	0–10	<2.0	60–200
soft limestone (e.g. chalk)	2.3–2.5	5–50	<4.0	20–50
dolomite	2.5–2.75	1–5	<1.5	30–200

2.2.5 Metamorphic rocks

If a rock is subjected to increased temperature or pressure, or both, to such a degree that it is altered by recrystallisation, then a new rock with a new texture and possibly a new mineral composition is produced. Rocks formed in this way belong to the third major category of rocks, the **metamorphic rocks**. The process of change is referred to as **metamorphism** of the original rock. In **dynamic metamorphism**, increased stress is the dominant agent, extra heat being relatively unimportant. It is characteristic of narrow belts of movement, where the rocks on one side are being displaced relative to those on the other. Whether the rocks are simply crushed, or whether there is some growth of new crystals, depends largely on the temperature in the mass affected by dynamic metamorphism.

In **thermal metamorphism**, increased temperature is the dominant agent producing change, and the degree of recrystallisation of the original rocks bears a simple relation to it. It is characteristic of the country rocks that lie at the margins of any large igneous intrusions and have been baked and altered by the hot magma. In large masses of granite, the changes extend outwards from the granite for distances of the order of 2 km. This zone of metamorphism is called the **thermal aureole** of the granite mass, and the type of new minerals formed at any part within it depends on the temperature attained and the distance from the granite. The changes that occur in clay or mud rocks within a thermal aureole are shown in Figure 2.37.

Temperature, load and directed pressure are important agents of **regional metamorphism**, which invariably affects wide areas rather than being related to an individual igneous mass or one zone of movement.

The degree of metamorphism is related to the conditions of temperature and pressure under which the new metamorphic rock has formed, and may be assessed by the appearance of certain new minerals. These **index minerals**, each of which indicates a particular temperature and

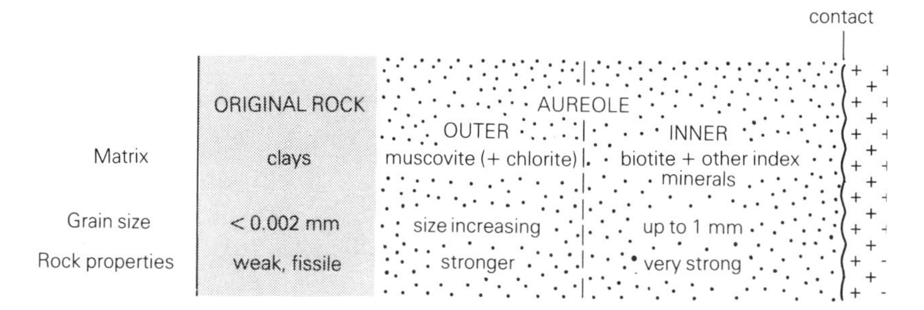

Figure 2.37 Schematic cross section across a thermal aureole in clay rocks. *Index minerals in the innermost aureole are not aligned but arranged in a random manner, and such rocks are called hornfelses, which are strong and make good engineering materials.

pressure at the time it formed, are used to define the **metamorphic grade** of the rock in which they occur. Their presence and interpretation are conditional on the composition of the original parental rocks as well as on the grade of metamorphism. Textural changes also occur as metamorphic grade increases. The progressive regional metamorphism of a clay rock is shown diagrammatically in Figure 2.38. As metamorphic grade increases, so does the grain size, from clay-sized particles (<0.02 mm) in the original clay rock and slate, up to about 1 mm in schists. Clays still occur in **slates** but are replaced by chlorite and muscovite in **phyllites** and biotites in **schists**. As metamorphic grade increases still further, index minerals (such as garnet) appear in the schist and the foliae become distorted around these minerals. At the highest metamorphic grades, the foliae become less distinct and often discontinuous. In these rocks the grain size is coarse (⩾1 mm) and the rock is called a **gneiss**. Regional (or thermal) metamorphism of a pure sandstone produces a metamorphic

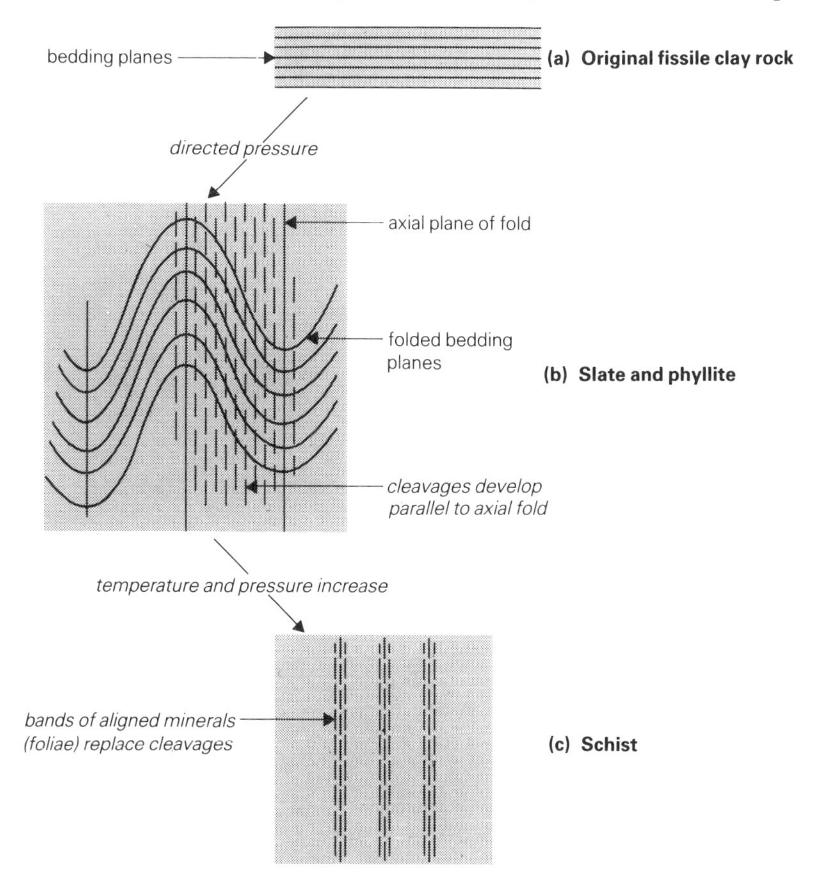

Figure 2.38 Regional metamorphism of a clay rock.

Table 2.19 Relationships between metamorphic grade, index minerals and parental rock types.

			Rock type undergoing regional metamorphism			
Grade	General rock name	Grain size	Mudstones or shales	Greywackes	Impure limestones	Basic igneous rocks
low	slate phyllite	fine	chlorite	chlorite	chlorite and Na-plagioclase	chlorite and Na-plagioclase
			biotite	biotite		
medium	schist	medium	Fe-garnet		Ca-garnet	Mg-garnet
				Na-plagioclase	hornblende	
			kyanite		Ca-plagioclase	Ca-plagioclase
high	gneiss	coarse	sillimanite		augite	
			minerals above represent index minerals			

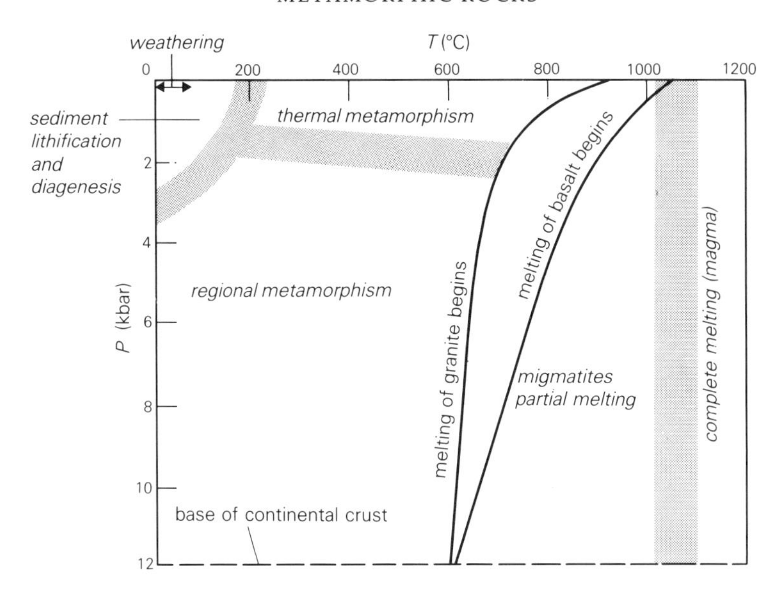

Figure 2.39 Temperature and pressure conditions of metamorphism. For comparison the temperatures and pressures of formation of igneous and sedimentary rocks are included.

quartzite, and of a limestone a **marble** (or a banded marble if the limestone is impure).

The relationships between index minerals, metamorphic grades and parental compositions are summarised in Table 2.19. Figure 2.39 shows temperature and pressure conditions of metamorphism, and relates these to formation temperatures (and pressures) of igneous and sedimentary rocks.

At extreme grades of regional metamorphism, partial melting of a rock may occur to give a metamorphic rock of variable texture in which bands of crystalline quartz and feldspar occur. These have been derived from melted material, leaving a finer grained matrix which is relatively rich in mica. The rock is called a **migmatite**. The layering formed by alternating bands of different texture is referred to as **metamorphic banding** or **foliation** (see Fig. 2.40). Other types of banding also occur in metamorphic rocks.

Since metamorphic rocks may be formed from any type of existing rock, their mineral composition ranges more widely than that of all other types of rock combined. They cannot be covered by any simple scheme of classification, but a simple textural classification is given in Table 2.20.

From Table 2.20, and from what has been written in this chapter, it will have become evident that most metamorphic rocks are strongly **anisotropic** – having different properties in different directions. Thus, for example, slate is very strong in compression with the cleavages perpendicular (or at a high angle) to the direction of compression, and much

Figure 2.40 Foliation in coarse-grained gneiss.

weaker when compressed in a direction parallel to the cleavages. All other foliated rocks behave in a similar fashion; thus the range of values for some tests may be very great. Some common engineering properties for metamorphic rocks are presented in Table 2.21.

Table 2.20 Textural classification of metamorphic rocks.

Texture	Rock type	Grain size (etc.)
granular or equidimensional	hornfels	fine to medium grained; index minerals in random arrangements
	quartzite	fine to medium grained
	marble	medium to coarse grained
foliated	slate, phyllite	fine grained; well defined cleavages
	schist	medium grained
indistinct foliation	gneiss	medium to coarse grained
banded	migmatite	compositional banding

Table 2.21 Engineering properties of some common metamorphic rocks.

	Density (sp. gr.)	Porosity (%)	Water absorption (%)	Unconfined compressive strength ($MN\ m^{-2}$)
slate	2.6–2.8	0.1–0.5	<0.5	70–200
schist	2.6–2.8	0.1–1.5	<1.5	50–150
gneiss	2.7–3.0	0.5–1.5	<1.0	50–200
quartzite	2.6–2.7	0.1–0.5	<0.5	150–300
marble	2.4–2.7	0.5–3.0	<1.0	70–150

References and selected reading

British Standards Institution 1973. *Aggregate from natural sources for concrete.* BS882, 1201 Part 2: 1973.
British Standards Institution 1974. *Methods of testing soils for civil engineering purposes.* BS 1377: 1974.
British Standards Institution 1975. *Methods of sampling and testing of mineral aggregates, sands and fillers.* BS 812 Part 1: 1975.
Military Engineering 1976. *Applied geology for engineers,* vol. 15. London: HMSO.
British Standards Institution 1981. *Cole of Practice for site investigations.* BS 5930: 1981 (formerly CP 2001).
Gillott, J. E. 1968. *Clay in engineering geology.* Amsterdam: Elsevier.
Ministry of Defence and the Institution of Civil Engineers 1976. *Applied geology for engineers.* London: HMSO.
Pettijohn, F. J. 1975. *Sedimentary rocks.* 3rd edn. New York: Harper & Row.
Press, F. and R. Siever 1982. *Earth,* 2nd edn. San Francisco: W. H. Freeman.
Read H. H. 1970. *Rutley's elements of mineralogy.* 26th edn. London: George Allen & Unwin.

3 *Superficial deposits*

3.1 Soils

The term **regolith** is used to include the mantle of surface deposits and soils overlying the bedrock. The regolith may vary in thickness from less than 1 m to over 30 m; its density, usually between 1.5 and 2, is less than that of rock.

Geologists and engineers give different meanings to the word 'soil'. Geologists use the term to refer to any rock waste, produced by the disintegration of rocks at the surface by weathering processes, which has formed *in situ*. These untransported surface deposits are called **sedentary** or **residual deposits** (Fig. 3.1).

In contrast, engineers use the term 'soil' more widely and more loosely, to describe any superficial or surficial deposit which can be excavated without blasting. This definition covers **transported sediments** as well as residual deposits (Fig. 3.2). Thus, engineers would regard 'soil' as including water-transported sediments (**alluvium**), wind-transported material (**dunes** and **loess**), sediment transported by glaciers or their meltwaters (**till** or **glacial drift**) and material moved downhill by gravity (**colluvium**), and these will be discussed in this chapter. Readers are also referred to the discussion of sedimentary rocks in Section 2.2.4. Some rocks (such as London Clay) may even be thought of as 'soil' by the engineer, since these can be easily excavated.

3.2 Weathering

3.2.1 Introduction

Three processes of weathering and erosion at, and near, the surface transform solid rock into unconsolidated rock waste (the regolith). These are as follows:

(a) the mechanical disintegration of a rock mass at the surface as water, wind, ice and the rock fragments carried by them buffet or press against it, or force it apart;
(b) chemical reactions between the original minerals of the rock, the near-surface water and the oxygen of the atmosphere to produce new minerals which are stable under the conditions at the Earth's surface and remove other more soluble constituents;

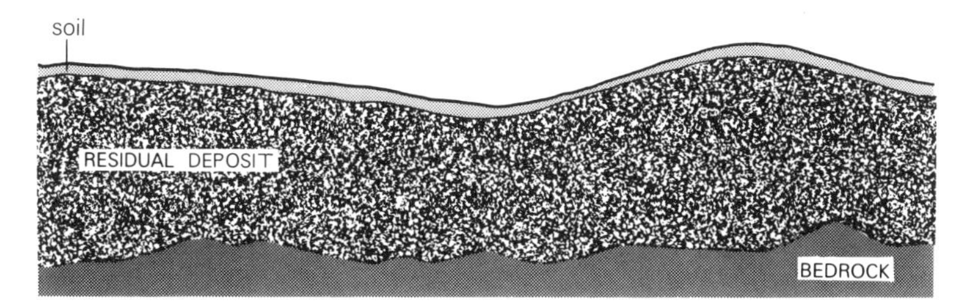

Figure 3.1 Recent layer of soil developed from, and lying above, an ancient 'soil' or residual deposit which has developed *in situ* from the underlying bedrock. The residual deposit may be tens of metres thick.

(c) biological activity, which produces organic acids, thus adding to the chemical reactions, and which may also be an agent assisting mechanical disintegration.

The effectiveness of these processes in destroying the rock is dependent on its constituent minerals and its texture. For example, a mineral property such as cleavage influences how readily water and air can enter the mineral grain and increase the rate of any reaction. The well developed cleavages of feldspar allow the change to clay minerals to take place within, as well as around, the mineral grain (see Section 2.1.3). In contrast, quartz possesses no cleavages, is chemically inert and is a particularly stable mineral. The principal effect of weathering and erosion on it is to reduce the size of each quartz grain by abrasion, though the hardness of quartz helps to preserve it from this type of attack as well.

Large cracks and fissures in the rock mass also facilitate the entry of water and air, and so assist weathering (Figs 3.3 a & b). The more fissures and other rock discontinuities that the rock possesses (Appendix H), the greater is the surface area exposed to chemical reaction, and hence the faster does weathering take place. Permeability also facilitates weathering. Figures 3.4a and b illustrate two permeable rock structures in which weathering will occur.

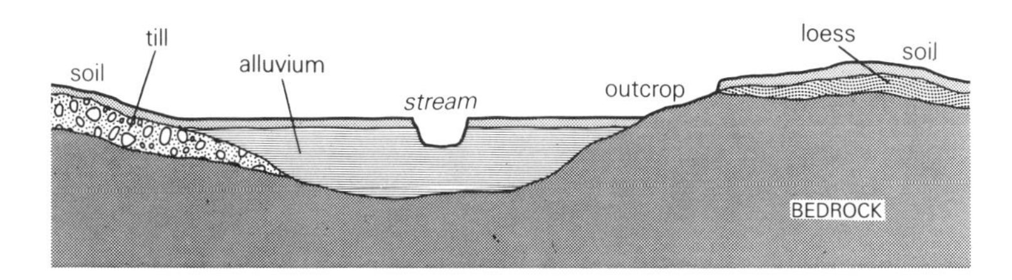

Figure 3.2 Thin (<1 m) soils developed on top of transported deposits, which include alluvium, glacial drift or till, and wind-deposited loess. Bedrock crops out on hillside. Even in this situation, residual deposits could be found occurring immediately above the bedrock and below the transported deposits.

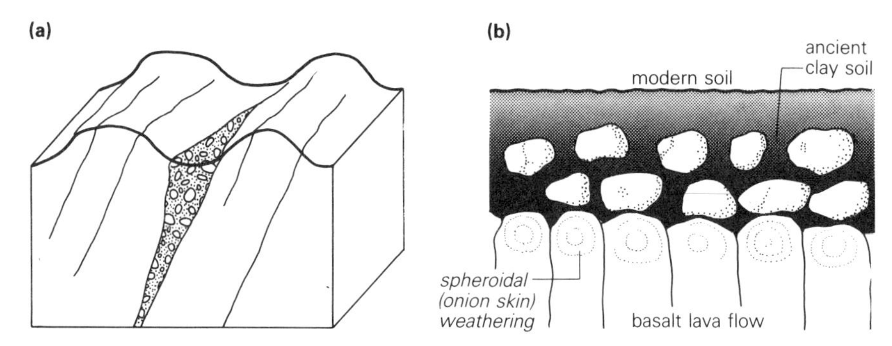

Figure 3.3 Rock structures assisting weathering. (a) Two joint planes (sets) in granite mass providing access for water to weather rock mass. Superficial deposits collect in open cracks. Granite forms rounded columnar masses. (b) Vertically jointed basalt lavas weather to form clay residual deposits. The basalt weathers in a spheroidal fashion; rounded masses separate and weather severely, becoming smaller upwards (away from the coherent rock mass).

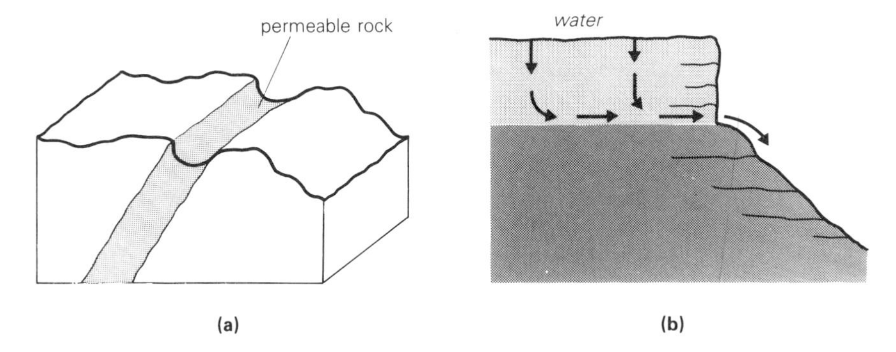

Figure 3.4 Permeable rock layers assisting weathering. (a) In steeply dipping rocks, permeable layers allow water to ingress and these layers weather more deeply than others. (b) In horizontally bedded rocks, the base of permeable beds can weather as a result of water seepage (shown by arrows).

3.2.2 *Mechanical weathering*

Mechanical weathering leads to a physical disaggregation of the original rock mass into smaller particles. This can be caused by any one of several natural agents. For example, freezing of water within a crack produces an expanded wedge of ice which forces the walls of the crack apart. If the process is repeated by alternate thawing and freezing, fragments from the outer surface of the rock eventually break off to form loose scree. The same mechanical effect may be produced locally in the rock by chemical reactions between certain minerals and water that has penetrated along cracks. The hydration of these minerals produces a local increase in volume, and local pressure causes disintegration of the rock. Similarly, entry of water into the minute void spaces in rocks may allow salts to crystallise there and press against the walls of the void, thus weakening the rock.

Other processes involving the action of water in its various forms include erosion by ice, wave action at coasts, river erosion, and slopes being made unstable by the presence of water in the ground. Some of these agents of mechanical weathering are discussed later.

Expansion and contraction of the outer skin of a rock mass as it was heated by the sun and cooled at night were formerly thought to be an important agent of weathering, but careful modern studies show that this process (called **exfoliation**) requires water to be present for it to work.

3.2.3 Chemical weathering

The process of chemical weathering involves the simultaneous operation within one rock of three things:

(a) the reaction of the original minerals of the rock with water, oxygen and organic acids at the Earth's surface so that they are broken down chemically with the production and release of new products, some of which are soluble;
(b) the dissolving of these soluble products by water present in cracks in the rock and their eventual removal from it;
(c) the continued reaction of the insoluble rock constituents with water, carbon dioxide, oxygen and other atmospheric gases to form stable new assemblages of minerals.

The principal reactions of the last type are called hydration, carbonation, oxidation and reduction. **Hydration** is a reaction in which water combines with a rock constituent producing a mineral that has hydroxyl groups (OH) in its structure, the hydroxyl group coming from the water. **Carbonation** is a reaction involving carbonic acid and limestone. Carbonic acid is formed when carbon dioxide from the atmosphere dissolves in water. It can then react with limestone (calcite) to produce soluble calcium or sodium bicarbonate. In the process of **oxidation**, oxygen combines with a rock constituent, and in two typical reactions produces new iron-bearing minerals:

magnetite + atmospheric oxygen = haematite
magnetite + atmospheric oxygen + water = hydrated haematite (goethite or limonite)

Reduction takes place in environments deficient in oxygen, and the products of such a reaction contain relatively little oxygen. For example, in oxidising conditions organic matter is converted to carbon dioxide (CO_2), but in reducing conditions methane (CH_4) is often formed.

An important factor governing the type of chemical reaction that will take place in weathering is the **pH value** of the local environment within the rock. The pH is related inversely to the concentration of hydrogen ions present in any water. A neutral environment has a pH of 7.0, whereas a pH of 6.0, with a hydrogen ion concentration ten times as great, is acidic. A pH of 8.0 has one-tenth the concentration of hydrogen ions as pH 7.0, and is alkaline.

The ease with which pH value may be changed depends on the chemical properties of the rock or soil, and particularly on the availability of hydrogen ions from chemical reactions. Thus a sandy soil and a clay soil with the same initial pH value react very differently when their pH values are increased by adding lime. The clay soil requires more than twice the amount of lime to effect a given change of pH compared with the sandy soil. Rainfall decreases the pH value of a soil as metallic ions are leached out and replaced by hydrogen ions. In regions of high annual rainfall, soils are usually acidic. A change of pH as rainfall leaches a soil may trigger further reactions within it, as the first clay minerals formed become unstable with decreasing pH. Chemical weathering of common rock-forming minerals to produce clay minerals has been discussed fully in Chapter 2, Section 2.1.3.

Non-silicate minerals are minor constituents of most rocks. Sulphates, nitrates, chlorides and carbonates are highly soluble, though the calcium-magnesium carbonate (dolomite) is much less soluble than calcite. Most oxides are unstable and react with water to produce hydrated minerals, but titanium oxides are stable and occur in soils in very small amounts. Sulphides are unstable and may give rise to small quantities of sulphuric acid.

3.2.4 *Biological weathering*

Organic matter in soils is broken down by micro-organisms to give water and either carbon dioxide or methane and small quantities of ammonia and nitric acid. Carbon dioxide is produced by aerobic micro-organisms, which require oxygen to survive. Methane is produced by anaerobic micro-organisms which flourish in environments that are deficient in oxygen, referred to as **reducing environments**. Decay and decomposition take place much more quickly where oxygen is present. An excessive accumulation of decaying organic matter will consume all the available oxygen and produce a reducing environment, for example in a poorly drained area such as a bog or in lakes where circulation is poor. In these circumstances sulphides form and pyrite is usually present.

Lichens are combinations of fungi and algae and contribute to weathering of rocks, although the precise role that they play in this is uncertain.

3.2.5 *Role of vegetation*

The macroflora of a region reflects its climatic conditions. Trees and bushes in particular may effect the physical conditions of rocks by splitting them after roots have penetrated along cracks. The chemical conditions of the ground may be affected both by extraction of chemical substances by roots and later by the addition of different types of organic matter when plants decay.

Geological conditions may be revealed by plants. Some species tend to grow in ground that is predominantly acid (for example, the pines and heathers that occur throughout the Scottish Highlands), whereas other types of vegetation require a more alkaline ground. In the Scottish Highlands unexposed limestone outcrops can often be predicted by the presence of grass growing on the overlying soil cover.

3.2.6 *Influence of climate*

The processes of weathering and erosion depend very much on climate, and there is a correlation between soil type and local climatic effects, such as ground temperature. Freezing slows down some processes of soil formation; thus arctic soils are poorly developed. In contrast, the process of soil formation may penetrate to great depths in the tropics. The mean temperature at the surface over periods of years rather than the daily or even annual variations is the factor that controls ground temperature below the top few metres. Rocks and soils are very poor conductors of heat, and annual fluctuations cancel out at depths of about 3 m. Daily variations have no effect below the top 1 m.

The relative importance of reactions involving water depends both on rainfall and on the temperature of the water. The reaction rate doubles for each 10°C rise. The effects of rainfall are demonstrated simply by describing two very different environments. In desert regions, which have very low rainfall, water on and in the rocks and soils is evaporated during most days. It is not available to leach soluble metallic ions from the rock and replace them with hydrogen ions. This, plus an abundance of oxygen in contact with dry mineral grains, as well as a lack of organic material, produces high pH values of 7.5 to 9.5. Reactions are slowed, and iron from the rocks is changed to the oxygen-rich ferric form. Montmorillonite, illite and chlorite are other new products that form.

In contrast, tropical rainforests have heavy rainfall, which in turn yields a profusion of vegetable matter and saturation of all cracks and other voids in the rocks and soils to within a short distance of the surface. Soil formation takes place in a reducing environment in which metallic ions are leached out and reactions are fast; pH values are typically between 3.5 and 5.5. The rapid, thorough weathering results in an accumulation of aluminium- and titanium-rich deposits, such as bauxite.

In limestone country, closed depressions are sometimes seen. These form as a result of the solution of limestone by meteoric waters (Section 5.2.1) that carve out underground caverns, the roofs of which collapse. This landscape feature is known as **karst topography**.

3.3 Modern residual soils

3.3.1 Soil development and engineering grades of weathering

Most modern soils are developed on top of other superficial deposits, which are either transported or residual. In the latter case, the present-day soil is developed from an older one. Only occasionally is the development of soils taking place immediately on top of solid bedrock, with a continuous gradation into it. A profile through a modern soil typically shows, from top to bottom, plant debris grading downwards into decaying organic matter, which in turn grades into the weathered parental material or **saprolite**.

Figure 3.5 shows a typical residual deposit developed by weathering *in situ* above bedrock. The bedrock shows a gradual change upwards from fresh rock into weathered rock showing discolouration, and then into more severely weathered material, which increases in amount until all the rock mass is decomposed. The rock mass may decompose to give either a continuous framework, in which structures originally occurring in the rock mass appear as ghost structures, or corestones, which gradually decompose nearer the surface.

To describe the degree of weathering of the rock mass from which an overlying residual soil has formed, engineers have devised a descriptive scheme for grading the progressive changes. This is shown in Table 3.1. All grades of weathering may be visible in a large rock mass, with variation in grade related to the relative ease with which water has penetrated at different points. This is likely to be determined by the presence of open cracks in the rock mass.

In Figure 3.5, the grade of weathering of the rock mass has been indicated at various levels in the residual deposit. Note that the thicknesses of the zones in the residual deposits shown in Figure 3.5 are highly variable. Zone II may vary up to 5 m but discolouration may extend along joints and cracks deep into the rock mass. Zone III is usually between 2 and 6 m thick and zone IV can be many tens of metres thick. Zone V is variable and zone VI usually thin, 2 m or less.

3.3.2 Soil description

The decomposition of the organic matter produces a variety of acids, the most important of which is carbonic acid. These acids move downwards

Table 3.1 Descriptive scheme for grading the degree of weathering of a rock mass.

Term	Description	Grade
fresh	no visible sign of rock material weathering; perhaps slight discoloration on major discontinuity surfaces	I
slightly weathered	discoloration indicates weathering of rock material and discontinuity surfaces; all the rock material may be discoloured by weathering	II
moderately weathered	less than half of the rock material is decomposed and/or disintegrated to a soil; fresh or discoloured rock is present either as a continuous framework or as corestones	III
highly weathered	more than half of the rock material is decomposed and/or disintegrated to a soil; fresh or discoloured rock is present either as a discontinuous framework or as corestones	IV
completely weathered	all rock material is decomposed and/or disintegrated to soil; the original mass structure is still largely intact	V
residual soil	all rock material is converted to soil; the mass structure and material fabric are destroyed; there is a large change in volume, but the soil has not been significantly transported	VI

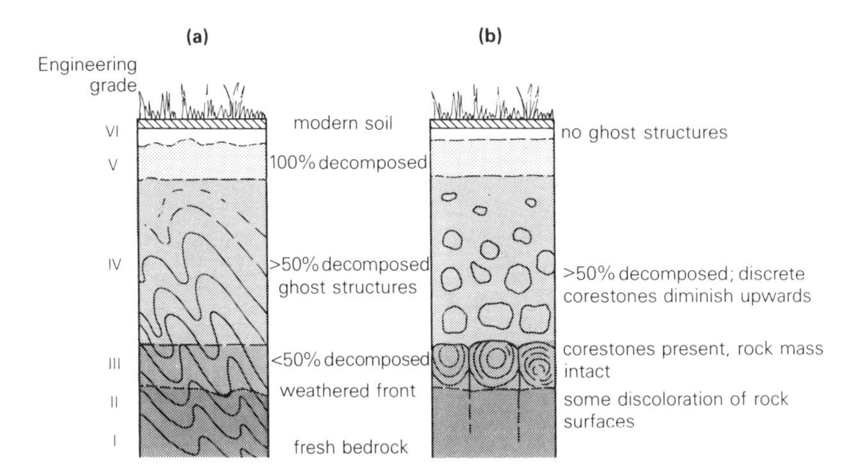

Figure 3.5 Two typical residual deposits overlying bedrock. (a) Continuous framework with ghost structures (a structured saprolite) appearing in the decomposed rock, with total decomposition (a massive saprolite) occurring just below the modern soil layer. (b) Corestone development from the original bedrock, with corestones decreasing upwards until total decomposition again occurs just below the modern soil layer.

and react with existing minerals at lower levels in the soil, producing soluble components which in turn move downwards. The soil layer in which these reactions occur and from which soluble compounds are removed is called the **leached** (or eluviated) **layer**, or the **A horizon**. The soluble components are deposited below it in the **illuviated layer** (the **B horizon**). These uppermost three layers (the organic debris, the A horizon and the B horizon) comprise the **solum**. Below it there is minimal solution and redeposition in the **C** and **R horizons**; C consists of weathered and R of fresh parental material (Fig. 3.6).

In humid regions, acidic soils called **pedalfers** occur. Aluminium and iron are removed from the leached layer and precipitated as oxide or hydrated oxide at greater depths, but usually within 0.5 m of the surface. This process of solution and precipitation is called **podzolisation**, and the soils are called **podzols**. The original alkali elements (sodium, potassium, barium and calcium) go into solution and are eventually carried away from the soil by drainage.

Only the upper layer of the soil in arid regions receives any moisture, and there is no seepage through it. Organic material and carbon dioxide are also scarce and, as a consequence, carbonates are precipitated within the soil. Alkaline soils of this type with carbonate layers are called **pedocals**. They are closed chemical systems from which nothing is removed, unlike pedalfers, which are open systems.

In arid regions where soils are enriched in lime, a layer of **caliche** may be formed. Such a deposit is formed on a peneplain in a climate that leads to sharply defined alternations of saturation and desiccation. Caliche is a 'hard cap' deposit produced by upward capillary migration of ground

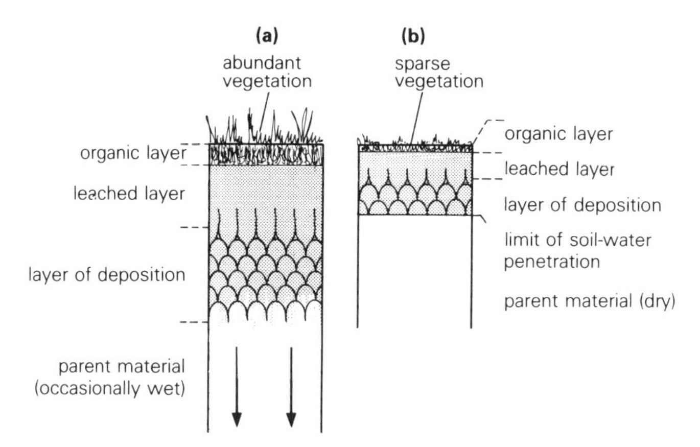

Figure 3.6 Profiles of two major kinds of soil. (a) Acid soils from which soluble constituents are removed by water draining from the bottom of the profile, characteristic of humid regions. (b) Alkaline soils from which water is insufficient to drain all the way through the ground, characteristic of arid or semi-arid regions.

waters during the arid period. Depending upon the composition of the underlying rock, and therefore the composition of these migrating solutions, precipitates of carbonate (**calcrete**), siliceous (**silcrete**) or ferruginous (**ferricrete**) materials may be deposited. The position of deposition of different chemical layers depends upon the solubility of the constituents of the migrating solutions, the most soluble salts being deposited nearest the surface (although rainwater action may produce a reversed sequence just below the surface).

3.3.3 *Soil terminology*

The boundaries between individual soil layers may be sharp or gradual. The terms used are given in Table 3.2. The layers are designated by the use of capital letters, with lower-case letters to identify subdivisions or to denote the special properties of a particular layer. For example, Bsa signifies an accumulation of soluble salts in the B layer, and BA_z indicates more precisely the layer of horizon B in which sulphate accumulation occurs. The symbols and their definitions are shown in Table 3.3.

3.3.4 *Classification of residual soils*

Most modern soils develop profiles related to the climatic and vegetational zones in which they occur, and for this reason they are referred to as **zonal soils**. In contrast, **azonal soils** have poorly defined profiles. **Intrazonal soils** are produced by unusual local climatic or geological conditions, and examples are **hydromorphic soils**, formed where there is an excess of water which saturates rock and soil, **halomorphic soils**, where there is an excess of alkalis because salt water is present or there has been much evaporation, and **calcimorphic soils**, where excess lime is present. Intrazonal soils also include gleys, bog soils, salt crusts and brown forest soils. For example, gleys occur where there are fluctuations in the level of water saturation in the soil. A classification of residual soils is given in Table 3.4 and descriptions of the important soil groups are given in Appendix A.

Table 3.2 Descriptive scheme for boundary widths between layers of soil.

Term	Width of boundary (mm)
abrupt	<25
clear	25–62.5
gradual	62.5–125
diffuse	>125

Table 3.3 (a) Designation of layers of soils by capital letters, with numbers to designate gradational layer (from Olsen 1984). (b) Letters used to denote special properties of a layer of soil. Subscripts not given in the table include f (frozen soil), which is applicable in tundra soils with permafrost; m (hard pan), where an indurated layer occurs, and b, which indicates a soil layer buried by a surface deposit.

(a) Layers

Symbol	Alternatives			Definition
solum	O	O	A horizon	organic layer
	A	A_1		organic-rich A layer
	A	A_2		layer of maximum leaching
	AB or EB	A_3		A layer gradational with B layer
	BA or BE	B_1	B horizon	B layer gradational with A layer
	B or BW	B_2		layer of maximum deposition
	BC or CB	B_3		B layer gradational with C layer
	C			weathered parental material
	R	D		parental material (substratum)

(b) Properties

Letters	Alternatives	Definition (all subscripts imply accumulations)
s	ir	sesquioxides (illuvial)
o		sesquioxides (residual)
t		clays
y	cs	calcium sulphate or gypsum
k	ca	calcium carbonate
c	cn	concretions or nodules
h		organic matter or humus
z	sa	soluble salts
g		gley (organic rich layer, usually water saturated and in a reduced condition)

Variations in physical conditions on a local scale may produce corresponding changes in the soils being formed, even though the parental material is identical. This relationship is called a soil **catena**. For example, soil on the side of a hill receives moisture from higher up the hill, whereas soil at the bottom of the slope receives additional moisture from a larger gathering area plus some dissolved salts from the same source.

3.4 Transported superficial deposits

3.4.1 Introduction

Transported superficial deposits are formed of rock debris which has

Table 3.4 Residual soil classification.

Order	Suborder	Great soil group
zonal soils	(1) soils of the cold zone	tundra soils
	(2) soils of arid regions	sierozem brown soils reddish-brown soils desert soils red desert soils
	(3) soil of semi-arid, subhumid and humid grasslands	chestnut soils reddish-chestnut soils chernozem soils prairie or brunizem soils reddish prairie soils
	(4) soils of the forest–grassland transition	
	(5) podzolised soils of the timbered regions	podzol soils grey podzol soils brown podzol soils sol brun acide red–yellow podzol soils
	(6) lateritic soils of forested warm–temperate and tropical regions	reddish-brown lateritic soils laterite soils
intrazonal soils	(1) halomorphic (saline and alkali) soils of imperfectly drained arid regions and littoral deposits	solonchak, or saline soils solonetz soils (partly leached solonchak) soloth soils
	(2) hydromorphic soils of marshes, swamps, seep areas and flats	humic gley soils alpine meadow soils bog and half-bog soils planosols groundwater podzol soils groundwater laterite soils
	(3) calcimorphic soils	brown forest soils rendzina soils
azonal soils		lithosols regosols alluvial soils

been carried by some natural agent from where it was formed by weathering and erosion to where it now occurs. They are 'soils' in the engineering sense but not in the geological usage of the term. Rock decay with little or no transport of the products is termed weathering; and when the rock or its products is simultaneously removed, this is termed erosion. The effects of burial and other processes will eventually transform these products into layers of sedimentary rock. For that reason, some of them (marine clays and sands, beach sands, estuarine and deltaic muds, lagoonal clays and lake deposits) are mentioned in Section 2.2.4, which discusses the formation of rocks. Transported superficial deposits are described and classified most simply by reference to the agent that moved them and the local conditions that controlled their deposition. Four important agents of erosion and transport (wind, rivers, the sea at coasts, and glaciers) and the deposits associated with them are described in the following sections.

3.4.2 *Aeolian (windborne) deposits*

A strong wind blowing across rock debris or soil can lift and carry fine material as dust, and can move the larger sand grains by rolling them and making them bounce across the surface. This windborne movement of material occurs in areas with little or no vegetation and is typical of hot desert regions, although the process also operates in cold deserts and some coastal areas. Wind both transports and sorts the material. The finer, silt-sized fraction is carried in suspension by the wind, and may travel great distances before it is eventually deposited as loess. The coarser material that remains forms sand dunes. Both loess and sand dunes are liable to further erosion by wind unless their surface is stabilised by vegetation or another binding agent. They are composed mainly of quartz with a smaller fraction of other stable minerals such as iron oxides. Near coasts calcite in the form of shell debris may occur occasionally. Clay minerals are virtually absent.

Erosion and transport by wind are most important in desert regions, which lack a protective cover of vegetation and a skin of surface water to bind the grains together. Conditions favouring the formation of aeolian deposits were, however, more widespread in the present temperate zones during and immediately after the Great Ice Age, which ended approximately 10000 years ago (see Section 3.5.1). For example, lowered sea levels at certain times left broad stretches of beach sand exposed to wind action, and assisted the formation of sand dunes along many British coasts. In other regions that were free, or freed, of ice and temporarily devoid of plants, the deposits left by the retreating ice, together with other soils, provided sources of loess.

3.4.3 *Alluvial (riverborne) deposits*

Lateral erosion by a river into its banks eventually produces a valley which is much wider than its course. Erosion at one point is matched by deposition at another and a wide valley is eventually filled with **alluvial deposits**. These include cross-bedded and evenly bedded sands, silts and gravels, plus spreads of fine silt and clay across the flat **floodplain** at the sides of a river (Fig. 3.7). These fine-grained layers are related to periods of spate, when flood waters spilling beyond the river channel suffer a sudden decrease in their rate of flow. Settling of the load of suspended sediment takes place in the slack water covering the floodplain, especially close to the banks. Large mounds form along the sides of a major river in this way, and are called **levees**.

Any sudden decrease in the rate of flow of a sediment-laden river affects its ability to transport sediment, and part of its load settles. This happens commonly when the river enters the sea or the still waters of a lake. The result is a shallow (or even emerged) area in the lake, fanning out from the mouth of the river, called a **delta**. Similarly, when an upland stream flowing down steep slopes enters more level country, deposition, usually of coarser detritus, takes place. At the change of slope, large

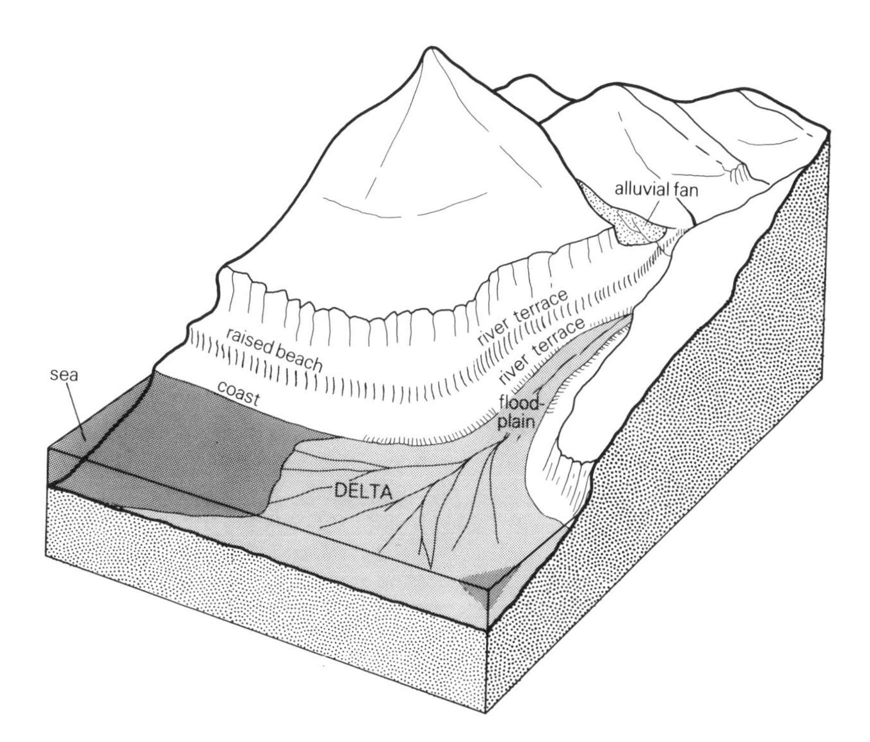

Figure 3.7 The mountain stream has deposited the coarser part of its load as an alluvial fan where the gradient changes suddenly at the plain. Deposition of finer sediment has taken place at the coast to form a delta. As the river winds towards the coast, it cuts a broad floodplain. The river terrace is an older floodplain, which is now well above the river bed.

stones settle near the apex of a fan-shaped mound, within which gravel and finer deposits are present down stream. The stream follows a channel cut through its own deposits. The magnitude, proportions and composition of the **alluvial fan** are dependent on the change of slope and the size of stream (see Fig. 3.8). A small stream a metre or two wide when not in spate may build an alluvial fan up to 2–3 m thick and about 100 m wide.

The **discharge of a river** (Q) can be estimated by the formula $Q = 750M^{0.67}$, where Q is discharge in *cusecs* (that is cubic feet per second) and

Figure 3.8 An alluvial cone (or fan) formed where the mountain stream descends into a flat-lying area in the Scottish Highlands. (British Geological Survey photograph, B773, published by permission of the Director; NERC copyright.)

M is the catchment area of the river in square miles. In the USA another formula (the Myers Formula) is used: Q (peak discharge in cusecs) $= qa = 10\,000\,(a)^{\frac{1}{2}}$, where a is catchment area in square miles. For a maximum flood, $Q_{max} = 100\,P(a)^{\frac{1}{2}}$, where P is a Myers rating for rivers of different catchment areas. In 100 rivers in the USA, P varied from 0.14 to 289. Catchment areas of less than 1000 square miles have P values of less than 50. In general, British rivers have a P rating of about 16.

If the river were to cut deeply into its bed (for example, in the process of adjusting to a lowered sea level), then the floodplain is left high and dry above the river as a **river terrace**. Each time the base level of the river changes, a new set of terraces may be formed. The oldest would be perched highest above the river if sea level were lowered progressively. The deposits of a river terrace are generally coarser than the neighbouring alluvium, since maturity of the river decreases the rate of erosion and the grade size carried by it at a particular locality. The difference in height between the flood plain and a terrace is typically a few metres, but may be as much as a kilometre in regions of rapid uplift such as the Andes.

3.4.4 *Erosion and deposition at the coast*

The sea erodes by the pounding action of waves on the base of cliffs, and also by compressing air present in crevices and so causing further explosive impact within them. The cutting action of its currents as they sweep across the foreshore is aided by the rubbing of pebbles dragged along by the moving water. Limestone cliffs may also suffer some erosion by solution of the rock in water. As a result of these agents, a notch is cut at the foot of cliffs (Figs 3.9 & 10) undermining them and causing them to collapse and retreat from the sea. This leaves a **wave-cut platform** between the cliffs and low-water mark. Where the cliffs are formed of poorly consolidated rock and soils, erosion may proceed at rates that average 1–2 m per year. This erosion usually takes the form of sporadic and alarming inroads by cliff falls after storms. For example, in 1953 the coincidence of a few adverse factors produced a disastrous surge of sea level in the southern North Sea to heights of $2\frac{1}{2}$ m above normal, and in the course of two hours some low cliffs in eastern England were cut back by 10 m. Along such coastlines, waves generated by wind action are more important than 'tidal waves'. The size of waves generated in the open sea is dependent upon fetch, which is the distance over which the wind operates. The **Stevenson Formula**, $H = 1.5\,(f)^{\frac{1}{2}}$, is used to calculate wave size, where H is trough-to-crest wave height in feet and f is fetch in miles. For example, the northeastern coast of England has a maximum fetch of some 300 miles to Denmark, thus the maximum value of H would be $1.5 \times \sqrt{300} = 26$ ft. Wind speed is obviously an important factor and the reader is referred to Bretschneider (1952), Darbyshire and Draper (1963)

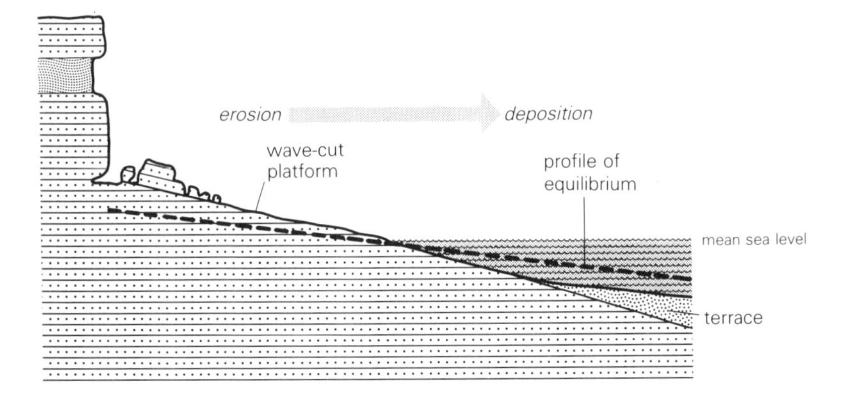

Figure 3.9 A profile drawn across a shoreline where marine erosion has cut a notch into the base of the cliffs and a rock platform in front of them. Debris from the cliffs and other material above the profile of equilibrium (broken line) are carried seawards and form a terrace off shore. Erosion and deposition tend to grade the slope of the beach to the ideal profile of equilibrium. The vertical scale is exaggerated in the order of ×10 for clarity.

Figure 3.10 A rock platform cut by waves into well bedded sandstones. Note the more intense erosion at the base of the cliffs. (British Geological Survey photograph, C1611, published by permission of the Director; NERC copyright).

and Wood (1969) for detailed information on this topic. Where the toes of cliffs are sited in deep water, these cliffs will have to withstand waves of Stevenson Formula height.

The debris from erosion is carried seawards across the platform to be added to a wave-built terrace of sediment below low-water mark. The platform and the terrace together produce an increasingly wide expanse of shallow, shelving water. This must be crossed by waves before they attack the cliffs, and their erosive sting is decreased progressively. With time, the shore profile (of platform and terrace) tends towards an equilibrium slope for the given sea level and local conditions, and seaward and landward transport of sediment across it are roughly in balance over a period of about a year. Winter storms tend to sweep beach sediment seawards, and beaches are more slowly restored in the summer. A beach may, however, take a few years to recover from the effects of one catastrophic storm. This movement of sediment can be investigated using radioactive tracers, or by means of dyed pebbles combined with skindiving.

A change in the profile of equilibrium, or interference with the normal movement of sediment, produces compensatory erosion and deposition. If the slope of the foreshore were to become more gentle, there would be some erosion of the terrace as sand and shingle are carried landwards on to the beach. Part of this sand may be added temporarily to dunes, and later returned to the sea by offshore winds. Any loss of sediment from the offshore terrace and consequent steepening of the profile can increase beach and cliff erosion dramatically. For example, dredging of the terrace sands at one site may produce relatively deeper water in shore (Fig. 3.11), and as a consequence waves are concentrated by refraction towards adjacent parts of the shore. The breakers have greater force and produce stronger backwash than at other parts of the same coast. As a result, the erosion of beach and cliffs and the movement of debris seawards to regrade the shore profile progress rapidly. The importance of dredged gravels as a source of aggregate in the UK (Section 7.2.4) has meant that changes to the coast are being induced where these dredging operations are occurring, such as off the south-east coast of England. In the 1960s, between 5 and 7% of the total sand and gravel production was marine dredged, whereas in the 1970s and 1980s, the percentage was between 11 and 16% (in 1980, 16%).

So far, movement of beach material has been described in terms of a simple two-dimensional model of a shore, where the striving for equilibrium moves sediment up or down a profile at right angles to the shore. In real shores, this is complicated by movement of sediment along the coast, that is, by **longshore drift**. The direction in which waves meet a shore is controlled partly by its trend, partly by wind direction, and partly by the direction of the advancing flood tide. When the latter two factors operate,

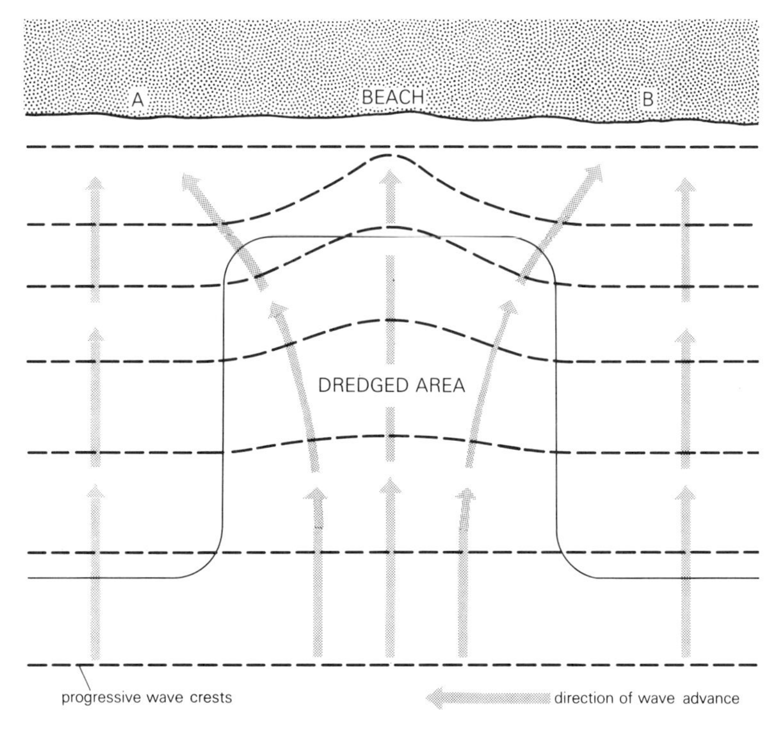

Figure 3.11 Diagrammatic map showing an offshore area where dredging has produced an area of deeper water (outlined by the −5 m contour) close to the beach. In the areas flanking the dredged region, a shelving of the sea bottom towards the beach progressively retards the forward velocity of the waves, and the spacing between adjacent crests becomes less. There is no similar retardation in the uniformly deep dredged area, and waves which were initially straight far from the shore become increasingly curved as they advance. The direction of advance to the beach is normal to the waves and the refraction of these in the dredged area focuses wave action on A and B. The lack of retardation in the dredged area also results in more wave energy reaching the beach between A and B.

waves approach the shore obliquely (Fig. 3.12), and the sediment they carry is swept up the beach in the same direction. Backwash, however, carries the grains straight down the slope of the beach, so that they suffer a net lateral displacement along the shoreline. Cumulative movement of sediment by this process is controlled by the balance of the dominant wind and the strength of the advancing flood tide. Where they are in harness (for example, in the English Channel, prevailing winds and flood tides both come from the south-west), longshore drift is appreciable. Tidal flow also produces movement of sediment in the shallow seas of the continental shelves. In narrower straits, tidal flow may reach values of over 10 km per hour, and may scour sediment from the bottom. Longshore drift may be controlled by groynes built normal to the shore, which hinder sediment movement. A beach builds up at the groyne, but

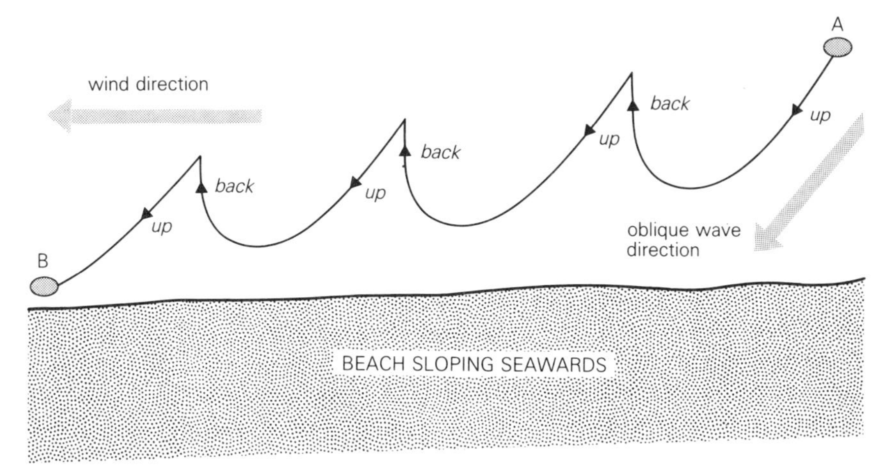

Figure 3.12 The path of a sand grain from A to B in an area where a longshore wind causes the waves to approach the beach obliquely. The uprush of the advancing wave carries the grain in the same direction (up), but the direction in which it is taken up by backwash (back) of the retreating wave is controlled by the slope of the beach.

at the expense of erosion of the beach's downdrift from it. Terminal erosion cannot be avoided unless the deficiency in longshore drift is counteracted by supplying material from off shore. For this reason, coastal protection should not be considered piecemeal. With planning it may be possible to arrange for terminal erosion to occur along an unimportant stretch of coast. Longshore drift is interfered with naturally at the mouths of rivers. On being swept into the deeper water of an estuary, the sand settles beyond the easy reach of wave action, and its accumulation produces a **spit** (Fig. 3.13).

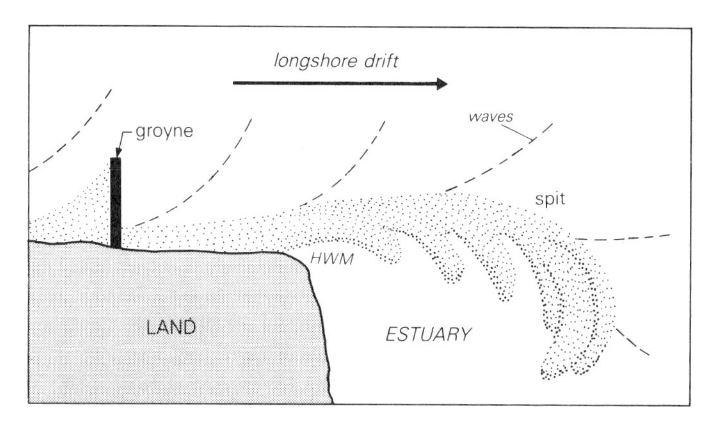

Figure 3.13 Longshore drift of sand is hindered by the groyne, and sand accumulates on the upstream side, which may be difficult to recognise since this changes with the tidal conditions and the seasons. Sand also accumulates in the deeper water of the inlet (in this case an estuary) to form a spit. The plumose form is moulded by refraction of the waves.

Coastal sediments vary according to the material being eroded. On a rocky coast, much of the shore will be bare rock, or in places cobbles and gravel. Where the sea is eating into a stretch of wind-blown sand, the shore will be a flat sandy beach. In confined waters, such as an estuary, clays, silts and fine sands cover some shorelines and grade into intertidal flats and saltmarsh. Marine clays and silts are interbedded with alluvium and peat as changing sea levels have produced the same range of environments at one spot as occur laterally at one time. The presence of plant roots and other organic debris may create very soft organic clays under these conditions. Sheltered areas with similar deposits may exist behind natural barriers such as dunes or storm beaches. These ridges of gravel are formed near high water mark during gales.

3.5 Glacial (iceborne) deposits

3.5.1 Introduction

Erosion by ice, and deposition of superficial deposits from it, are processes limited geographically at the present day to arctic regions and to very high mountains. Quite recently on a geological scale of time, however, wide areas of the present temperate zones, including most of Britain, were affected by glaciation and many of their superficial deposits were laid down directly from ice, or from melt waters flowing from the glaciers. This **Great Ice Age**, which ended approximately 10 000 years ago and spanned a period of about two million years, occurred during the division of geological time called the **Pleistocene**. During it, not only were superficial deposits called **glacial drift** laid down over wide areas of Britain and North America, but glacial erosion shaped the solid rock surface (**rock head**), and fluctuations of mean sea level left coastal deposits well above present high water mark. For these reasons, an account is given of other processes active during and after the Pleistocene, and of the erosional and other phenomena related to glaciation, as well as of glacial deposits.

3.5.2 The Pleistocene glaciation

The distribution of land, sea and mountains in Pleistocene times was broadly similar to that of the present day. The main changes that have occurred are in mean sea level, the scars left by glaciers, and the uneven deposition of glacial drift. In some places this forms small hills; in others it chokes and obscures older valleys.

The Pleistocene epoch is a short span of time compared with the hundreds of millions of years that have elapsed since many of the solid sedimentary rocks of Britain were formed. During it, worldwide temperature dropped on a few occasions (the **glacial periods**) when ice sheets advanced to cover wide areas of the northern hemisphere, including Britain as far south as the Bristol Channel. Beyond the ice front, **periglacial effects** comparable to phenomena in present-day tundra can be recognised. During the **interglacial periods** the climate was sometimes warmer than in present-day Britain. Much of this Pleistocene history of northern regions has been scraped away and destroyed by the last advance of the glaciers. There has been no steady increase of world temperatures since the last glacial period, but instead there have been short-period fluctuations of temperature superimposed on longer-period changes.

The optimal climate, with the highest worldwide temperatures, occurred around 5000 BC. The prime cause of the changes appears to be fluctuations in solar radiation reaching the Earth; this itself is partly related to sunspot activity and to long-term astronomical cycles. The Pleistocene period has ended by arbitrary scientific definition, but the Earth is not yet definitely clear of the Ice Age.

3.5.3 Glaciers and their flow

Large accumulations of ice are of two main types: **ice sheets**, such as exist in Antarctica and Greenland, and **valley glaciers**, such as are seen in present-day Switzerland. Ice sheets accumulate in extensive areas of high ground. These may be depressed progressively under an increasing load of ice. For example, in Antarctica the ice sheet is about 3000 m thick near the centre of the continent, where the solid rock surface is depressed to near sea level. The rise from the fringing mountains to the South Pole is across a great lens-shaped mass of ice. Flow within an ice sheet is radially outwards from the centre where the ice is thickest. Major topographic features have only a secondary control. The ice sheet may follow the general trend of a large valley, but may also breach a watershed which obstructs its radial flow. Smaller topographic features may be obliterated, either by erosion or by being plastered over with glacial deposits.

Movement of the ice sheet 'down hill' (in relation to its upper surface) is accompanied by melting once it passes below the snowline. Eventually the rate of advance is balanced by the rate of melting along an **ice front**, where meltwater streams are discharged from caves in a wall of ice, which is now too thin to flow.

In contrast, the movement of the valley glaciers is constrained by pre-existing major river valleys in upland areas. Valley glaciers are present at the early and late stages of a glacial period, that is, before an ice sheet has formed or after it has wasted. Permanent snow gathers on peaks, often on the shady northeastern side, and the rock below it is eroded preferentially by, for example, frost expanding in cracks and shattering the rock. A large hollow, resembling a huge amphitheatre (called a **corrie**), is formed high on the mountainside, and provides a gathering place for more snow. (At the present day the snowline, above which snow lies the whole year round, almost brushes the tip of Ben Nevis in Scotland.) These corries are often occupied by small lakes, which may be partly infilled with peat, clay from rotted granite, or more rarely by diatomaceous earth (a pure siliceous deposit consisting of minute skeletons of simple plants). Corries are used as sites for high-level reservoirs in some pump-storage schemes, where a dam has been built across the 'lip' of the corrie. Flow from the ice field in a corrie starts after the snow has compacted to ice. Once this ice has achieved a certain thickness it produces instability, and pressure changes cause melting and refreezing. Progress is guided by the steep, confining sides of a highland valley, but after leaving the valley and entering flat ground the glacier may spread out in a broad arc before further movement is stopped by the rate of melting.

The flow of ice differs in mechanism and effect from that of water. A river channel slopes persistently in one direction, apart from where fast turbulent flow has produced pools and potholes. The floor of a glaciated valley may rise and fall by several metres. *En masse* the glacier is a rigid solid tending to grind into the floor of the valley especially where the rocks are soft. It is unable to follow the sinuous course taken by water. Having gouged into a weaker part of its floor to form a **rock basin**, it tends to be deflected upwards over the next hard outcrop, such as an igneous intrusion (rather like a chisel meeting a knot in wood), and leaves it as a **rock barrier** (see Fig. 3.18). The glacier cannot wend its way around the projecting spurs of a river valley, and as it punches a near-straight path it leaves them as **truncated spurs**, and produces a **through valley**, which is **U-shaped** in profile. The enormous power of the glacier is concentrated at its base. Glaciation, unlike river action, does not widen a valley to any great extent. As a result, most glaciated valleys are **overdeepened** in relation to their tributary valleys, where ice was too thin to flow, and to mean sea level. The tributaries are left as **hanging valleys** from which waterfalls now tumble into the glaciated valley. Near the coast, river valleys, graded to sea level, are overdeepened by glacial erosion to give fjords (or freshwater lochs) with depths which locally reach as much as 300 m below sea level. Similar deep ice-gouged hollows are present in the sea floor, off glaciated mountainous coasts.

The presence of rock basins can have important consequences for engineering projects sited on them. For example, a traditional way of disposing of dangerous waste or noxious effluent has been to dump or discharge it into the deepest water readily available, at sea or in a sea loch. In many cases this coincides with a rock basin, which acts as a reservoir and limits the dispersal and dilution of the waste. On land, glacial hollows become ponds and lakes, which fill progressively with deposits such as weak, bedded clays and thick peat. They present difficult sites for foundations and are likely to add to costs. The presence of lake deposits in a U-shaped valley may well indicate the existence of a rock barrier, which acted as a natural dam, further down stream. The siting of a dam on the shallower solid rock there would lead to substantial savings on the costs of the foundations.

3.5.4 *Erosion by ice*

There are two important ways by which moving ice can erode. (a) It adheres to rock surfaces, and if a block can be detached easily along minor fractures in the rock, the moving ice plucks it from the outcrop, especially if the outcrop is on the downstream side of a rocky obstruction. The blocks, which are incorporated into the base of the glacier, are usually less than 1 m across, but there are cases where plucking has detached great slices of bedrock sufficiently large to appear to be still in place, until deeper boring showed boulder clay below them. Plucking by ice does not always detach blocks, but may simply open up minor fractures near the surface. (b) The rock fragments embedded in the base of the glacier become cutting tools as it grinds forwards. They make scratches on the solid rock surface called **glacial striae**, and may polish parts of the bedrock surface to a high glaze (see Fig. 3.14). Any minor projection of solid rock in the ice's path is ground away, and larger obstructions become streamlined in the direction of ice flow. In the process, the rock fragments themselves become abraded, and part is eventually crushed to fine rock powder, consisting of unweathered minerals.

A combination of scraping and plucking by ice produces the shape of the solid rock surface shown in Figure 3.15. This may be hidden under a cover of drift, but as a land form, visible in upland areas of glacial erosion, it is referred to as a **roche moutonnée**.

These processes are very effective, and nearly all the pre-existing soils and the mantle of weathered rock on outcrops have usually been swept away. Fresh rock is present in exposures or immediately under the cover of drift. This is in great contrast to conditions in many unglaciated regions, and the thick mantle of rotted rock in some tropical areas can be surprising to anyone whose sense of 'normality' is based on a glaciated area.

3.5.5 *Deposition from ice*

On high ground, glaciers suffer no appreciable wastage and can add to their load of **moraine**. In these areas of glacial erosion, the cover of drift left when the ice melts is thin and consists mainly of glacial sands, gravels

Figure 3.14 A glaciated rock surface which has been smoothed and moulded into a streamlined form by ice. The superficial scratches on the rock surface are glacial striae. The deeper, more prominent, cracks along which the rock is broken are small joints. (British Geological Survey photograph, C2096, published by permission of the Director; NERC copyright.)

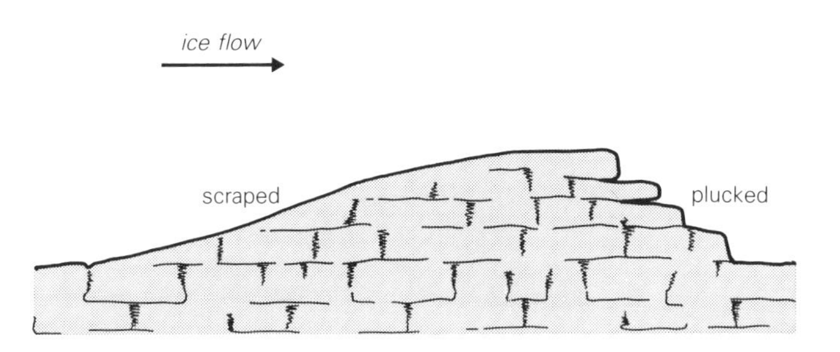

Figure 3.15 A vertical section across a *roche moutonnée*, which has been carved from solid rock by glacial erosion. The ice flow has scraped and streamlined the original exposure, and has plucked blocks from the lee side to leave a stepped profile. Jointing and other planes of weakness are shown. The shape in plan view is like a whale's back. The length is most commonly between a few metres and tens of metres, and the height is of the order of metres.

and layered clays. At lower altitudes, a glacier is burdened increasingly with moraine, yet, as the ice wastes, it is less able to transport it. Most **till** is deposited from the bottom of the ice sheet while it is still flowing. The movement sometimes moulds the till into streamlined low hills called **drumlins** (Fig. 3.16). The composition of the till is determined by the nature of the rocks cropping out up stream from where it was deposited. For example, where an ice sheet has crossed Carboniferous strata, the till is often a **boulder clay**, consisting of an unsorted, unstratified mixture of rock fragments in clay, which has been pressed and sometimes overconsolidated by the load of ice. The fragments are usually only a few centimetres across, but may be boulders over a metre in diameter. They have been orientated by the ice flow to give the till a directional texture. Large streaks of gravel may be present as an integral part of the sheet of till, and its structural relationship to the boulder clay may be difficult to determine. The shearing caused by the flow of ice produces a rough horizontal fissility, like the flaky quality of pastry, in some boulder clays.

Erosion and deposition may occur in close proximity, not only at the margin of an area of erosion but also elsewhere on local topographic irregularities. A small hill formed by an igneous intrusion, called a **plug hill**, may be eroded by ice until a steep craggy face is left to confront the ice flow; but, protected by it, a streamlined tail of softer rock or thick till is preserved on the lee side. The resultant landform is called a **crag and tail** (Fig. 3.17).

Thick till is also deposited in valleys which trend transverse to the flow. The infill conceals most, or all, of the older valley, so that under a shallow present-day valley, or even a featureless plain, there may be a deep preglacial **buried channel** cut into the bedrock. Since the solid rock surface lies below it, at greater than average depth for the area, the presence of a buried channel could lead to increased costs if heavy structures were founded on it. In some larger buried valleys, the infill may consist of churned-up preglacial alluvium and late-glacial meltwater deposits, as well as till.

Boulder clays (or tills) comprise a clay matrix containing boulders of

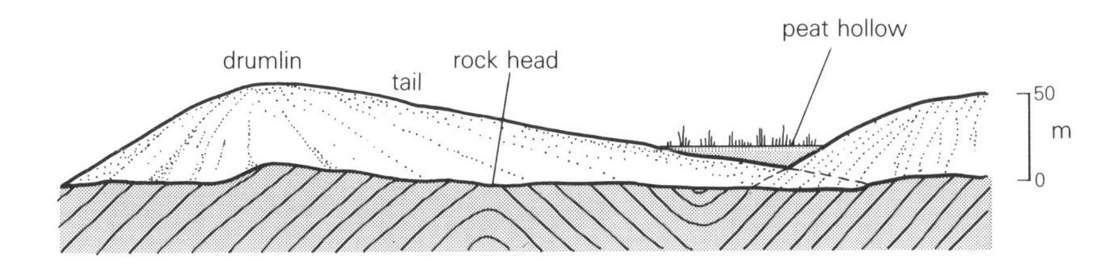

Figure 3.16 The drumlins are hills of till resting on the solid rock. The badly drained hollow encircled by drumlins has become filled with peat, and a flat area is superimposed on the slopes of the drumlins.

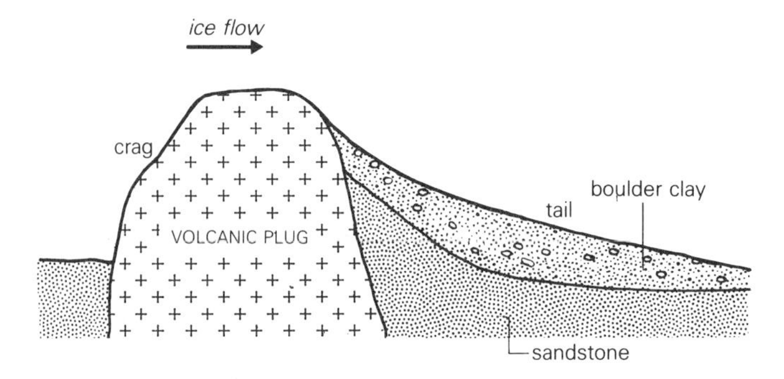

Figure 3.17 A vertical section across a crag and tail. On the left, a rock relatively resistant to glacial erosion (in this case the igneous rocks of a volcanic plug) is left protruding as a crag. It shelters softer sandstones in its lee from erosion by ice and they are left forming part of a streamlined 'tail' pointing in the direction of ice flow. The change of pressure at the base of the ice has left a thicker cover of boulder clay on the tail.

various sizes and shapes. Sand and gravel layers deposited by subglacial streams may also occur within the boulder clay. Grain size distributions in boulder clay and morainic soils are variable, and McGown (1971) recognised a split or gap in the grading around the sand-size fraction. An investigation of boulder-clay cliffs (Attewell & Farmer 1976) showed that a number of mechanisms contributed to their erosion, in particular surface weathering, creep, rotational and planar sliding, and mud flow. Courchée (1970) gives a detailed account of the properties of boulder-clay slopes.

3.5.6 *Deposition from melt waters*

A glacier deposits its remaining load of moraine at the ice front. If this is static, the coarse gravel and boulders form an irregular ridge called a **terminal moraine**. A series of recessional moraines, of similar character, may be formed at successive halting stages as the ice front retreats. Finer fractions of the moraine are transported by melt waters issuing from tunnels in the glacier and from wasting ice at the front of the glacier. The position of such a tunnel mouth may be marked as the ice front retreats by an elongated hummock of cross-bedded sand and gravel. The rate of flow of the meltwater stream decreases as it spreads, free from the confinement of the tunnel walls, and much of its load is dropped. A sinuous ridge called an **esker** may form as the tunnel mouth changes position while the ice retreats, if there is a steady supply of sand. A more complicated retreat, with large masses of stagnant ice present, may produce lines of hummocks called **kames**. When a large block of ice melts, it leaves a depression in the drift (commonly a few metres across and one or two metres deep) called a **kettle hole**. Sand and fine gravel may be carried for kilometres from the ice front, and be deposited as a flat spread of well bedded, well sorted sediments. Other terms, such as kame terrace and fluvioglacial fan, are used to describe landforms produced by spreads of gravel and sand.

Deposits from melt waters are said to be **fluvioglacial**. If large enough to be delineated, they are normally shown on Geological Survey maps as 'glacial sand and gravel', without further differentiation, although they range in composition from clean, well bedded, poorly graded sand, to unstratified mixtures of coarse gravel, boulders and clay. They also vary widely in compactness. For example, a deposit that contained dead ice is less compact than one laid down in water, and may be recognised by its steep margins and lack of regular bedding.

3.5.7 Late-glacial and postglacial phenomena

At the end of the last glaciation, as temperatures rose, the ice sheets wasted, leaving a patchy cover of fluvioglacial sand and gravel on the glacial till. The disruption of the preglacial drainage system, with some rivers choked with drift and others blocked by stagnant ice or moraines, produced ponding on an extensive scale. Many of these small shallow lakes are now filled with peat or bedded clays (Fig. 3.18), and others remain to give northern landscapes a distinctive character. The layering in some clays consists of alternating light and dark bands, each a few millimetres thick. The dark sediment is clay and the light is silt or fine sand. This **varved clay** is often formed in lakes which were fed seasonally by waters from melting glaciers. Spring floods dumped a mixture of clay and silt in them. The silt settled quickly but the clay particles stayed in suspension longer. A complete rhythm of silt and clay constitutes a **varve** and usually represents the deposition of a single year. Occasionally in upland valleys the level of a late-glacial lake is marked by erosion of the valley sides. Wave action at the shores cuts narrow terraces, which ring the valley.

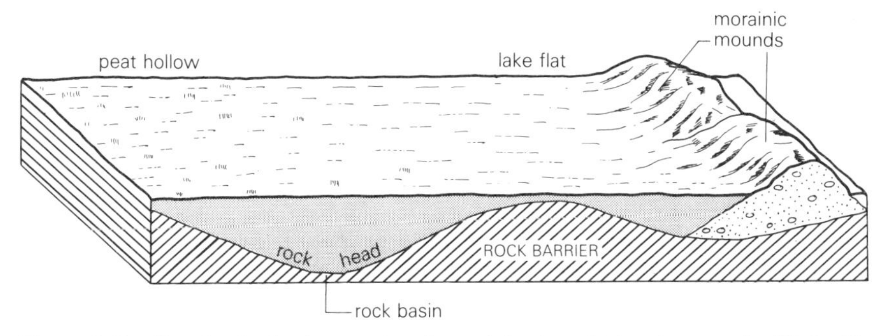

Figure 3.18 Gouging by ice has produced a rock basin, which subsequently filled with flat-lying lake clays and peat. Natural ponding has also occurred behind the morainic mounds of sand and gravel. The lake flat is the surface expression of this recent sedimentation.

In postglacial times there have been marked fluctuations of mean sea level around Britain, partly because the melted ice added to the volume of the oceans and partly because, in glaciated regions where ice sheets were thick, there was a response of the crust to the removal of the load of ice. Freed of ice the plate started to rise in the denser viscous asthenosphere (see Fig. 3.5) to attain a new hydrostatic equilibrium. The process is referred to as **isostatic recovery**. It is still continuing in northern Britain and other glaciated regions, since the high viscosity of the material that flows in as the plate rises makes it a slow process. Beyond the ice front, that is, south of the Severn–Thames line in Britain, the adjustments involve relative subsidence of the land, and produce drowned coastlines and submerged forests. Regional subsidence is a matter of long-term concern in The Netherlands: there was a disastrous influx of the sea in the 12th century. Not all of the subsidence is attributable, however, to isostasy.

The different sea levels, if they have persisted long enough, have produced terraces and raised beaches in coastal areas. These are gently sloping strips, often backed by sea cliffs, which are covered by typical beach deposits. Their mean elevation is usually between a few and fifty metres. In estuaries, the deposits of raised beaches are likely to be bedded clays with lateral changes to sand or gravel near their shorelines. Similar terracing and strandlines, which were formed at times of lower mean sea level, exist off shore.

3.5.8 Periglacial phenomena

Periglacial soils are formed under associated climatic conditions but are not formed directly by ice action. They may occur as wind-blown soils (loess) or as residual soils resulting from freeze-and-thaw weathering conditions. Generally speaking, such periglacial soils are found to the south of where the Pleistocene ice sheet reached – in the UK, south of a line from the Severn to the Humber.

Beyond the ice front lay a broad zone of tundra, the soils of which were frozen for long periods (permafrost). This affected their stability, resulting in movements both within the soil, and of the mass of the soil, by **solifluxion**. During the last glacial period, southern England lay in a periglacial zone, and as the ice front retreated periglacial conditions moved with it.

If periglacial soils occur on low-angle slopes, slope failure by **mudslides** resulting from sheet movements may result. Such mudslides take place because of the formation of ice lenses and of coarse and fine layers parallel to the ground surface, which are caused by freezing and thawing.

Head deposits are superficial accumulations which have usually been moved by solifluxion, and they are a type of drift often formed in periglacial conditions. Although their formation is favoured by frost action, it is not totally dependent on it, and head is still being formed by soil creep and downwash, for example, as rock screes. The label 'head, undifferentiated' on Geological Survey maps covers a variety of soil compositions. In general, head consists of poorly sorted, locally derived, angular fragments, and is massive or poorly bedded. Since it is formed by downhill movement, it normally lies on, or close to, a slope. The outcrops of solid rock below it may, especially if they are well bedded, show a **terminal curvature**, produced by the drag of the superficial movements. Head is thickest (in places it may be over 15 m) where the hill slope is concave, and at the valley bottoms. There it may grade into river deposits in a way that makes any boundary quite arbitrary.

Where a near-horizontal, strong, rigid layer, such as a bed of sandstone or limestone, rests on clays, superficial movement may produce **cambering**. In this, the limestone or sandstone layer moves down slope in large units called **gulls**, which are separated along joints, particularly those infilled with younger soils. During this cambering process the underlying clay may be deformed. The lines of gulls in a cambered valley tend to follow the contours. Conditions that favour cambering also tend to produce **valley bulges**. The stresses generated by the topographic loads of the sides of the valley may produce a superficial anticline in the weaker rocks near the central line of the valley. Periglacial conditions have been the critical factor in creating bulges in some valleys, but they have also formed wherever steep slopes and weak sediments are adjacent to one another.

A widespread periglacial soil in the chalk areas of southern England is **clay with flints**. It is a reddish-brown clay rich in flints, and in places contains large blocks of quartzite called sarsens. The clay and sarsens are derived by solifluxion from deposits on the chalk that persist after softer deposits have been eroded away. The distribution of clay with flints resembles that of a till. It forms a mantle of variable thickness on the solid rock, with the additional complication that it may have collapsed into cavities dissolved by ground water in the chalk. Pipes infilled with clay with flints may extend down for over 30 m.

Other periglacial phenomena produced by the freezing and partial thawing of the soil include fossil ice wedges filled with later sediment, polygonal ground, and disturbance of earlier layering in soils (cryoturbation). Some of these effects are important in site investigation. For example, periglacial weathering of the chalk on the floor of the English Channel along the proposed tunnel line has increased the flow of water through it locally; and 'fossil' mud flows which were active under periglacial conditions remain as one of the causes of instability in some present-day escarpments and cliffs on the south coast of England.

3.6 Landforms

3.6.1 Erosional landforms

Rivers, glaciers and other agents slowly wear away the land surface, at rates determined at any locality by the balance between the intensity of the erosive forces and the resistance to erosion of the rocks exposed to this assault. In general, resistant rocks are left protruding and forming high ground, and weaker rocks are etched out so that their outcrops match depressions in the ground surface. At a geological boundary where hard and soft rocks abut, there is likely to be a break of slope (a **scarp**) and the boundary may be traced by following the surface feature. The outcrop of the resistant rock will extend approximately to the lower edge of the scarp. The nature of the boundary (that is, whether it is a bedding plane, a fault or an intrusive contact), can only be determined from other observations. The types of rock can only be inferred by linking the boundary to exposures where the resistant and the weak rocks on either side of it can be seen.

If erosive forces attacked the land surface everywhere with uniform intensity, then a simple relationship between erosional landforms and solid geology would be near-universal where superficial deposits were thin or absent. On the real Earth, erosive agents are often concentrated at certain localities, and they cut deeply into the land surface at these places with an intensity that is conditioned only by the different hardnesses of the rocks. This is particularly the case with glacial erosion. Most glacial, erosional landforms, such as U-shaped valleys or *roches moutonnées*, indicate the direction of ice flow and suggest the presence and orientation of other glacial phenomena, but they reveal little about the distribution of rocks in the area. Glacial erosion usually confuses rather than accentuates landforms such as scarps or erosion hollows, since the gouging by ice often cuts across the grain of the country. In contrast, many depositional landforms of glacial or fluvioglacial origin provide a useful indication of the nature of the drift deposits forming them.

SCARP AND DIP SLOPE

As shown in Figure 3.19, a scarp and dip slope indicates the presence of a relatively resistant layer, such as a bed of sandstone in shales or a dolerite sill in sediments, dipping at a low angle in the direction of the dip slope. The scarp trends in the strike direction. Note that the controlling property is resistance to erosion relative to adjacent beds, and not necessarily hardness. Thus the porous chalk of southern England forms scarps. The magnitude of a scarp and dip slope is dependent on the thickness of the resistant layer, and the scarp may be up to a few hundred metres in height.

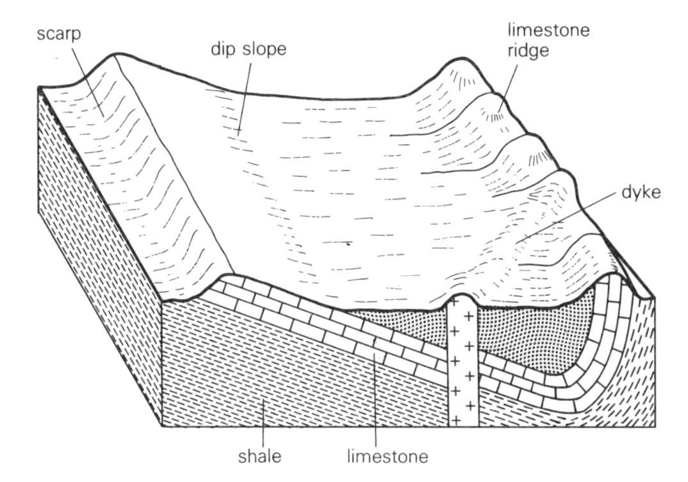

Figure 3.19 The outcrop of a resistant limestone, surrounded by softer rock, produces a steep scarp, with a gentler slope in the direction of dip. Where the limestone is vertical, scarp and dip slope are equally steep and produce a ridge. A smaller ridge is related to the outcrop of a dyke.

RIDGE (or 'hogsback')

Where a resistant layer dips steeply, the dip slope is as steep as the scarp, producing a ridge (Fig. 3.19). In an area of gently dipping strata, a ridge may indicate the presence of a dyke.

STEP FEATURING

The uppermost part (about 1 m) and the base of many individual basaltic lava flows are comparatively weak because of the concentration of vesicles. In the Tertiary and Carboniferous lavas of the British Isles, contemporaneous weathering may have altered, and further weakened, the top of some flows by changing the vesicular lava to a lateritic soil. Present-day weathering etches out these weaker zones preferentially within each lava flow, undermining the flow above. Rock falls away from the firmer basalt part of the overlying flow, breaking along the existing vertical, columnar joints. The result is a stepped profile (Fig. 3.20) of successive scarps, each corresponding to a flow. The dip slopes occur where the softer tops crop out. The size of each step corresponds to the thickness of the flow, and typically is within the range 4–25 m.

EROSION HOLLOWS

A steeply inclined weak layer, when etched out by erosion, produces a straight shallow depression. If it trends parallel to adjacent ridges, it is probably produced by a bed within the local sequence of strata. If the hollow cuts discordantly across the grain of the country, it may be related to a dyke, but more probably to a fault zone and the fault breccia within

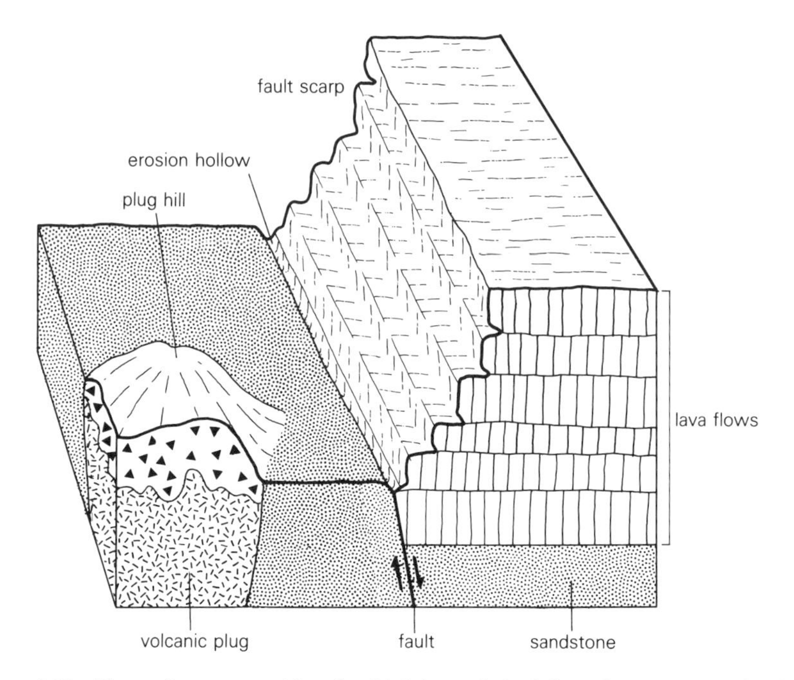

Figure 3.20 The agglomerate and basalt which have choked the volcanic vent are harder than the surrounding sandstone and give rise to a plug hill. The lava flows overlie the sandstones and have been faulted down to their present position. An erosion hollow is etched out along the fault zone, and there is a fault scarp at the boundary of hard lava flows and softer sandstone. Note that the slope of the scarp is not controlled by which side is upthrown, but by the relative hardness. Preferential erosion of the top of each lava flow has undermined the flow above it, and a rock fall is about to take place from the top flow. A falling block will come away along a cooling joint. Step featuring is superimposed on the fault scarp.

it. This interpretation would be reinforced if scarps or ridges died out close to the erosion hollow, thus suggesting the truncation of strata. Strike slip faults characteristically give rise to wide zones of breccia and marked erosion hollows (Fig. 3.20). The largest landform of this type in Britain is the Great Glen, which is a few kilometres wide along most of its length and has been gouged to a depth of a few hundred metres by ice. The smaller erosion hollows that would be more of service to a geological surveyor and of interest to the engineer are typically a few metres wide.

COLLAPSE STRUCTURES AND SWALLOW HOLES

Since limestone is soluble in fresh water, a steady trickle along a major joint or fault will leach away the rock until a tunnel or line of caverns is formed. Streams flowing across thick limestones may disappear dramatically down swallow holes (colloquially and confusingly referred to as potholes) to follow a subterranean course. A cavern at depth may ultimately collapse and be marked by a shallow depression at the surface.

The depression is typically a few metres across, and roughly circular in plan. It resembles a crown hole (see Section 6.2.3) produced by the artificial 'erosion' of a coal seam, but if it occurs in a context of dry river valleys, there is confirmation that thick limestone is present and that it is a solution collapse structure.

3.6.2 *Depositional landforms*

In many parts of the world, superficial deposits are formed *in situ* by the weathering of rocks into soils. The land surface indicates the distribution of such soils only in so far as it still reflects the distribution of parent rocks by erosional landforms that are older than the soils. Such residual soils are uncommon in Britain, where most superficial deposits have been transported by ice, rivers, wind or gravity (acting on deposits on a slope) before deposition. The local environment and the mode of deposition produce a characteristic landform, which may be used to infer the probable nature of the deposits forming it. Transported superficial deposits, the natural agents that moved them, and several of the landforms related to them, have already been discussed (Section 3.4), so only a few examples are described here.

DRUMLINS

The formation of drumlins (see Fig. 3.16) is described elsewhere (Section 3.5.5). They typically occur in a swarm, and any ground between individual drumlins is left poorly drained and possibly ponded. A typical drumlin is 0.5–1.0 km in length in the direction of ice flow, and half that in maximum width. Its height lies in the range 50–100 m. Its blunter, steeper end was formed on the upstream side of the ice flow, and its tail in the lee. The hill is formed mainly, or entirely, of till. Some drumlins have a small rock core, seldom more than a fraction of the size of the drumlin, which may have encouraged deposition from the ice. Others are simply mounds of till on an irregular glaciated surface, and the height of the drumlin is a measure of the thickness of the superficial deposit. There is no certain way of discriminating between these two extreme possibilities from the form of the drumlin.

MORAINIC MOUNDS

These deposits from glacial melt waters (Section 3.5.6) are laid down close to the ice front to form low mounds (Fig. 3.18) typically 5–15 m high and about 10–30 m wide. The length and regularity of a particular mound depend on the persistence of flow of melt water from which they were deposited. A marked accumulation which produces a continuous belt of coalescing mounds is formed along a static ice front and is usually composed of well graded sand and gravel with large boulders. The sinuous ridges formed from a single large stream normally contain better sorted, cross-bedded sands.

LAKE FLATS AND PEAT HOLLOWS

Flat, featureless, often poorly drained ground usually indicates that the area has been ponded and that deposits have been laid down in standing waters (see Figs 3.16 & 18). Sometimes the natural dam which has blocked the hollow is destroyed by overflow at its lowest point cutting a deep spillway. The dam may have been morainic mounds, ice or the coalescence of drumlins. Other natural lakes occur in rock basins (Section 3.5.3) which are obliterated by being filled in completely with sediment and peat. The flat area is typically underlain by well bedded clays, which may be varved. Part or all of the infill, especially in areas of glacial erosion, may be peat.

References and selected reading

Attewell, P. B. and I. W. Farmer 1976. *Principles of engineering geology*. London: Chapman and Hall.

Bretschneider, C. L. 1952. The generation and decay of wind in deep water. *Trans Am. Geoph. Union* **33**, 381–3.

Casagrande, A. 1948. Classification and identification of soils. *Trans Am. Soc. Civ. Engrs* **113**, 901–92.

Courchée, R. 1970. *Properties of boulder clay slopes and their influence on coastal erosion at Robin Hood's Bay, Yorkshire.* Unpublished M.Sc. Thesis, Durham University.

Darbyshire, M. and L. Draper 1963. Forecasting wind-generated sea waves. *Engineering* **195**, 482–4.

Dixon, H. H. and R. H. S. Robertson 1970. Some engineering experience in tropical soils. *Q. J. Engng Geol.* **3**, 137–50.

Fookes, P. G., W. R. Dearman and J. A. Franklin 1971. Some engineering aspects of rock weathering with field examples from Dartmoor and elsewhere. *Q. J. Engng Geol.* **4**, 139–86.

Gillott, J. E. 1968. *Clay in engineering geology*. Amsterdam: Elsevier.

Higginbottom, I. E. and P. G. Fookes 1970. Engineering aspects of periglacial features in Britain. *Q. J. Engng Geol.* **3**, 85–118.

Holmes, A. 1965. *Principles of physical geology*, 2nd edn. London: Nelson.

Hunt, C. B. 1972. *Geology of soils*. San Francisco: W. H. Freeman.

Loughan, F. C. 1969. *Chemical weathering of the silicate minerals.* Amsterdam: Elsevier.

McGown, A. 1971. The classification for engineering purposes of tills from moraines and associated landforms. *Q. J. Engng Geol.* **4**, 115–30.

McGown, A. and E. Derbyshire 1977. Genetic influences on the properties of tills. *Q. J. Engng Geol.* **10**, 389–410.

Millot, G. 1970. *Geology of clays*. London: Chapman and Hall.

Morgan, A. V. 1971. Engineering problems caused by fossil permafrost features in the English Midlands. *Q. J. Engng Geol.* **4**, 111–14.

Olsen, G. W. 1984. *Field guide to soils and the environment.* New York: Chapman and Hall.

Schroeder, W. L. 1976. *Soils in construction*. Chichester: Wiley.

Wood, A. N. M. 1969. *Coastal hydraulics*. London: Macmillan.

4 *Distribution of rocks at and below the surface*

4.1 Introduction

It is necessary to know what patterns of rock distribution occur naturally, and what processes give rise to them, in order to understand why particular rocks are present at certain depths in a borehole, and to predict from such fragmentary evidence the probable distribution of these rocks over a wider area. The distribution of rocks at and below the surface is the result of two factors: (a) the original distribution of these rocks at the time they were formed, and (b) the change produced by later deformation, as the rocks are buckled and fractured within the Earth. Deformation may also change the character of the rock mass and its engineering properties by producing fractures or even by causing recrystallisation to occur.

The shapes of igneous rock bodies, including the range of dimensions commonly encountered and the variation of rock types within the igneous mass, have been described in Section 2.2.3 as igneous structures.

The distributions of sedimentary rocks, involving both vertical and lateral change of rock type, are described as **stratigraphic models**, using selected examples. These models reflect the changes in depositional environment with place and time within one area during which sediment was accumulated. The redistribution of rock layers by deforming stresses produces characteristic forms, described here as **structural models**.

A large part of a geologist's training is in the interpretation of three-dimensional structures of rocks from evidence collected at the surface and possibly, but not necessarily, from boreholes. A combined stratigraphic and structural model of the rocks in an area is then proposed, which is continuously refined as more evidence from surface rocks and boreholes becomes available.

4.2 Geological maps and sections

The outcrop of a body of rock, such as a layer of shale, is the area that emerges at the surface where the rock is present, or lies immediately below a cover of vegetation, soil, or other superficial deposit. Those parts of the outcrop where rock is visible at the surface, and where observations may be made, are called **exposures**. Exposures are usually found in the banks of streams, in road cuttings, on cliff or scarp faces, and in quarries.

A **geological map** is an Ordnance Survey map, showing locations and topography, on which is superimposed geological information. This includes exposures, inferred outcrops and the nature of the boundaries between them, any deformation of the rocks such as fracturing or other observations, and a **geological legend** at the side of the map, explaining the symbols and colours used. Large geological departments in commercial companies and government scientific bodies, such as the British Geological Survey (BGS; Appendix C), have a prescribed set of geological symbols for use in their reports and maps. Each is known as the **standard legend** of the particular organisation. Maps in this book conform to the standard legend of the BGS. An arrow is used to indicate the precise location of an exposure and the inclination (**dip**) of any original layering (**stratification**). The arrow points in the direction of dip, and the amount of dip, measured from the horizontal in degrees, is given as a figure close to the tip of the arrow, which marks the precise location of the exposure; the observations refer only to what is seen there and not to the entire area straddled by the arrow. Other common symbols in use are listed in Table 4.1. For horizontal and vertical layers, the position of the exposure is where the two bars of the symbol cross.

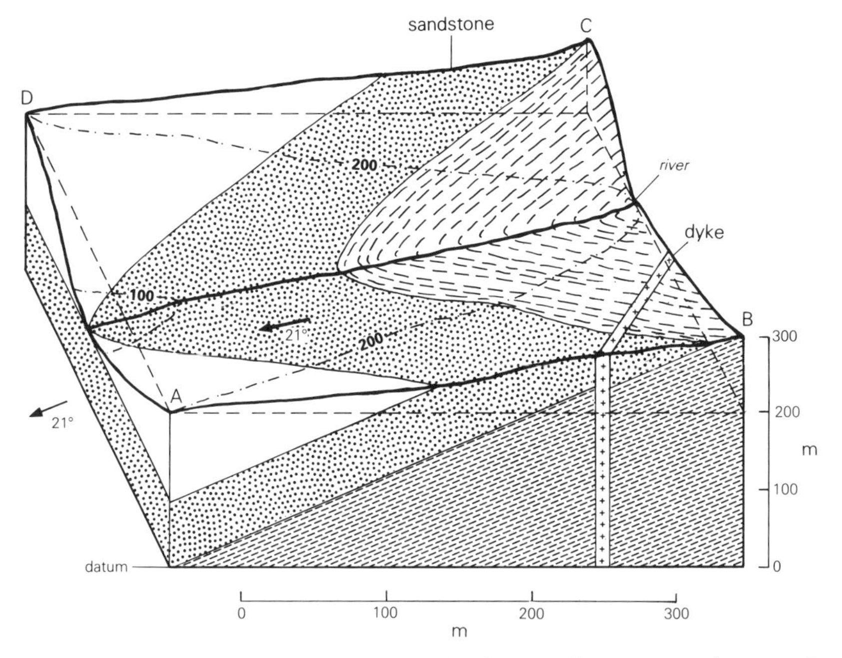

Figure 4.1 A block model showing the relationship between the outcrops and topography on the upper surface ABCD and the section AB. Contours at +100 and +200 m are drawn on the map. An exposure of the sandstone layer near the river is indicated by the dip arrow, and the figure 21 beside it gives the angle of dip in degrees.

Table 4.1 Some symbols for geotechnical maps and plans.

Symbol	Feature
	General features
LWMT	low water mean tide
HWMT	high water mean tide
6	stream with arrow showing direction of flow, and waterfall (height in metres)
	mine shaft in use
	mine shaft abandoned
	borehole
	trial pit
10	surface slope (up to 55°)
	undulating slope
	glacial striae with direction of ice movement
	glacial striae (direction of ice movement unknown)
	raised glacial feature surrounded by formline and identified by name (DRUMLIN, MORAINE, etc.)
	Geological features
	horizontal strata
20 or 20	inclined strata (dip given in degrees; arrow or small line indicates direction of dip of strata)
	gently inclined strata
	moderately inclined strata
	steeply inclined strata
	vertical strata
70	joint (horizontal, inclined or vertical)
60	shear zone (horizontal, inclined or vertical)
	certain geological boundary between rock types
	approximate geological boundary between rock types
75 T 10	fault dip and throw T (in metres) on downthrow side
	axial trace of anticline
	axial trace of syncline

An outcrop of younger strata completely surrounded by older strata (the older beds being those occurring below the younger in the vertical succession of strata) is called an **outlier**. An outcrop of older strata surrounded by younger is called an **inlier**.

The distribution of rocks on a vertical plane (comparable to an 'elevation' in engineering drawing) below a particular line at the surface is called a **vertical geological section**, or more simply a **section**. A geological model of an area ABCD showing outcrops, and the section AB across it, are presented in Figure 4.1. This shows, within a scaled frame of a datum plus vertical and horizontal scales, what would be seen if a vertical cut were made through the Earth along the line AB to reveal the side facing the viewer. The topographic profile of the ground surface shows the rises and falls along the same line. Sections are a conventional way of presenting information about the distribution of rocks below the surface. It is common practice to assist comprehension of a geological map by drawing one or more representative sections below it, and to present geological information needed for planning and costing excavations, such as tunnels, by drawing a section along the proposed line.

The relationships between the shape of the ground surface (that is, the topography), the outcrops, and the structure in idealised models in which layers are constant in rock type and in thickness over the area of the map, are discussed in this chapter. The relationship between surface geology and a section is also illustrated, and the way in which the section may be drawn is discussed. The construction of more elaborate sections, which combine structural deformation with stratigraphic change laterally and are based on real geological maps and reports, is discussed in Section 6.2.

4.3 Nature and uses of stratigraphy

4.3.1 Rock type and past environment

A guiding principle of geology is that 'the present is the key to the past'. This concept is expressed more formally, and more fully, as the **Doctrine** (or **Principle**) **of Uniformitarianism**, which states that the processes that created and moulded the rocks of the Earth in its past are by and large the same as those active today. Volcanic activity, which produces igneous rocks, and the deposition of eroded fragments to form sedimentary layers, proceeded in similar ways, if at different rates, in the past as they do at the present time. The Earth's surface had environments – shallow seas, river deltas, deserts – similar to those found now, though not necessarily in the same places. The geography of our planet has changed continually, but the general physical conditions that may be found have remained qualitatively more or less the same for thousands of millions of years. The relative areas of seas, mountains, deserts and ice sheets have varied considerably, however, as time passed. The Principle of Uniformitarianism is a particular version of one of the foundation stones of scientific thought; that is, the premise that the same laws can be applied to understanding the distant parts of our Universe, the frontiers of both space and time, as those derived from observations on Earth at the present time. There is the qualification that human experience of limited space and time must be inadequate as a basis on which to conceive an infinite Universe, just as the terms of everyday experience are an inadequate language in which to express some modern cosmologies.

The processes that *have* produced rocks in the Earth's past may be understood by studying how many rock types *are* being formed at the present. The cooling and solidification of a modern lava flow can be seen to create a layer of fine-grained igneous rock, such as basalt. It may be inferred that similar, but older, layers of basalt are ancient lava flows, and are a record of volcanic activity in the past. Each is an event in the **geological history of the area**.

Each different rock layer is a record of a past event, and the sequence of events (that is, the geological history of the locality where they are observed) may be inferred using the **Principle of Superposition**. This states that if one layer B lies above another A, then B was formed after A and is younger than it – unless it can be shown that the layers have been inverted from their original position by later earth movements. It may be possible to identify the original top and bottom of a layer by a way-up criterion (see also Section 2.2.4) such as graded or cross bedding (see Fig. 4.2).

The relative ages of rocks and of the structures that deform them may also be inferred from other observations. For example, the presence of pebbles of granite in a conglomerate shows that the granite is older. Other examples are given in Figure 4.3.

In some layers of sedimentary rock there are traces of the animals and

Figure 4.2 Cross bedding in layers of sandstone. The point of the pencil indicates the original way up of the beds, as shown by truncation of the finer layering on that side of each unit. (Reproduced by kind permission of Mr David Campbell.)

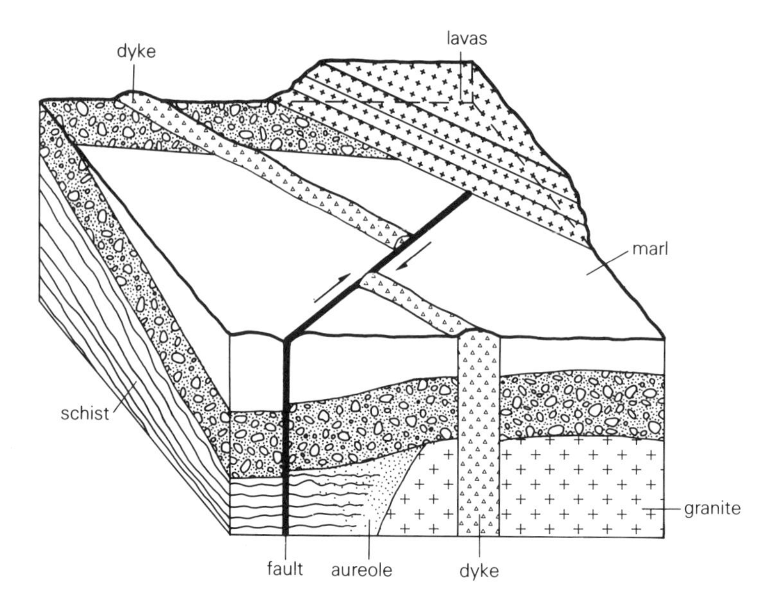

Figure 4.3 The relative ages of rocks and geological events may be inferred from geometrical relationships and from the nature of some rocks. An example of the former is shown in the figure, where the superposition of the lavas on all other beds can be seen: the fault demonstrates that the lavas are the youngest strata present; also, the emplacement of the dyke in the marl shows that the dyke is younger than the marl, and the displacement of the dyke by the fault shows that the fault is younger. Pebbles of the granite in the conglomerate confirm that the granite is older, and the thermal aureole in the schist shows that the granite is younger than the schist.

plants that were living at the time the sediment was being deposited and were covered by sediment after their deaths. These remains are called **fossils**, and their study is **palaeontology**. The most common fossils are the hard parts of organisms – bones, shells, coral skeletons, scales and teeth – and the imprints of leaves with the original vegetable matter transformed to a thin, black film of coal. Hollows in the ground produced by decayed plant stalks and roots serve as moulds for later sediment, and sandstone casts of roots are not uncommon in certain layers. Fossils are of importance as the source of what we know of past life on our planet, and of how it developed through geological time. Different groups of animals and plants are found in layers of different age, and the types of fossil found in a particular layer are characteristic and diagnostic of its age. Thus fossils, if present, provide a means of dating a sedimentary layer, and of **correlating** rock layers in different areas by showing that they are of the same age. As a result, it may be possible to recognise a **lateral change of facies**, that is, a change of rock type within one layer, produced at the time of formation by a change of environment across the area. This idea is illustrated by the simple example shown in Figure 4.4.

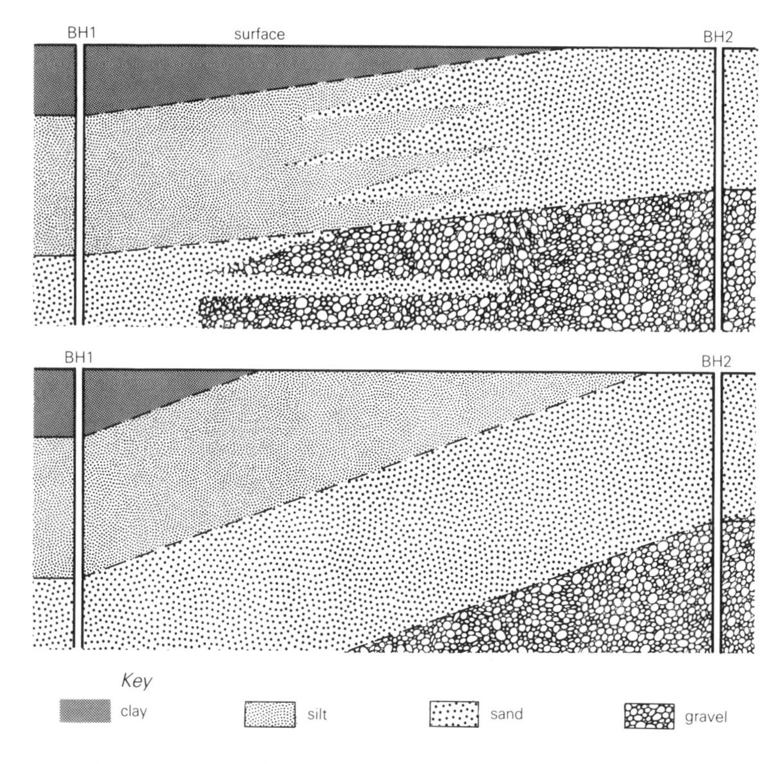

Figure 4.4 Strata penetrated by two boreholes, BH1 and BH2, are shown on a vertical section. If the presence of fossils were to prove that the silt in BH1 and the sand in BH2 were of the same age, and that the sand in BH1 and the gravel in BH2 could also be correlated, then lateral changes of facies would be present in both layers. In (a), the intervening vertical section between BH1 and BH2 is drawn. It shows a gentle inclination of the layering from BH2 to BH1 and a coarsening of sediment type in the opposite direction. The change from silt to sand and from sand to gravel is shown diagrammatically as interdigitation of the rock types.

All fossils are of some use in inferring the age of sedimentary rocks, but those of most use are relatively plentiful, even in a core sample, and flourished for only a relatively short spell of geological time. Many one-celled animals and plants meet these criteria and are used extensively, particularly by specialist geologists in the oil industry.

Life developed at an early stage in the history of the Earth, but few fossils are preserved in the older rocks. In the layers overlying these oldest rocks, fossils of more complex plants and invertebrate animals are present; then above these the first land plants and early fish are found. It is only in the uppermost, comparatively young, layers that the first traces of man and the immediate ancestors of present-day flora and fauna appear. Occasionally, changes in the appearance of a fossil community can be followed in detail over periods of a few million years, but typically the fossil record is fragmentary and disjointed. One general feature, however, stands out clearly: life on Earth has changed persistently and gradually with time, becoming more specialised in its adaptations to different environments and more complex in the organisation of individual animals and plants. These changes are described by the word **evolution**.

The idea that rocks and fossils can reveal the past history of the Earth and that rock layers may be understood and interpreted in terms of past environments is an important and central concept in geology. This branch of the science, which studies rocks – especially layered rocks – as Earth history, is called **stratigraphy**. It makes an important contribution to our knowledge of the universe, as well as of the planet Earth, since it shares with astronomy the ability to provide evidence of events in the distant past. These include, for example, the behaviour of the Earth's magnetic and gravitational fields, which are of wider scientific significance than the past geography of one area.

The teaching of an appreciation of stratigraphic research, especially as it relates to the long timespans involved and to the development of life and the emergence of man, should be part of a general education. Here, attention is focused on the practical knowledge of stratigraphy relevant to site investigation. The purpose of this is (a) to understand how environmental factors and ancient geographies control the original distribution of rocks and how they are related to the stratigraphic models used in helping to predict what rocks are present under a site; and (b) to understand the presentation of data on geological maps, and in memoirs and reports, in which rocks are classified according to their age. Some knowledge of stratigraphy and a grasp of its concepts are necessary in order to be able to cope with the stratigraphic terms and to comprehend the arguments and information presented.

4.3.2 The geological timescale

The strata at one locality represent only a *local* history of the Earth's surface. In order to compare events and layers at two different localities and assign relative ages (that is, to correlate the strata), it is necessary to recognise similar events in both places. A timescale on which local events can be dated must be based on happenings and changes that can be recognised in widely separated places. The fossil record provides such a history, as the assemblage of fossils in a particular environment is characteristic of a period of time and occurs contemporaneously over most of the Earth. Some time does elapse as a new species of animal spreads over wider regions from the place where it first evolved, but most successful groups spread rapidly compared with the rates of other geological processes.

The planet Earth was formed 4600 Ma (million years) ago from debris in space that coalesced. Life appeared comparatively early in its history, and fossils have been found in rocks which other evidence shows to be more than 3000 Ma old. The evolution of many species of animals or plants with calcareous (shelly) and chitinous (horny) hard parts, which survive as fossils and are suitable for dating rocks, happened much later, about 590 Ma ago. The interval of Earth history after 590 Ma, which can be divided into time units by fossils, is known as the **Phanerozoic eon** of time. It is divided into three great **eras** of time: the **Palaeozoic**, when 'old life' such as the first primitive fish existed; the **Mesozoic**, when 'middle life' forms such as the dinosaurs were dominant; and the **Cenozoic** (or Cainozoic), when 'new life' forms such as mammals and grasses flourished. Each era is divided into **periods** of time and these in turn are subdivided into smaller time units. The periods are listed in the geological timescale of Table 4.2.

The pre-Phanerozoic interval of Earth history from its formation up to *c*. 590 Ma is known as the **Precambrian**. This enormous interval of time is divided into three eons: the **Priscoan** – from the Earth's origin (*c*. 4600 Ma) to 4000 Ma; the **Archaean** – from 4000 Ma to 2500 Ma; and the **Proterozoic** – from 2500 Ma to the beginning of the Phanerozoic eon (at *c*. 590 Ma). Each of the three Precambrian eons is further divided into eras, as is the Phanerozoic. The earliest rocks discovered on the Earth are gneisses of early Archaean age from Amitsoq, West Greenland, which have a **radiometric age** (explained later in this section) of 3700 Ma. Other early Archaean rocks are known from Central and South Africa, and represent igneous and sedimentary as well as metamorphic rocks. In the UK the earliest rocks are late Archaean gneisses found at Scourie in the North-west Highlands of Scotland. A fuller account of the Precambrian rocks of the UK is given in Section 4.6.2.

There have been differences of opinion about the most convenient stratigraphic boundaries to take when dividing geological time, particularly the Phanerozoic, into periods and **epochs**. These are being resolved by international agreement between stratigraphers, and time units are now defined by designating certain sequences of rocks as standards (that is, by using type localities), the unit being the time elapsed between the events represented by two horizons (much as the 'imperial yard' used to be defined as the length of a piece of platinum stored in a government office). Correlation with, and consequential dating of, strata other than at the type localities is done by comparing fossil assemblages found at the type locality with fossils found in the strata to be dated. In practice a more complex argument involving correlation with other localities must usually be employed, so that the equivalence in age of fossils which have not been recorded at the type locality may be inferred.

Dating a geological event with reference to the geological (strati-

Table 4.2 The geological timescale. The age of the *beginning* of each period is given in million years before present. Notable events and features of economic importance in the British area are also listed.

Era	Period (System)		Epoch (Series)	Age to beginning of period (Ma)	Major orogenic events (Europe)	Some rock units
Cenozoic						
	Quaternary		Holocene	0.01		fen silts
			Pleistocene	1.5–2.0		boulder clay, fluvioglacial sands
	Tertiary	Neogene	Pliocene Miocene	25	Alpine Orogeny	
		Palaeogene	Oligocene Eocene Palaeocene	65		London Clay volcanic rocks of Inner Hebrides
Mesozoic	Cretaceous			144		chalk deposits: thick sedimentary deposits in North Sea containing oil
	Jurassic			213		Portland Stone, Oxford Clay
	Triassic			248		Cheshire salt, European evaporite deposits
Palaeozoic	Permian			286	Variscan or Hercynian Orogeny	New Red Sandstone, gas deposits in southern North Sea
	Carboniferous	Pennsylvanian*		320		British, European, USA and USSR Coal Measures, extensive cover in Midland Valley of Scotland and northern England, good quality limestones
		Mississippian*		360		
	Devonian			408		Old Red Sandstone (non-marine) in Scotland, many Scottish granites
	Silurian			438	Caledonian Orogeny	shales and greywackes in Scottish borders and Wales volcanic rocks of Lake District and South-west Scotland
	Ordovician			505		
	Cambrian			590		limestones in North-west Scotland, Welsh slates
Precambrian						

*Pennsylvanian and Mississippian are names of sub-periods of Carboniferous in North America.

graphic) timescale is like dating human history by dynasties (for example, speaking of 'in early Victorian times' without reference to the date in years). It is the commonly used practical method in Phanerozoic sedimentary strata, but is inconvenient or even useless *for certain purposes*, such as working out the rate at which processes operated in the past. Furthermore, its application is limited to Phanerozoic sedimentary rocks.

Any method of obtaining an age in years of a rock specimen must depend on assessing the cumulative effects of some geological process over a long period of geological time, the quantitative effects of which are well known from studying its action at the present time. The rate per year must be determined and must be constant, or able to be corrected reliably for past fluctuations, during the span of geological time.

The methods used in geology depend on the process of radioactive decay, that is, the disintegration of certain **unstable isotopes**. Most elements in their chemically pure form consist of more than one type of atom. Each variety (that is, each **isotope**) is distinguished by minor differences of atomic structure and of atomic weight, but *not* of chemical properties. The rate of decay of an unstable isotope is expressed by its **half-life**. This is the time taken for half of the original **parent isotope** to disintegrate and produce energy and **daughter products**.

In general, a radiometric age is obtained from the equation:

$$t = (1/\lambda)\log_e[(D/P) + 1]$$

where t is the age, λ is the decay constant, D is the concentration of daughter atoms in the mineral due to radioactive decay, and P is the concentration of parent atoms in the mineral. Isotopes used in determining radiometric ages of rocks should be present in certain rock-forming minerals. Several unstable isotopes with different half-lives are used to date rocks of widely different ages – from early Precambrian rocks to recent Pleistocene raised-beach deposits. The isotopes commonly used are given in Table 4.3. For old Precambrian rocks, Rb–Sr ages are usually obtained, these isotopes being concentrated in feldspars; U–Pb ages are also obtained for such rocks, as well as for old rocks with the isotopes concentrated in the mineral zircon. For rocks with a wide spread of ages, K–Ar ages can be used, particularly if the rocks contain appreciable amounts of K-bearing minerals such as micas or hornblendes.

Radiocarbon (^{14}C) dating is used to date recent deposits, particularly glacial or raised-beach deposits formed during the last ice age. It is also a method commonly used to date artefacts obtained from various archaeological sites.

A few laboratories in every country specialise in the radiometric dating of rocks and minerals, and these laboratories have produced tens of

Table 4.3 Unstable isotopes of the most important elements used in the radiometric dating of minerals and rocks.

Unstable isotope	Daughter product	Half-life
rubidium ^{87}Rb	strontium ^{87}Sr	50 000 Ma
thorium ^{232}Th	lead ^{208}Pb	4 510 Ma
uranium ^{238}U	lead ^{206}Pb	4 510 Ma
uranium ^{235}U	lead ^{207}Pb	713 Ma
potassium ^{40}K	argon ^{40}Ar	11 850 Ma
potassium ^{40}K	calcium ^{40}Ca	1 470 Ma
carbon ^{14}C	nitrogen ^{14}N	5 730 years

thousands of such radiometric dates. Since the rocks dated come from every geological era, from earliest Precambrian to late Cenozoic, a framework of timed events can be produced – a radiometric timescale from which other events may be dated relatively, using the simple methods of traditional stratigraphy. Engineers should understand the way in which absolute ages are determined. However, it is the simple methods of inferring the relative ages of the rocks invariably encountered in site investigation that are important, and to convey an impression that radiometric dating is a standard part of geological surveying would be misleading.

4.3.3 Stratigraphic presentation of rocks in maps and reports

The boundaries on a geological map sometimes define the outcrop of a single, highly distinctive rock mass such as a layer of coal or limestone or a dolerite dyke. More commonly, sedimentary rock layers define a **formation**, which is a bed or assemblage of beds with well defined upper and lower boundaries that can be traced and mapped over considerable distances. A formation is simply a rock unit, which can be dated if it contains the appropriate fossils that define a particular subdivision (**age**) of geological time. Such a rock unit may be defined by its general properties, such as the nature of the pebbles and the colour of the cement in conglomerates within it. The significant differences between formations reflect significant differences in environments of deposition. Changes of environment can also occur laterally across a region at any one time, with corresponding lateral changes of rock type.

In the legend of a geological map, the age of formations (or of several formations, called a **group**) is shown. The description of strata in reports, for example BGS District Memoirs (Appendix C), is also arranged according to stratigraphic order, with the oldest group of rocks described first. Descriptions of the strata present at different exposures within the area considered are given within the chapter on the particular group or formation, and the available evidence on lateral changes of facies is presented.

4.4 Deformation of rocks (structural geology)

4.4.1 Outcrops and subsurface structure

If the structure of an area is known, then so also is the distribution of rocks on any two-dimensional section across it. A geological map is, generally speaking, such a section, and the distribution of rocks in this case is the pattern of outcrops. The principal factor that may make a map more complicated than a vertical section is that the latter is a planar surface, whereas the ground surface is likely to be irregular. In the paragraphs that follow, the qualitative relationships between structures, vertical sections and the corresponding geological map are illustrated for different structural models. In order to avoid complication and to show geometric relationships clearly, both the stratigraphy and topography are kept simple. The rocks are layers of constant thickness and are without any lateral change of rock type. The ground surface approximates to a level plain except where it is dissected by river valleys.

By a converse argument, the general distribution of rocks in the subsurface may be inferred by studying the outcrops, and by relating a particular pattern to the structural model that would produce a similar one. The direction of dip, the nature of folding of the strata and the amount of displacement across faults may be recognised quickly by inspection of a geological map of the area and used as a first step in analysing the geology.

Inferring the nature of a structural model from the evidence of outcrops, or from scattered exposures, involves inductive reasoning (see Section 6.1.2) and makes the implicit assumption that the most probable answer is the simplest one consistent with the observations. For example, failing evidence to the contrary, it is assumed that dip and strike stay constant between adjacent exposures where they are identical, and the interpolated outcrops are constructed on this premise. The probability of being correct increases, of course, with the amount of information available.

The structure of strata is usually shown on a map by symbols, such as dip arrows. A more convenient way for some purposes, where the structure is simple, is to draw contours relative to a datum and so represent the ups and downs of a particular geological interface (just as topographic contours depict the shape of the ground surface). The interface is usually a bedding plane separating two different layers. A common use of contours to depict geological structure in engineering geology is a **contour map of rock head**, which shows the depth to solid rock and the shape of the bedrock surface. To distinguish them from topographic contours they are referred to as **stratum contours** or, when they are used in certain interpretative constructions, as **strike lines** (see next section for a fuller explanation). When irregularities of the ground surface are represented on a geological map by topographic contours, and when the structure of a geological interface is represented by stratum contours drawn at the same vertical interval (for example, at elevation difference of 25 m), then it is possible to construct the outcrop of the interface of the map. This is illustrated in the next section.

4.4.2 Layers of uniform dip (horizontal, vertical and inclined strata)

Original layering in sediments is called stratification or bedding. (Different terms are used for the layering produced by other igneous and metamorphic processes.) Each layer is referred to as a **stratum** (plural, **strata**) or **bed**, though the latter term is normally applied only to sedimentary rocks and volcanic ash layers. It is never applied to the layer formed by a basaltic or rhyolitic lava flow. The interfaces between beds are called **bedding planes**. The outcrops of major bedding planes that separate thick beds of different rock type are depicted on geological maps. Bedding planes may also occur within a thick layer of sandstone, where minor changes of composition or texture, or even a break in deposition, are present. These are planes of weakness, or potential planes of failure, and their presence produces a change of mechanical properties of the rock with the direction in which they are measured (that is, the bulk properties of the rock are anisotropic). Minor bedding planes are abundant in micaceous shale, because mica flakes lie in the bedding direction as they grow in the rock or settle in water. Rocks that are devoid of bedding and other planes of weakness such as joints (see Section 4.4.4) are said to be **massive**.

At the time of deposition, most beds are nearly horizontal. Later, tilting or buckling of the Earth's crust may leave them inclined at an angle to the horizontal. The dip is easily measured with a **clinometer**, consisting of two hinged arms. One rests on the bedding plane and the other has a spirit level and is made to lie horizontally. The angle between the arms is shown by a protractor. In Figure 4.5 a layer of sandstone is seen to dip at 30° to the north.

A clinometer aligned north–south would give a reading of 30°. A measurement made in any other direction on the bed would give a smaller reading of dip, with the dip decreasing progressively as the angle B increases, until at $B = 90°$ the dip measured is zero. In this example, north is the **direction of maximum** or **true dip** and 30° is the **true dip**. The direction of 90° to it (that is, east–west) is the **strike** of the bed. A line

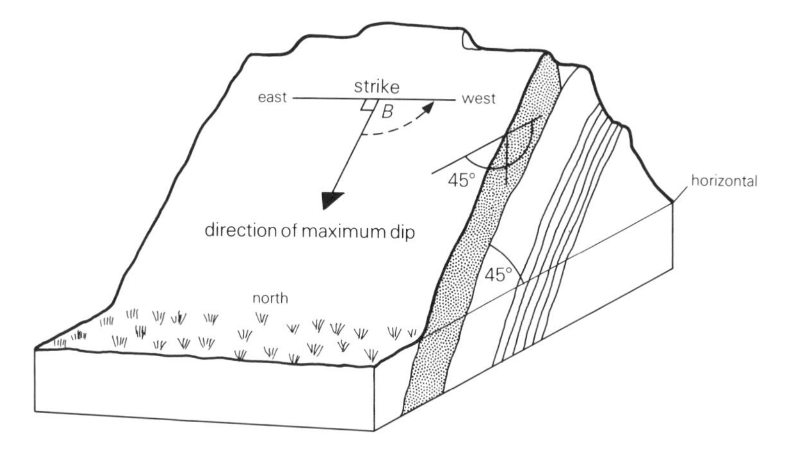

Figure 4.5 The layer of sandstone is tilted to the north at an angle of 45° to the horizontal. A clinometer has been placed on a bedding plane and orientated in a north–south direction, that is, it lies in the direction of maximum (true) dip. The dip observed is 45°. When an angle of dip is read in any other direction on the bedding plane (that is, at some angle B with the direction of maximum dip), the apparent dip observed is less than 45°. In the limiting case where $B = 90°$, the apparent dip is zero. This direction is called the strike of the dipping layer.

drawn in this direction on the bed is horizontal, and since it is thus a line joining points at the same height on the bed, it is a stratum contour or strike line. In any of the directions between true dip and strike, the angle of dip observed between bedding plane and horizontal depends on the true dip and the angle B. A value of this type is referred to as an **apparent dip**. Most observations of sections in cuttings, quarries or cliff faces reveal apparent dips, but where a bedding plane is exposed, the true dip can be measured. The relationship of one to the other is given by the equation

$$\tan(\text{apparent dip}) = \tan(\text{true dip}) \times \cos B.$$

If the structure of a bedding plane is represented by strike lines and the topography by contours, then the outcrop may be constructed by joining all points on the map where the values of both sets of contours are identical. At such points the bedding plane coincides with the ground surface (that is, it crops out). Points on the outcrop where contours cross are readily identifiable by inspection. The others, needed to construct a continuous line of outcrop, are determined by intrapolating between both sets of contours.

The converse problem of inferring the structure from the relationship of outcrops to topography is more common. The altitude of the bedding plane along its outcrop may be read directly from the map using the topographic contours that cross the outcrop. For example, in Figure 4.6

the bedding plane separating the sandstone and marl is at an altitude of 400 m at A and C, and at 350 m at B. Strike lines are drawn by joining points of equal altitude, for example A and C. A straight line is drawn as the simplest geometry of the bedding plane consistent with the observations. In practice, there is a partial check on the assumption that no fold (that is, change of dip and strike) is present between A and C. The apparent dips seen on the sides of the valley (that is, along AB and CD) are constant, and barring the presence of a very local, very complex and very unlikely structural peculiarity restricted within the area ACD, the assumption that the true dip does not change is valid. Nevertheless, the strike lines, and hence the structure, are *inferred* and not *deduced* (see Section 6.1.2).

If strata are horizontal, then the entire surface of each bedding plane lies at the same altitude, and outcrops are parallel to topographic contours (Fig. 4.7). In this case the position of the outcrop of any bed is *dependent entirely* on the topography. Conversely, the presence of horizontal beds on a map may be recognised by inspection, from the control that topography has on their outcrops.

If a layer is vertical, its outcrop is parallel to the strike of the layer but is completely *independent* of topography. For example, a vertical dyke (Fig. 4.7) crosses hill and valley without any deflection of the trend of its outcrop.

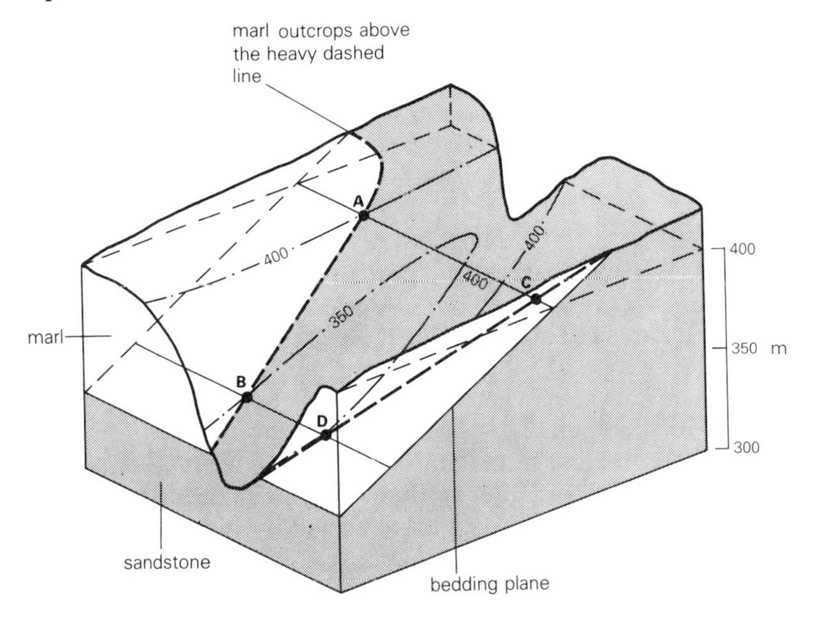

Figure 4.6 A block model showing a layer of marl overlying sandstone: the outcrops of the bedding plane between marl and sandstone on the valley sides are shown, together with strike lines at +400 m from A to C, and at +350 m from B to D.

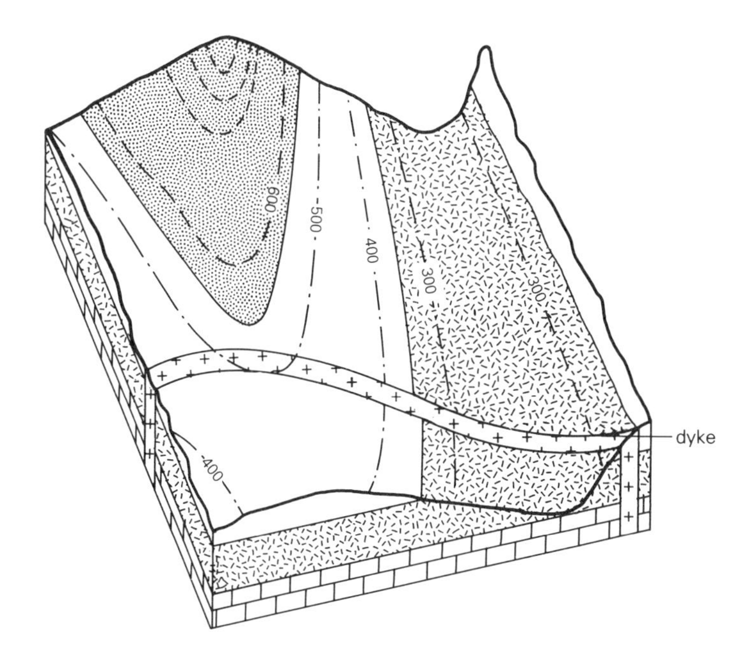

Figure 4.7 A block model showing a sequence of *horizontal* strata cut by a *vertical* dyke, and their relationship to the irregular upper surface of the block. The hill and valley are shown in profile, and are also defined by topographic contours drawn on a 100 m interval. The distribution of outcrops on a map is similar to that portrayed on the upper surface. The horizontal layers are parallel to the contours, but the outcrop of the vertical dyke is a straight line and is completely independent of the contours.

In the general case of dipping layers, the outcrops are controlled partly by dip and strike and partly by topography, with the control by topography increasing as dip decreases. On a flat, level plain, the outcrop of a dipping layer forms a strip parallel to the strike. The width of the strip (W) is related to the thickness (D) and to the amount of dip (a) of the bed by the equation $W = D/\sin a$. D is the true thickness of the layer, measured at right angles to the bedding planes. The vertical thickness (Z), such as would be observed in a borehole, is related to D by the angle of dip (a) by the equation $Z = D/\cos a$.

Where there are topographic irregularities, specific relationships can be deduced using strike lines. One distinctive relationship (Fig. 4.6) useful in reading maps is the V-shaped deflection of the outcrop, which is present where a dipping layer crosses a valley. The V, if nearly symmetrical, points roughly in the direction of dip, if the dip of the bed is greater than the slope of the valley. This is a common state of affairs. The phenomenon provides an indication of the dip of a fault or dyke, as well as the dip of beds.

4.4.3 Folds

If subjected to stresses beyond those that their strength can resist, strata are permanently deformed by either buckling or fracturing. The type of deformation depends on the mechanical properties of the rocks and the nature of the stresses. In general, stresses which are applied slowly, either deep within the Earth where the confining pressure produced by overburden is high, or to rocks that are not brittle, tend to produce **folds** by buckling or plastic flow. The essential characteristic of a fold is a *change of dip*. This occurs at the **hinge** of the fold. On either side of the hinge is a **limb**, where the dip is comparatively steady. In real folds the change of dip is seldom restricted to a line, but is concentrated in a liner zone. The plane of symmetry (Section 2.1.1) that bisects the angle between the limbs is the **axial surface** (or **axial plane**). The intersection of the axial surface with each bedding plane is an **axis of the fold**.

A fold where the limbs diverge downwards (that is, where the dips of the limb are away from the hinge) is an **anticline**; a fold where dips are towards the hinge is a **syncline** (see Figs 4.8 & 9).

The symmetry of a fold about its axial plane is reflected in its outcrop

Figure 4.8 Aerial view of an eroded anticline in Iran, which plunges at a low angle (towards the top left of the photograph). An arch of massive limestone, which is intact in the distance, has been eroded away in the foreground to reveal the pattern of outcrops characteristic of a plunging fold. (Photograph by Huntings Surveys, reproduced by kind permission of British Petroleum Co.)

Figure 4.9 An anticline and syncline affecting greywacke and shale layers in the Southern Uplands. Note the tension joints near the hinges of the folds. (British Geological Survey photograph, C2070, published by permission of the Director; NERC copyright.)

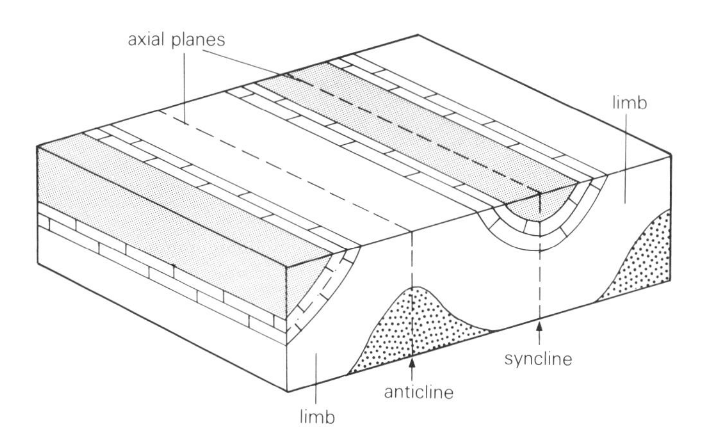

Figure 4.10 A block model showing the distribution of folded strata on a map (the upper surface) and on two vertical sections. The axial planes of an anticline and syncline are shown by broken lines. The outcrops on the map are symmetrical about the traces of the axial planes.

pattern (Fig. 4.10). The sequence of beds cropping out on one side of the trace of the axial plane on the ground surface is repeated, but *in reverse order*, on the other side. The oldest strata that crop out are present near the axis of an anticline, and the youngest strata at the axis of a syncline. These patterns, plus the dips, may be used to recognise folds on a map or in the field, where a fold is too large to be visible in one exposure.

If the axis is not horizontal, the fold is said to **plunge**, with the amount of plunge being the angle between the axis and the horizontal line vertically above it. The U-shaped outcrop of a plunging fold is shown in Figure 4.11. As a result of the plunge, the map is an oblique section across the fold hinge and reveals its shape, including the nature and width of the hinge zone. If the plunge were 90° and the area of the map were flat and level, the map would show the true thicknesses of the beds. As the plunge decreases from this maximum value, the U-shaped outcrop is progressively elongated in the axial direction.

Topographic irregularities make the outcrops of folds more complicated, in ways that can be deduced using strike lines. A simple fold approximates to two layers inclined towards, or away from, the hinge. In

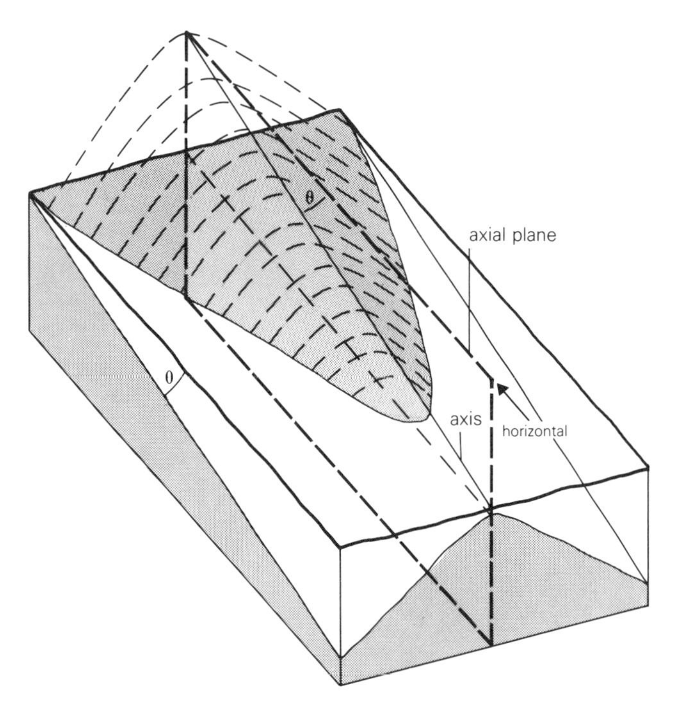

Figure 4.11 A block model showing the outcrop of an anticline plunging at angle θ. The axial plane, an axis, and the former continuation of a bedding plane above the map are represented.

the converse problem of determining the strike and dip from the outcrops, it is best to define the area of outcrop of each limb by first tracing in the axes of the fold by inspection.

Folds are described and classified by certain features of their geometry which are diagnostic of how they were formed. The criteria and the terms used to define folds are illustrated in Figure 4.12; they are as follows:

(a) The *dip of the axial plane:* where it is vertical the dip of each limb is the same in magnitude, but opposite in direction, and the width of outcrop of any bed is the same on both sides of the axis. In inclined folds, one limb is steeper and its outcrop is correspondingly narrower than that of the other. Overfolding occurs where one limb is overturned, such that both limbs dip the same way. The rotation of the axial plane, seen from left to right in the diagram, is probably produced by transport of the fold with relative drag at its base.
(b) The size of the *angle between the limbs:* this reflects the intensity of compression of the fold.
(c) The *relative lengths of both limbs.*
(d) The *style of folding:* that is, the manners in which different types of rock have behaved mechanically, particularly whether they are **competent** (that is, comparatively strong) or **incompetent**. Under the stresses of folding, an incompetent layer tends to flow and change thickness, sometimes to a stage where it is discontinuous. Arenaceous rocks are usually competent and clay rocks are incompetent.

Concentric folds are produced by simple buckling, and the beds are curved concentrically by the same amount. The beds remain constant in *true* thickness and parallel to each other after folding. Slip has taken place along the bedding planes. A concentric fold becomes progressively less open as the 'centre' of the concentricity is approached. Open folds affecting competent strata are usually concentric.

In **similar (shear) folds** all layers are deformed to a similar shape, and the bed thicknesses are constant in a direction parallel to the axial plane (see Fig. 4.13). This produces an increase of true thickness at the hinges and a decrease on the limbs. Similar folds are produced by differential movements along minor shear planes in a direction parallel to the axial plane. Intense folding of metamorphic rocks at a time when temperature and pressure were high is usually of this type.

Flow folding is produced when incompetent rocks flow in a complex pattern of movement to accommodate either the local stresses related to the buckling of competent beds or lateral changes in the loads of overburden compressing them. The pattern of folds which results is usually complicated, and such that the shape of one folded bed may bear

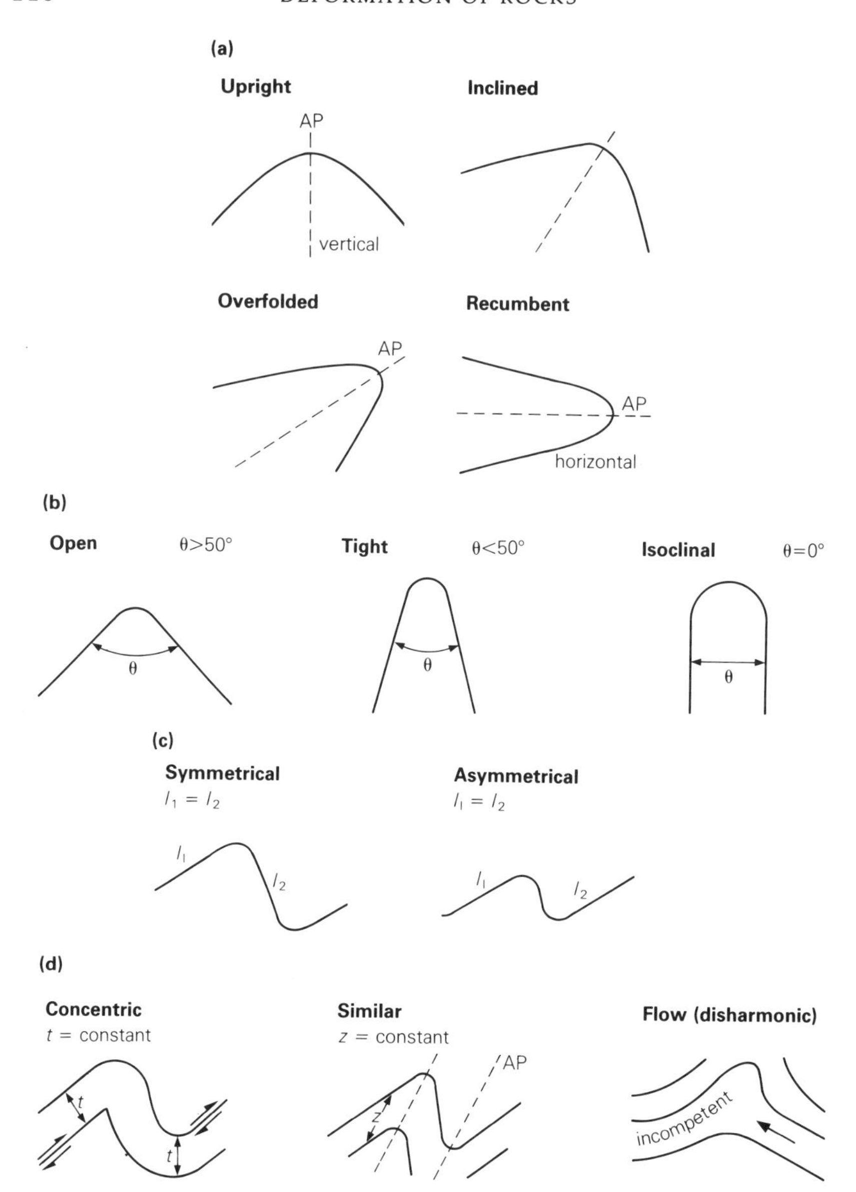

Figure 4.12 Classification and nomenclature of folds according to (a) dip of the axial plane (AP), (b) size of angle between limbs, (c) relative lengths of both limbs, and (d) style of folding.

little obvious relationship to that of another. Folding is then **disharmonic**. Folds involving a thick layer of salt are of this type. The salt, which is highly incompetent, rises locally as **salt pillows** and **salt plugs**, folding the beds above them to form **salt domes** (Fig. 4.14).

Figure 4.13 Similar folds in metamorphic rocks. (Reproduced by kind permission of Mr David Campbell.)

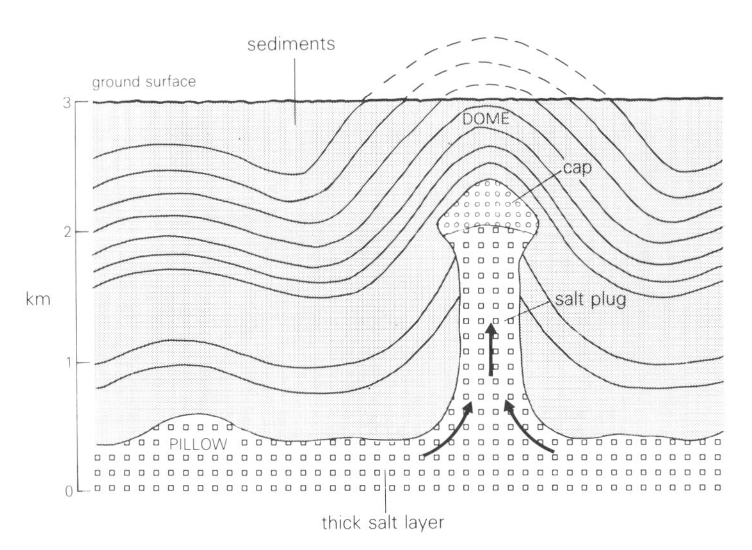

Figure 4.14 A thick layer of salt (and other evaporites), present at a depth of a few kilometres, has flowed laterally to form a salt pillow and a salt plug. As the salt plug rose towards the surface it has truncated the sedimentary layers it pierced, and has folded the beds above it into a salt dome, the top of which has been eroded. The salt at the top of the plug has dissolved in ground water in the sediments to leave an insoluble residue of gypsum and other minerals as a cap on top of the plug.

Most folds only approximate to one of these structural models and may, for example, be predominantly concentric but with some flow folding where a bed of mudstone is present.

4.4.4 Fractures in rocks

Brittle rocks deform by fracturing, especially if stress is applied rapidly. Whether or not a given rock is brittle depends not only on its texture and composition but also on its temperature, on the confining pressure around it, and on any fluids present within it. In general, fracturing takes place at shallower levels in the Earth than folding, though both may occur in the same place at the same time.

Where displacement, usually of a few centimetres, of the broken rock has taken place at right angles to the plane of fracture (Fig. 4.15), it produces a **tension fracture**. If the gap between the walls of the fracture is elliptical, the term **tension gash** may be used. These are often filled with later minerals precipitated from ground water, such as calcite or quartz.The nature of the displacement may be shown by the matching of some disrupted feature, say a sedimentary structure, on opposite sides of the fracture. Tension fractures form where **absolute tension** (a negative stress) has operated normal to the fracture plane and pulled the rock apart. Since the effects of overburden (and of the corresponding confining pressure) increase with depth, absolute tension is replaced, at a few hundred metres, by **relative tension** in the same direction. At these greater depths, stress is positive in all directions, but of minimal value in one – the direction of relative tension. **Extension fractures** (a type of

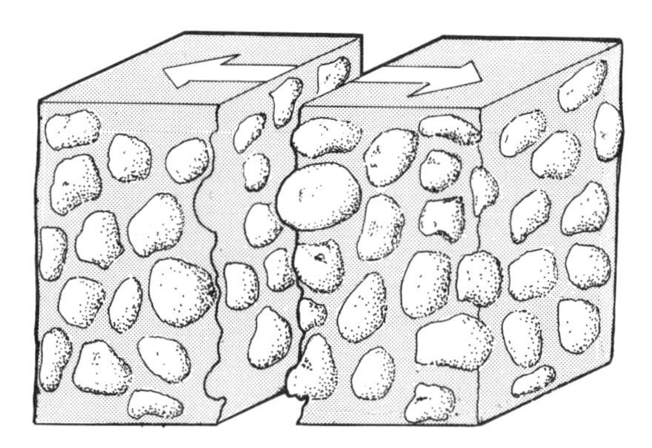

Figure 4.15 A fracture has been produced by tension in the conglomerate. The two walls of the fracture have moved apart in a direction at right angles to it. The plane of failure has not broken across pebbles but has followed the weaker cement of the rock to leave matching irregularities on both walls. Any space between the walls is usually filled with minerals precipitated from ground water.

tension fracture) may form in planes normal to this direction, in extremely brittle rocks, if stress is applied quickly.

More commonly, at depths where tension is only relative and where rocks are brittle or semi-ductile, failure takes place along **shear fractures**. In these, movement has taken place parallel to the plane of fracture, and the strata on either side are displaced by anything from a few millimetres to 100 km. The rubbing of one side of the fracture against the other results in polishing, and may produce grooves and ridges called **slickensides**, stretched in the direction of relative movement. Shears are formed normal to the plane in which the directions of maximum pressure and minimum pressure (that is, relative tension) lie, and at an angle of approximately 30° to the former (Fig. 4.16). Two sets of shears, lying symmetrically about the direction of maximum pressure (at 60° one to the other), should be produced. Both sets are commonly present, but one set is usually much better developed.

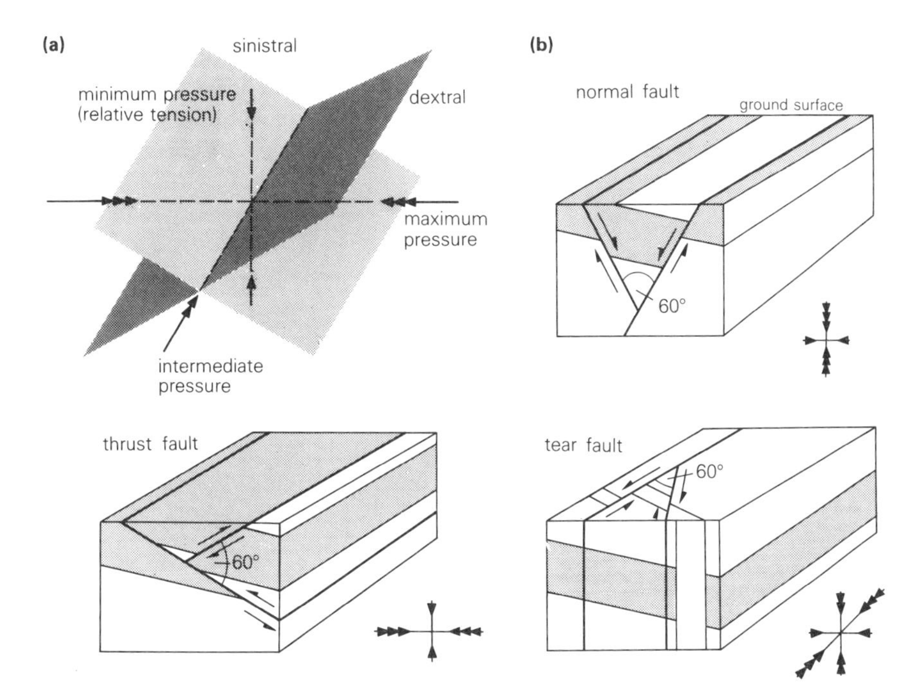

Figure 4.16 (a) The geometrical relationship between the directions of principal stress within a material and the shears that will appear once the shear strength is exceeded. Sinistral movement on a shear appears to an observer as relative movement of the far wall to his *left*, looking across the fault; dextral appears as relative movement to the *right*. (b) The geometrical relationship between the orientation of the directions of principal stress with respect to the vertical direction, and the type of fault formed by them. For example, the maximum pressure is vertical and the minimum pressure is horizontal when normal faults develop.

Minor fractures in rocks, both tensional and shear, are called **joints**. Larger shear fractures, where there are significant displacements of rock bodies across the plane, are called **faults** (see Fig. 4.17). (The distinction between a 'shear joint' and a 'small fault' is arbitrary, depending on circumstances and significance, but the former term usually suggests a displacement of a few centimetres.)

The presence of faults and joints is important to nearly all fields of economic geology. Joints, if frequent, have a considerable effect on the bulk properties of a rock mass. There is more difference in mechanical properties between a massive granite and a well jointed granite than between a massive granite and a massive gabbro. Joints affect the strength and stability of the rock mass, and the voids associated with their presence allow increased circulation of ground water through them. This may be relevant in water supply, in drainage of a deep excavation or in leakage through the sides or floor of a reservoir. Faults have similar effects on rocks, but are concentrated zones of weakness and of percolation, which may receive local remedial treatment in engineering works. Analysis of the geometry of fractures (Appendix H) is advisable in most construction projects, for example: where excavation can take advantage of planes of weakness; where rock bolts or other strengthening devices

Figure 4.17 Layers of sandstone and conglomerate cut by small faults and joints (some of which are small shears) similar, apart from magnitude of displacement, to the faults. (Reproduced by kind permission of Dr Brian Bluck.)

can be placed and orientated to be most effective; where design of major structures, such as dams, should be modified to avoid placing the maximum stress parallel to a plane of weakness; or where grouting has to be done to seal fractures against leakage. This latter is best achieved by drilling holes perpendicular to the fractures.

4.4.5 Faults

As with all structural surfaces, the orientation of a fault is expressed by its strike and dip. The displacement across a fault is called its **slip**, and the terms used to describe the components of slip (**strike slip**, **dip slip** and **throw**) are illustrated in Figure 4.18.

Only very small faults have a simple, clean-cut plane of movement. The majority are zones of shearing, usually a few metres across, but major faults may be more than 1 km wide. Within the **fault zone** there may be a number of planes of movement displacing the strata by **distributive slicing**, or the rock may be completely fragmented to form a fault **breccia** where competent rocks are affected, or **clay gouge** where incompetent rocks have been smeared out by shearing. (Gouge often acts as a seal to make the fault zone locally impervious to groundwater circulation.) At the **walls** (Fig. 4.19) of the fault the strata may show **terminal drag** in the direction of relative movement.

A fault is described by its geometry in two different ways: (a) by the

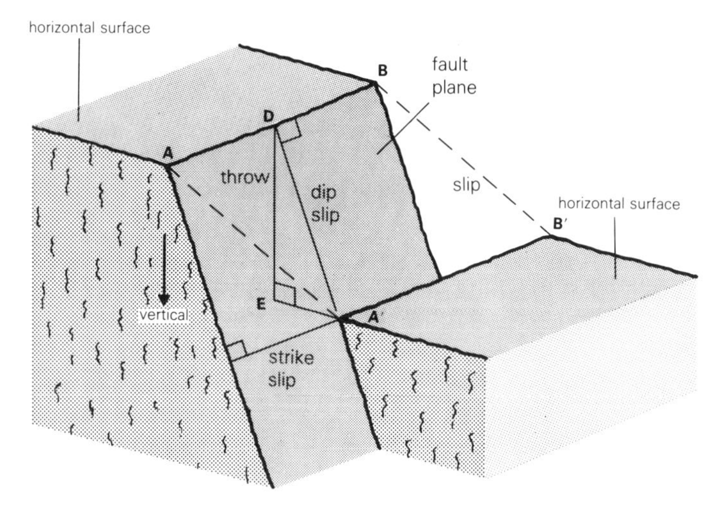

Figure 4.18 Before fault movement A and A′ and B and B′ were adjacent to each other. The total displacement A–A′ measured along the fault plane is called the slip of the fault. It may be resolved into strike slip and dip slip, which are respectively horizontal and vertical components, in the plane of the fault. Dip slip may also be resolved into a horizontal component and into a vertical component (the throw of the fault).

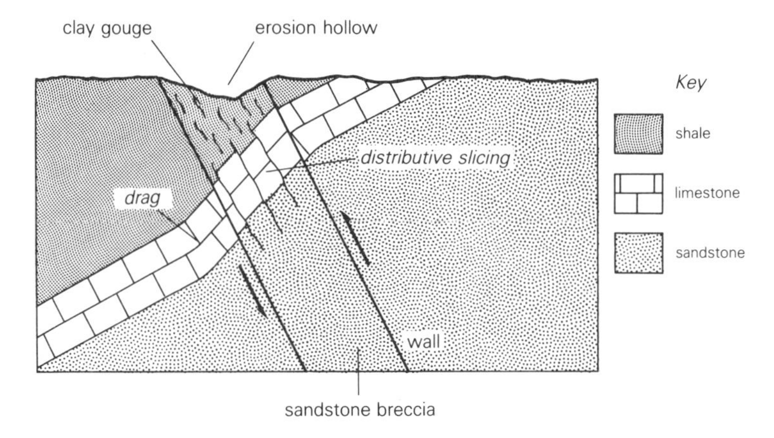

Figure 4.19 The section shows a reverse fault cutting a sequence of strata, sandstone–limestone–shale. Shearing within the fault zone has produced clay gouge from the shale, and fault breccia from the sandstone. The softer gouge in the fault has weathered out as an erosion hollow. The limestone is displaced in steps (that is, distributive slicing) by minor shears within the fault zone. The zone of shearing is bounded by the walls of the major fault, but some deformation (terminal drag in the directions of movement) is seen just beyond them.

relationship of its strike to the strike of the strata it affects, and (b) by the relative magnitudes of its components of slip.

(a) If the strikes are nearly parallel, the fault is a **strike fault**; if they are nearly at right angles, it is a **dip fault**; and if neither, it is **oblique**.
(b) Where slip is predominantly horizontal, producing a large strike-slip component and little throw or heave, the fault is described as **strike slip**. (The terms **wrench** and **tear** are used to describe certain varieties of strike-slip faults.) In faults of this type, the fault plane is near-vertical, and there is usually a comparatively wide fault zone in relation to displacement.

Where relative movement is up and down and strike slip is almost zero, a fault is described as **dip slip**. If its dip is towards the lowered side (that is, the **downthrown side**), the fault is **normal** (see Fig. 4.20). If the dip is towards the **upthrown side**, the fault is **reverse**. Normal faults dip at about 60°, and the common type of reverse fault, called a **thrust**, dips at about 30°.

This classification of faults into strike slip, normal and thrust is a genetic one that reflects the nature of the stresses which created them. The relationships may be inferred from what has been said in the previous section, and are shown in Figure 4.19.

Faults can be recognised from their effects on the pattern of outcrops. Dip faults and oblique faults truncate and displace the outcrops

Figure 4.20 A small normal fault affecting Coal Measures (sandstone and shales) revealed in an excavation in Central Scotland. The sense of movement along the fault may be inferred from the terminal drag of the layers. The fault zone of broken rock is very narrow. (British Geological Survey photograph, C2868, published by permission of the Director; NERC copyright.)

(Fig. 4.21b). Strike faults (Fig. 4.21a) produce a repetition of the succession of strata (but unlike a fold, repetition is in the same order) if the downthrow across the fault is in the opposite direction from the dip of the strata; they produce an omission of part of the normal succession of strata (which do not crop out because of the fault) if the downthrow is in the same direction as the dip.

4.4.6 Joints

The presence and orientation of joints have significant effects on the bulk properties of rocks, and their description (Appendix H) is an important stage in most site investigations. Joints may be non-systematic, but often occur as a set of parallel planes. If two or more sets are present, they form a system of joints (see Fig. 4.22). Prominent joints which are continuous as a single plane surface for hundreds of metres are called **master joints.** **Major joints** cross only a few beds before dying out. **Minor joints** are confined to one layer. A classification of joint spacing is given in Appendix H. In folded strata, joints are described as **longitudinal** if their strike is parallel to the fold axes, and otherwise as **cross** or **diagonal joints**.

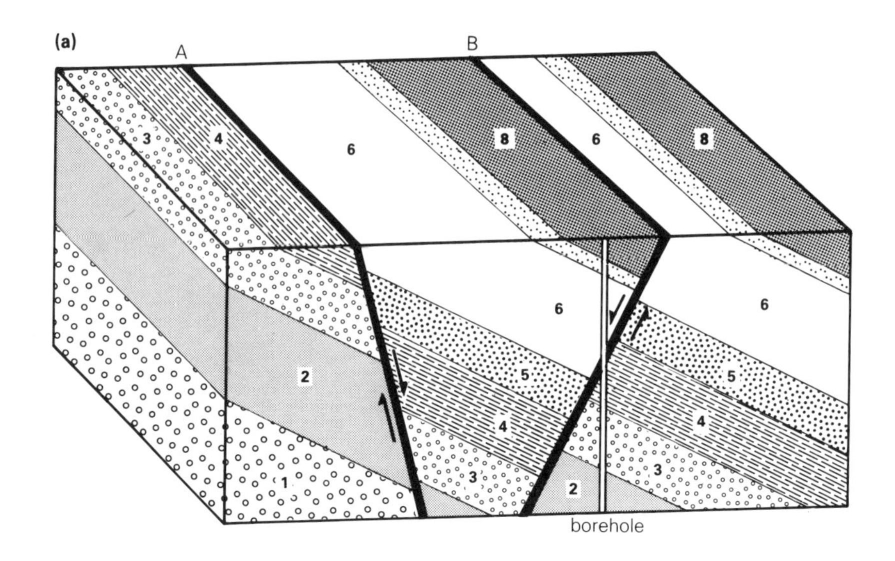
(a)
A
B
3
4
6
8
6
8
2
1
6
5
4
3
2
6
5
4
3
borehole

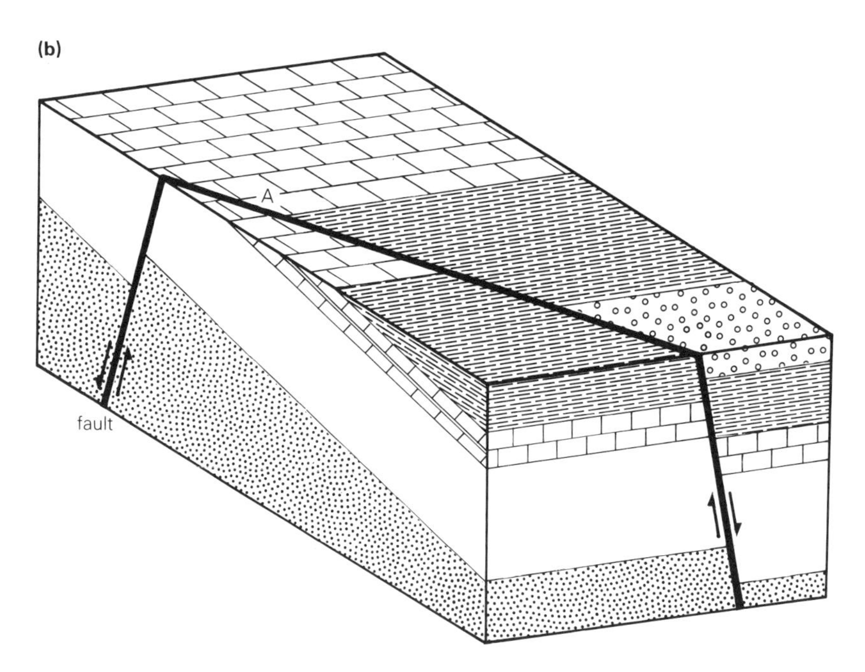
(b)
A
fault

Figure 4.22 Two sets of joints cutting a layer of sandstone, which is shown in plan view. Some joints are filled with a later mineral deposit, which has cemented the two faces together. (Reproduced by kind permission of Dr Brian Bluck.)

Joints may be tensional fractures or small shears. Their mode of formation may be shown by the correlation of disrupted structures across the joint, by slickensides or by crenulations on the joint surface. For example, plumose structure (which resembles large feathers chiselled on the rock) indicates that it failed by sudden rupture, and also shows the direction in which the fracture spread. Different geological conditions produce different patterns of joints. It is not possible to predict from general theory the position of individual joints, or the precise degree of

---

Figure 4.21 The patterns of outcrops related to faults. (a) Two normal strike faults, A and B, cut a sequence of strata, 1 to 8. The downthrown side of each fault is indicated on the map by arrows. Fault A throws down beds in the direction of dip, and as a result layer 5 does not crop out at the surface. There is *omission* of part of the sequence of strata as the fault is crossed. Fault B throws down beds in the opposite direction and produces *repetition* of the outcrops of layers 6, 7 and 8. Note that a borehole through the fault plane does not necessarily pass through a complete sequence of beds – layer 5 and parts of 4 and 6 are missed. (b) An oblique fault throws down the strata on the far side of the block and produces an offset of the outcrops, each of which is truncated by the fault. The amount of throw is equal to the thickness of the limestone layer; and top and bottom of that layer are aligned both in vertical section and on the map. The amount of throw of a fault can be read directly from the map, if the sequence and thicknesses are known, by seeing which beds are brought in contact and using this relationship.

development of any set in an unexplored site, but models of joint patterns and generalities about their probable development are of use in guiding further exploration.

In igneous rocks, a simple system of **cooling** or **contraction joints** is usually formed in lava flows, sills, dykes and other minor intrusions shortly after they are emplaced. Heat flows from the hot molten rock into the surrounding rocks or air and leaves the margins cooler than the centre. As the temperature of the intrusion drops to the melting point, solidification starts at random points at the margins. These cooling centres, by chance, are fractionally cooler than their surroundings. As these solid nuclei grow, thermal contraction of the igneous body is also taking place and this shrinkage produces small gaps between adjacent cooling centres (Fig. 4.23). As a result, a polygonal system of cracks develops around each centre, and these grow in towards the middle of the igneous body as cooling proceeds. Their orientation is controlled by the paths along which the temperature surface (isotherm) of value equal to the melting-point temperature retreats from the margin. This is usually at right angles to the margin, so that in lava flows and sills the joints are nearly vertical and in dykes they are near-horizontal (see Fig. 4.24). When well developed, as at Giant's Causeway in Ireland, they are referred to as **columnar joints**. Where columnar joints growing from opposite margins meet near the middle of a dyke or sill, a **median joint** should develop, but is not always present.

In plutonic masses of granite which have been forcibly emplaced, a

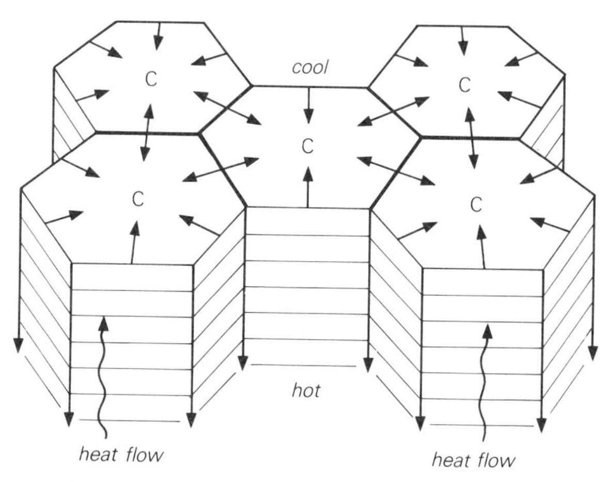

Figure 4.23 The diagram shows part of the upper contact surface of a sill and the distribution of cooling centres (C), where the temperature is marginally lower than elsewhere on the surface. Solidification of the magma started at these points. Further cooling and contraction has produced a polygonal system of cooling (tensional) joints, each lying midway between adjacent centres. The joints also grow downwards into the sill as it cools, to form columns.

Figure 4.24 Regular columnar jointing in a lava flow on the island of Staffa, Scotland. The jointing in the top flow is irregular, inferentially because of a peculiar pattern of cooling. (British Geological Survey photograph, D2218, published by permission of the Director; NERC copyright.)

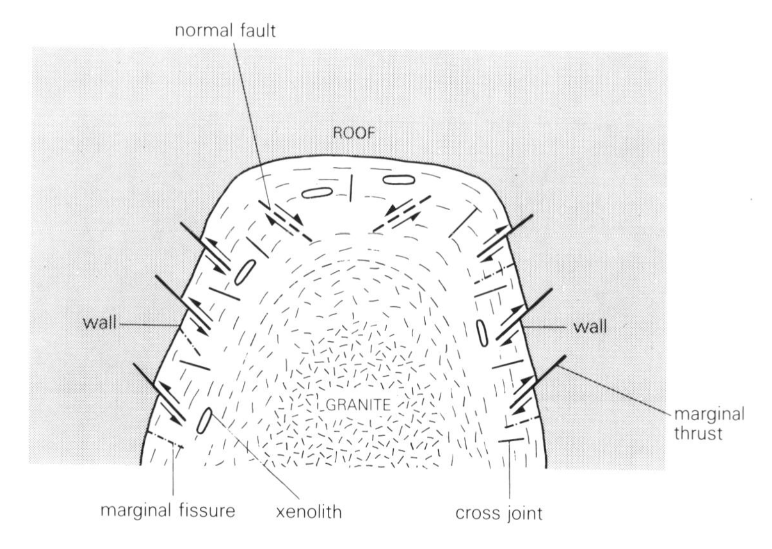

Figure 4.25 The section shows a typical pattern of fracturing (joints and faults) developed in a pluton which has been forcibly emplaced, and has steep walls and a broad roof. The granite of the pluton develops a flow texture, with parallel orientation of platy minerals towards its margins. Cross joints are produced by stretching of the rock in the direction of the flow texture; flat-lying normal faults are related to the stretching of the roof. Other features shown are marginal fissures, marginal thrusts and inclusions or xenoliths. The longitudinal joints lie in the plane of the paper and are not shown.

pattern of joints has been described and offered as a possible model (Fig. 4.25). In addition, a simple commonplace type of jointing (**sheet jointing**) is found close to the surface in large granite masses. The joints are usually near-horizontal and the dips are generally parallel to the principal topographic feature. For example, in a broad area of high ground related to the pluton, sheet jointing would be gently domed, but would not be parallel to minor valleys or spurs. When well developed, the granite characteristically shows a large-scale exfoliation with sheets of the order of 1 m thick. Sheet jointing may be produced by residual stresses in the granite arising from its earlier history, combined with the effects of unloading as erosion strips away the overlying rocks and reduces the confining pressure.

In sedimentary strata, a joint system may be present over a wide region and be related in whole, or in part, to the same stresses that produced folds and faults in the same rocks. Other mechanisms which can create regional joint sets are the very minor, but daily, distortion of the solid Earth by tidal stresses, and the relief of pressure produced as the load of overburden is stripped away by erosion. The frequency of joints in a layer is related to the rock type and the thickness. In general, the space between joints is less in brittle rocks with good elastic properties, and increases in proportion to the thickness. A thin layer of crystalline

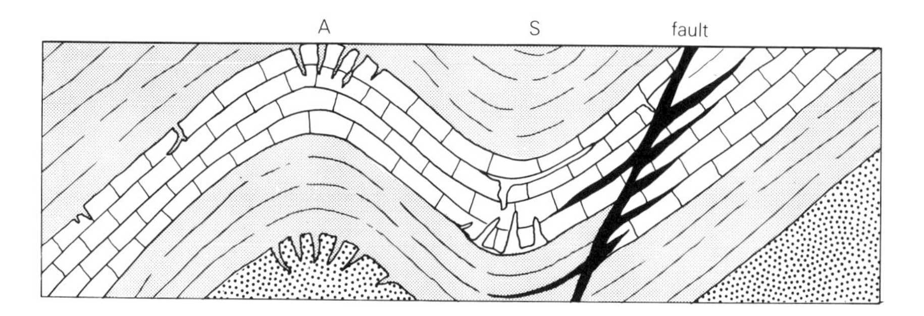

Figure 4.26 The increased curvature of the competent limestone and sandstone layers at the axial planes of the folds (A and S) produces the radiating joints shown. Minor shears, some parallel to the fault and others belonging to the complementary set of shears, are present as joints near to the fault zone.

limestone is usually well jointed. Joint frequency appears also to be increased by weathering, so that joints are more pronounced at the surface than underground.

In addition to any regional joint system, there are likely to be local concentrations of radiating tensional cracks caused by stretching at the hinges of folds, and formed close to fault zones as minor shears associated with the fault movements (Fig. 4.26).

4.4.7 Unconformity, overstep and onlap (overlap)

The geological history of an area is preserved in the strata. If there is a break in this record, because there has been no deposition of sediment and probably erosion for much of the time, the gap is called an **unconformity** between the lower and the upper series of strata (see Fig. 4.27). The surface corresponding to this time gap is the **plane of unconformity**. If there have been no Earth movements other than a vertical rise and fall relative to sea level, then the dip is the same above and below the unconformity. If tilting, folding or faulting of the lower formation (E–A in Fig. 4.28) before the upper series (K–M) is deposited on them has produced structural discordance, then the geometrical relationship at the plane of unconformity is an **angular unconformity**. The lowest bed (K) of the upper formation rests directly on progressively older beds of the lower formation and is said to **overstep** from E to B. The pattern of **subcrops** of E–A under bed K is identical to the outcrops once produced by the lower structure before K was deposited. A line of boreholes would show a systematic increase of the gap in the geological record, in the up-dip directions of the lower formation.

If subsidence has increased the area of deposition of the upper formation, then in the fringe of the area, where bed K was not deposited, bed L

Figure 4.27 Well bedded sandstones (Torridonian sandstone) resting unconformably on jointed metamorphic rocks (Lewisian gneiss). At the plane of unconformity that separates the two groups of rocks, there is a local gap in Earth history of hundreds of millions of years when no rocks were formed *and preserved* in this part of the North-west Highlands of Scotland. Note the slight overlap of the sandstone. (Reproduced by kind permission of Prof. Donald Bowes.)

rests directly on the older formation E–A, and is said to **onlap** or **overlap** K (Fig. 4.28). Onlap is always accompanied by overstep. The variable gap (with respect to the fullest sequence of strata for the area) at the unconformity is a combination of the absence of the top beds of the lower formation because of overstep and the absence of the bottom beds of the upper formation because of onlap.

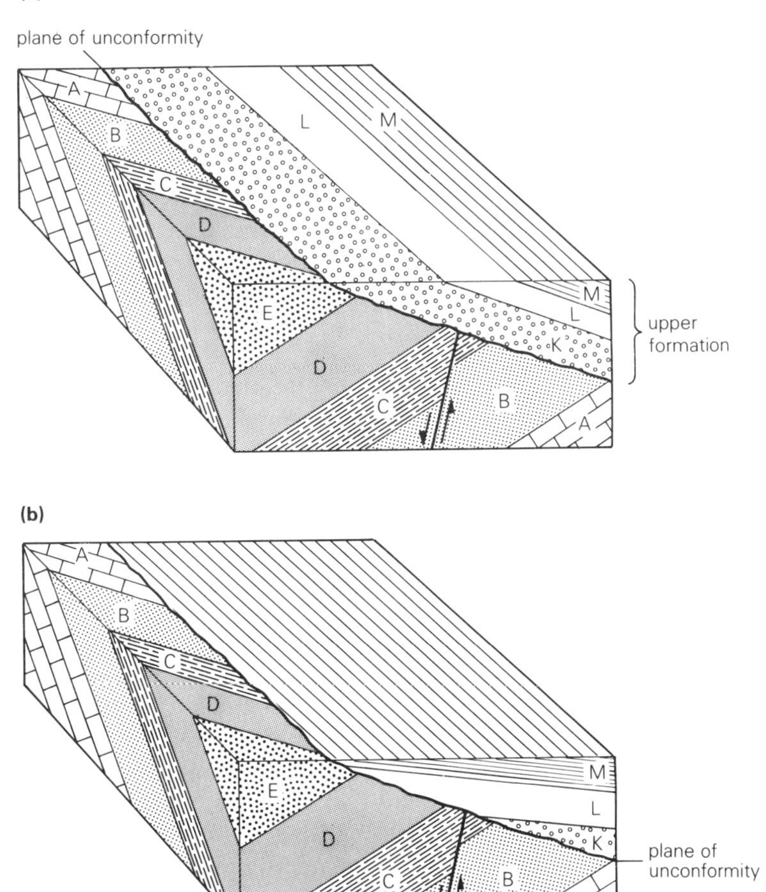

Figure 4.28 Overstep and onlap (overlap). (a) The upper formation K–L–M rests with angular unconformity on a lower formation A–E. The overstep of layer K which brings it above successive beds of the lower series is seen both in vertical section and on the map. (b) This diagram combines the overstep of the upper formation with onlap (overlap) within the upper formation.

4.5 Plate tectonics and seismicity

4.5.1 Earth's layers and plates

Seismic studies have shown that the Earth is comprised of three main concentric layers (Fig. 4.29):

(1) An outer **crust**, ranging in thickness from 5 to 10 km beneath the oceans and more than 40 km beneath the continents.
(2) The **mantle**, 2900 km thick, which is separated from the crust by the **Mohorovičić discontinuity** (called **Moho** or M discontinuity). The mantle contains several layers.
(3) The **core** is separated from the mantle by the **Gutenberg discontinuity**, and is also layered with an inner and outer core, the whole core being about 3500 km in thickness (Fig. 4.29).

The crust, mantle and core and any internal layers they may possess are distinguished from each other by their different seismic velocities.

Concentric layering in the Earth can be defined using another set of terms based on strength and viscosity. Again there are three layers or shells:

(1) The **lithosphere** is the outermost shell and is about 100 km thick (thus incorporating the crust and uppermost mantle). It can support large surface loads, like volcanoes, without yielding and is therefore rigid.
(2) The **asthenosphere** underlies the lithosphere and is 700 km thick. It is at a high temperature (near its melting point), has little strength, and can flow when stress is applied over a period of time. The upper asthenosphere is the zone where magma is generated.
(3) The innermost shell is the **mesosphere**, which is by far the thickest, including most of the mantle and extending to the core. It is more rigid than the asthenosphere and more viscous than the lithosphere.

The face of the Earth shows continual change throughout its long history. Some changes, particularly those produced by weathering and erosion of the land surface, are powered by heat from the Sun and the action of gravity. Other forces that change the Earth's surface, especially those acting upon the seas, are a result of complex reactions between the attractive forces of the Moon and the Sun. The relative positions of land and sea may alter during geological time, by either rises or falls in the level of the sea or of the land.

More important changes may occur because parts of the rigid outer shell of the Earth (the lithosphere) move on top of the asthenosphere. The

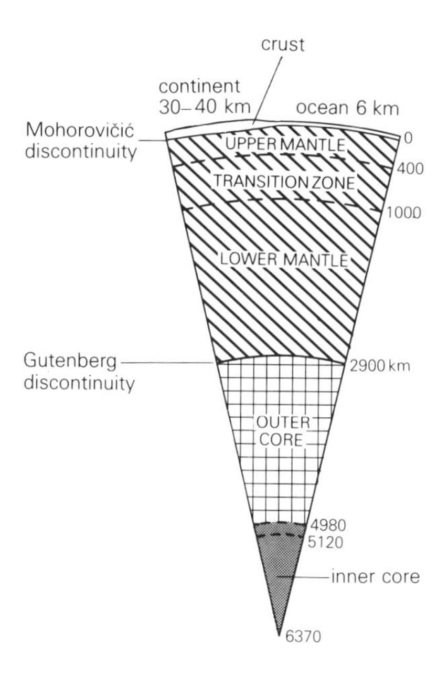

Figure 4.29 Layers of the Earth and major discontinuities.

study of the movement of these rigid shell parts (or plates) is known as **plate tectonics**. The essential concept of plate tectonics is that the entire surface of the Earth is composed of a series of rigid, undeformable, thin (<150 km thick) plates, and seven major plates cover most of the Earth's surface (Fig. 4.30a), namely the Pacific Plate, the African Plate, the Indian Plate, the Eurasian Plate, the Antarctic Plate, the North and South American Plate, and the Nazca Plate. Many smaller plates exist in addition to these, such as the Caribbean Plate, which is sited at the Caribbean Sea. The plates are continuously in motion, both in relation to each other and to the Earth's axis of rotation.

Plates are bounded by **active ridges** and **trenches**. Ridges or **mid-ocean ridges** are the centres of divergence or of spreading-apart of two plates. At such a mid-ocean ridge, new oceanic basalt is extruded and added to each plate. Such a process was recognised by examining the magnetic properties of the basalts at each side of a mid-ocean ridge over a distance of several hundred kilometres. Within igneous rocks, small crystals of magnetite act like magnets and align themselves in the Earth's magnetic field as the rock cools, eventually retaining a strong permanent magnetisation indicating the position of the Earth's magnetic poles at that time. Studies in this branch of the subject (called **palaeomagnetism**) have shown that the Earth's magnetic poles have moved throughout geological time and that the polarity of the Earth's magnetic field is not constant but that the north and south poles alternate at regular intervals, in the order

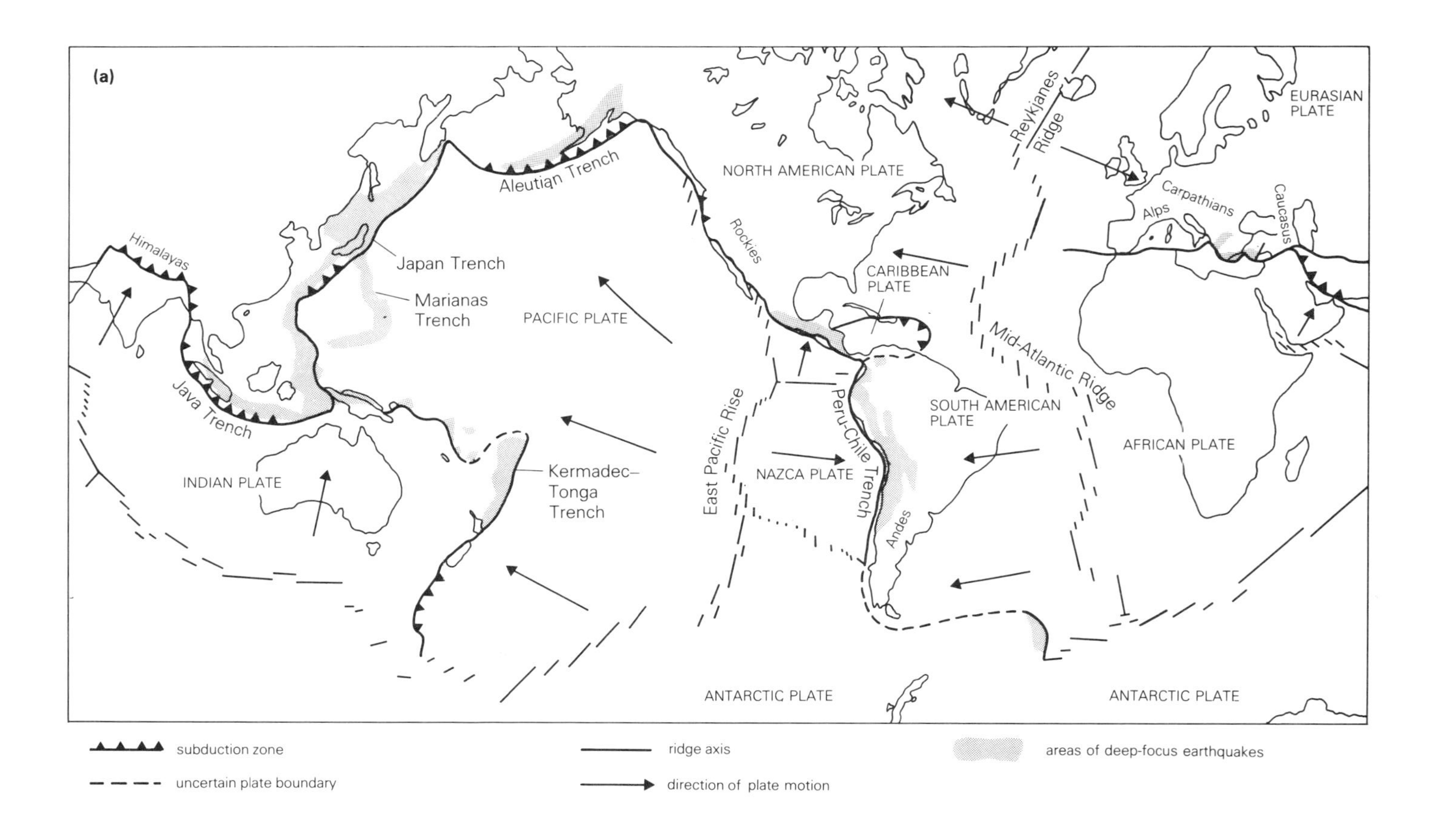
(a)
EURASIAN PLATE
Reykjanes Ridge
NORTH AMERICAN PLATE
Carpathians
Alps
Caucasus
Aleutian Trench
Himalayas
Japan Trench
Marianas Trench
PACIFIC PLATE
Rockies
CARIBBEAN PLATE
Mid-Atlantic Ridge
Java Trench
SOUTH AMERICAN PLATE
AFRICAN PLATE
East Pacific Rise
Peru-Chile Trench
Kermadec–Tonga Trench
NAZCA PLATE
INDIAN PLATE
Andes
ANTARCTIC PLATE
ANTARCTIC PLATE
subduction zone
uncertain plate boundary
ridge axis
direction of plate motion
areas of deep-focus earthquakes

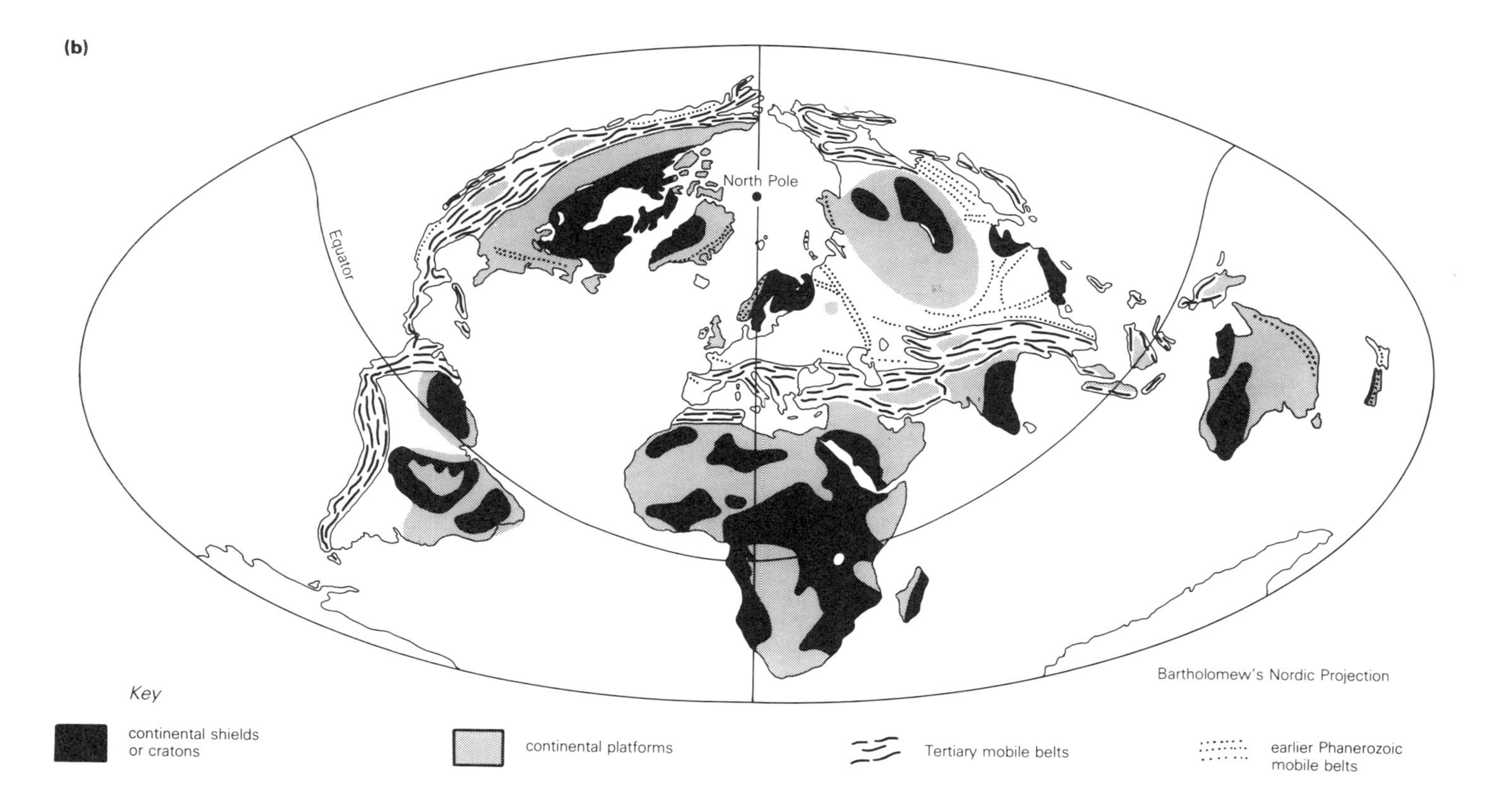

Figure 4.30 (a) Location and boundaries of the major lithospheric plates of the Earth, showing relative motion by means of arrows. (b) Major structural units of the world.

of half a million years. That is, at regular intervals, the Earth's magnetism is reversed.

These changes in polarity can be recognised in the ocean basalts forming at mid-ocean ridges. When several hundred kilometres of basalt (ocean-floor basalt) were examined at each side of a mid-ocean ridge, it was found that the basalts showed linear magnetic anomalies, with stripes of alternating normal and reversed magnetism, and that the mid-ocean ridge acted as a mirror plane, the magnetically striped basalts on one side of the ridge being the mirror images of those on the other side of the ridge (see Fig. 4.31).

These mid-ocean ridges are rarely continuous, but have a stepped appearance owing to **transverse fractures** or **transform faults** affecting the plates. No plate is destroyed or new material added along these fracture zones.

Trenches occur where the plates converge and one plate is thrust under the other, being 'consumed' at such a boundary. Plate boundaries at trenches are usually curved, the convex side pointing towards the plate being consumed. Trenches indicate **destructive plate margins**, like the Kermadec–Tonga Trench and the Peru–Chile Trench (Fig. 4.30a).

The underthrust plate is destroyed (consumed) in a **subduction zone**, which lies on the concave side of ocean trenches, and the plate is destroyed at the same rate as new material is added to it at a mid-ocean ridge. Because plates are poor conductors of heat, they must descend several hundred kilometres into the mantle before being consumed (Fig. 4.31).

As two plates collide along a destructive plate margin, **island arcs**, such as the East Indies may form on the surface, or great mountain ranges, such as the Andes or Himalayas, may form in the upper crust.

The global distribution patterns of **earthquakes** and their depths of origin were known before ideas on plate tectonics were formulated. Earthquake activity patterns were determined by **seismology** studies, and major earthquake zones were identified around the Pacific Ocean, along the Alpine–Himalayan belt, on mid-ocean ridges, at trenches and in young fold mountain belts. A seismicity map for the Earth in the 1960s is shown in Figure 4.32, which demonstrates that seismicity is concentrated along plate boundaries (compare with Fig. 4.30a). Seismic activity is of two types: shallow and deep earthquakes. The former are concentrated at ocean ridges and transform faults, and occur at depths of less than 30 km. Deep earthquakes occur in zones beneath the oceanic trenches, with the foci (see next section) located at varying depths along the subduction zone, which extends downwards to depths of more than 250 km. The deepest earthquakes have been recorded from active island-arc systems.

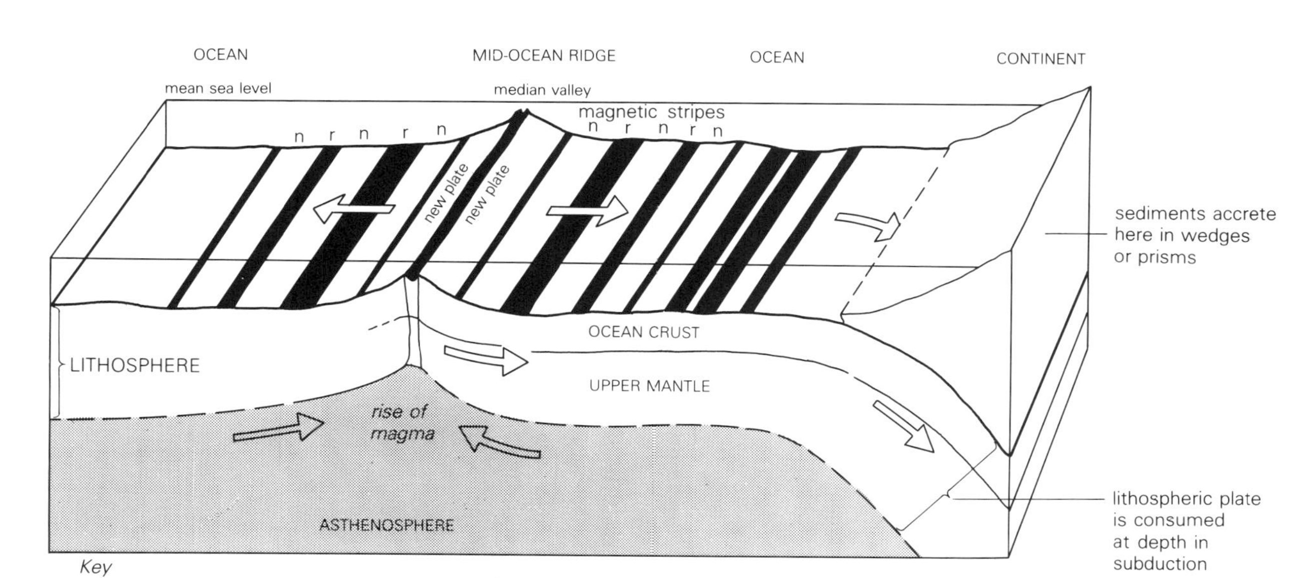

Figure 4.31 The diagram illustrates (with vertical exaggeration for clarity) ocean-floor spreading about a mid-ocean ridge. New oceanic crust is formed in the median valley where the crust is pulled apart and hot magma is intruded from the upper mantle. The upper mantle below the ridge is partly molten. It forms part of the lithospheric plates resting on a viscous asthenosphere which lies at depths of 70 km below sea level, and over which the plates can glide. The new oceanic crust is magnetised as it cools, and takes on normal or reversed polarity, depending on the geomagnetic field at the time of its formation.

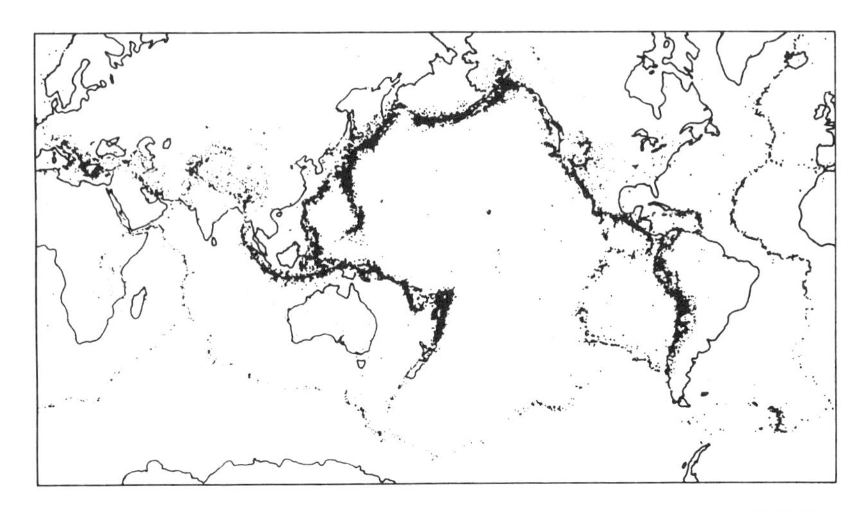

Figure 4.32 Distribution of earthquake epicentres on the Earth for the period 1961–67.

4.5.2 *Magnitude and intensity of earthquakes*

Seismic waves are generated when there is a sudden release of energy because rocks which have been strained elastically suddenly fail and move. The centre at which this happens, and from which waves are transmitted in all directions, is called the **focus**. The position on the ground surface vertically above it is the **epicentre**. The **magnitude** (M) of an earthquake is a measure of the energy generated at its focus. A scale of magnitude is a way of classifying earthquakes according to their *potential* destructive power. There is more than one specific definition of magnitude, and more than one scale of magnitude in international use. The **Richter Magnitude Scale** is commonly referred to, and defines magnitude by the equation:

$$M = \log_{10} A - \log_{10} A_0$$

where A is the amplitude of a given earthquake on the recorded trace of a standard instrument, at a given distance, and A_0 is that for a particular earthquake selected as a standard. The reference level A ($M = 1$) can be fixed by stating its value at a particular distance from the epicentre, for example 0.001 mm at 100 km. In that case, an earthquake recording a trace amplitude of 1.0 mm on a similar seismograph at 100 km from the epicentre would have a magnitude of 3. Magnitudes greater than 8 are rare and special events, and they would usually be perceptible to people more than 600 km away from the epicentre.

Earthquakes with magnitudes greater than 5.5 on the Richter Scale are large enough to be recorded on seismographs over the entire Earth. The first motion of the ground at such a recording station may be either *compressional*, with the first ground motion away from the source, or *dilational*, with the first ground motion towards the source. Elastic waves generated by an earthquake can be a mixture of both types.

Earthquake intensity is a measure of the amplitude of ground vibration *at one locality*. It is related to the square of the distance of the locality from the epicentre and to the amount of damping of the seismic waves produced by geological conditions along their path to the locality. It would be fortuitous to have a recording instrument at the locality when waves arrive, and assessment of intensity is based conventionally on other phenomena related to ground motion. The **Modified Mercalli Scale** is shown in Table 4.4, together with the *order of magnitude* of the maximum acceleration of the ground as it vibrates. These figures are meant as no more than an indication of a parameter of special interest to the engineer, since intensity is affected by local conditions. **Isoseismal maps** of an earthquake may be constructed by plotting the intensities at different localities and contouring equal values. Isoseismal lines are usually elongated along the fault which has moved, and are distorted by other geological structures.

4.5.3 *Relationship of earthquakes to faults*

The commonest sources of shallow earthquakes are active faults, and many epicentres are located close to the outcrop of a fault (see also

Table 4.4 The Modified Mercalli Scale (1931) of earthquake intensity.

Scale		Approximate maximum acceleration ($mm\,s^{-2}$)	Description
I	instrumental	10	felt only by instruments
II	very feeble	10	felt by a few people at rest
III	slight	25	like a passing lorry
IV	moderate	50	generally perceptible
V	rather strong	100	dishes broken, many people wakened
VI	strong	250	some plaster cracked, felt by all
VII	very strong	500	damage to poor construction
VIII	destructive	1000	much damage to normal buildings, chimneys fall
IX	ruinous	2500	ground cracked, great damage
X	disastrous	5000	many buildings destroyed
XI	very disastrous	7500	few structures left standing
XII	catastrophic	9800	total destruction

Section 4.5.5). The relationship is explained by the **Elastic Rebound Theory** (Fig. 4.33). There is slow differential movement of the blocks on both sides of the fault, but this initially does not cause slip along the fault plane. It is locked by friction. Instead, there is measurable elastic distortion of the rocks in a zone on both sides up to several hundred metres wide. Then, as movement proceeds and as elastic strain builds up to a critical value, friction is overcome and the fault moves. The elastic energy stored in the distorted rock is released suddenly as a pulse of earthquake waves. For example, the San Francisco earthquake of 1906 was accompanied by visible surface displacement along 430 km of the San Andreas Fault, with a maximum strike-slip displacement of 6.5 m.

When faulting occurs, a pattern of compressional (extension) and dilational (contraction) waves are generated, and the pattern can be used to define the movement along the fault plane. A network of seismic stations will determine the changing character of waves associated with an earthquake. From first motion studies a seismologist can determine the orientation of faults and the slip directions of earthquakes anywhere on Earth.

Britain is a region of low seismicity, but a few weak shocks are felt locally each year, and at intervals of several years an earthquake of greater magnitude (say Richter 5) causes minor damage to old masonry, loosens slates, breaks ornaments on shelves, gets coverage by the Press, and

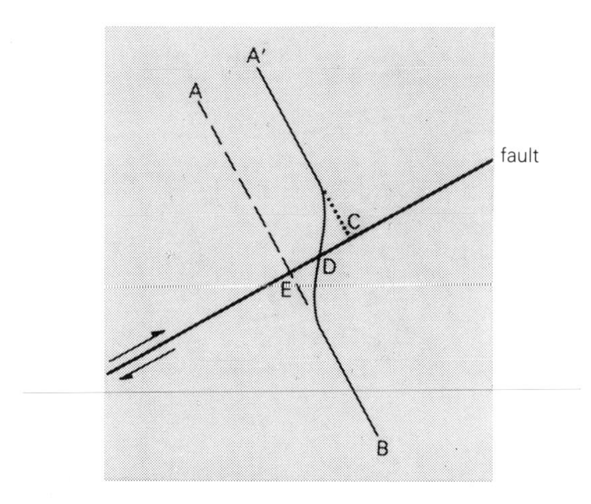

Figure 4.33 Diagrammatic map showing a fault along which horizontal dextral displacement is slowly taking place as indicated. An originally straight line A–B across the fault becomes deformed with movement in A′–D–B. At this stage of deformation the elastic distortion and the stresses generated close to the fault are sufficient to overcome frictional forces along the fault plane. Movement takes place suddenly and vigorously from D to C on one side and from D to E on the other, to produce two discontinuous lines A′C and EB, displaced across the fault. The corresponding release of elastic energy generates an earthquake.

makes insurance companies uneasy and policy holders read the small print. Many small earthquakes in Britain are associated with NE–SW trending faults such as the Highland Boundary Fault. No significant rupture of the surface or measurable displacement definitely associated with the shocks is clearly recorded, although there is mention of minor fissuring in a popular account of one earthquake associated with the Great Glen Fault. The stresses that stirred these ancient faults are related, in part, to geological movements affecting the British area at present (that is, to uneven isostatic recovery of the crust from the last ice sheet, and to tilting towards the subsiding North Sea Basin).

4.5.4 *Prediction of earthquakes*

Prediction of seismic risk can be assessed from general theory and from past records of an area. It is used for planning and for laying down building codes. Prediction of a specific, destructive earthquake is of greatest value only if it allows evacuation of a threatened city and deployment of emergency services a few days before the event. Prediction approaching this accuracy has been achieved in a few cases since 1973, and the Chinese claim to have successfully evacuated one of their towns in 1975. As elastic strain builds up within rocks close to the focus, over periods as long as decades, minute cracks in the rocks (**microfractures**) open and affect their physical properties. This dilatancy of the rocks increases their volume, and results in minor tilting of the ground surface from the active fault. Small quantities of the gas radon, trapped in some rocks, are released and can be detected by analysis of well water. The velocity of seismic waves (Section 6.3.2) changes, and V_p/V_s decreases by about 20%. This change can be monitored by local seismic surveys at daily intervals. A day or two before the earthquake, the seismic velocities revert to normal as water seeps into the microfractures. The fluid pressure within the voids of the rock (**pore pressure**) is increased, and by reducing effective pressures allows movement along the fault to take place.

The relationship between pore pressure and earthquakes was first observed when disposal of fluid toxic waste down a deep boring which traversed an active fault accidentally triggered off weak shocks. Control of the potentially destructive San Andreas Fault in California has been proposed by using this triggering mechanism to release as a series of weak, predictable shocks the enormous elastic strain energy that is already stored. Each stage of the programme would require three boreholes, each about 5 km deep and spaced 0.5 km apart along a segment of the fault. Water would be pumped from the outer holes to lock these points on the fault, and would then be injected into the middle hole to reduce friction across the fault plane to give a controlled tremor. This would be repeated along the the entire fault. The scheme would cost thousands of millions of dollars for 500 boreholes, plus a less predictable amount in subsequent civil damages suits. As in many imaginative engineering projects, there are political difficulties as well as economic ones.

4.5.5 *Seismic risk and problems for the engineer*

An earthquake may be of very small magnitude and only detectable by instruments, but may be powerful enough to cause annoyance or alarm to people and to damage buildings. Seismic risk from ground movements must be foreseen, understood and dealt with in planning and design. Regions of high seismicity (see Section 4.5.1) are often covered by building laws specified in terms of earthquake parameters, and even in countries of low seismicity, such as Britain, the probability of an earthquake of given power occurring within so many years must be considered in siting potentially dangerous structures like nuclear power generating stations.

Information about seismicity and seismic risk is obtainable by referring to the Global Seismology Unit of the BGS (Appendix C) in Edinburgh. A catalogue of earthquakes that have been recorded over the previous year is produced by the International Seismological Centre, Newbury, Berkshire. Older catalogues based on questionnaires to the local public and on historical records have been produced by individuals. The historical records have sometimes needed interpretation. For example, the area affected by one earthquake in Scotland at the beginning of the 17th century can be surmised from the records of Aberdeen. The town council decided to enforce a ban on playing golf on the Sabbath shortly after the shock!

Seismic activity must be considered when certain engineering structures, particularly dams, are designed. Dams in earthquake zones will be designed with a capacity for resistance to the dynamic forces that can be applied during an earthquake. Before construction begins, past seismological records of the dam site or nearby areas should be checked and major fault lines located. An instrumentation programme should be initiated on the actual site as soon as its location has been determined, particularly with regard to long-term seismic monitoring. The type of ground movements likely to occur should be predicted, and whether any earthquakes that may occur are likely to be of shallow depth, such as by slip along a fault plane, or of deep focus. Structural damage is likely to result from ground displacement, acceleration at low seismic frequencies and velocity from blast vibrations. Ambraseys and Sarma (1967) considered many of these factors in dam design. Earthquakes may cause damage in unconsolidated deposits. Filled ground may consolidate drastically, and studies have been made about the possibility of **liquefaction** development, either in the saturated fill material of an earth dam or rockfill dam or in the foundation material.

Earth or rockfill dams may fail in several ways during an earthquake. Circular shear failure or rotational failure of the embankment, and planar base failure along the embankment–foundation interface (see Section 8.1.3) are the three main types. Concrete dams may be subjected to failure by either the second or third of these modes.

Other important sources of small local tremors are created by human agency. The impounding of great masses of water in very large reservoirs may result in uneven and spasmodic settlement under the new load. More importantly, the extraction of coal causes collapse of the strata above the seam into the abandoned workings, and often produces small seismic foci. Instead of broad gentle subsidence, slippage may occur locally along a fault, with each jerky movement producing a small shock. Contemporaneously, **rock bursts** may explode from any massive rocks in the passageways as they collapse, with each burst acting as the focus of a minor shock. There is disagreement on the details of the mechanism linking seismicity with abandoned workings, and thus about the legal liability for damages associated with them in particular cases, although a general correlation has been demonstrated in more than one area.

4.6 The continental crust

4.6.1 Major crustal features

The continental crust consists of rocks formed at various times throughout the entire history of the Earth. These rocks can be divided into three main groups (Fig. 4.30b):

(a) **continental shields** or **cratons** of Precambrian age including igneous and high-grade metamorphic rocks;
(b) **continental platforms**, where a cover of younger sedimentary or metamorphosed sedimentary rocks overlies Precambrian basement;
(c) younger fold mountain belts (**mobile belts**) of several ages (of which Tertiary belts are most strongly defined). These mobile belts contain rocks of different ages, including younger igneous rocks. The position of these mobile belts should be compared with the plate margins in Fig. 4.30a.

4.6.2 *The rocks of Britain*

A brief general account of some of the rocks of Britain is given here to put some descriptive flesh on the bare bones of the general concepts of stratigraphy and structure. It should be read as a description of some stratigraphic models rather than as an account of local rocks and regional stratigraphy.

PRECAMBRIAN AND LOWER PALAEOZOIC ROCKS OF BRITAIN

Precambrian rocks crop out over most of the Scottish Highlands, in North and South Wales and in the Welsh Borders, and they are present at depth everywhere. For example, in the North-west Highlands, several hundred metres of Cambrian quartzite and limestone rest directly on two major groups of rocks, the Torridonian sandstone and the Lewisian gneiss (Fig. 4.34), both of which inferentially must be Precambrian. The Lewisian gneiss consists of a variety of metamorphic rocks, some of which are altered sediments and others altered igneous rocks. The commonest rock types are hornblende-gneiss, granulite and granite. They are cut by dykes and by broad shear zones where the rocks have been altered and weakened. The events stamped on these rocks include recrystallisation under changed temperature and pressure, intrusion of dykes and episodes of stress. That is to say, they preserve a history largely of happenings at depths of kilometres below the surface, rather than in the environments at the surface. The presence of metamorphic quartzite and marble (that is, altered sedimentary rocks) does, however, indicate that

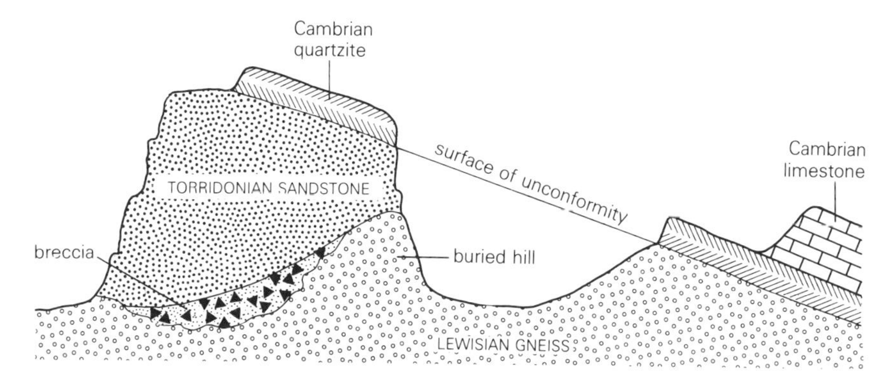

Figure 4.34 This diagrammatic section, drawn with an exaggerated vertical scale, shows the relationship between the Lewisian gneiss, Torridonian sandstone and Cambrian quartzite and limestone in the North-west Highlands of Scotland. The surface of unconformity at the base of the Cambrian strata is a plane dipping steadily in the same direction as the bedding, whereas the unconformity at the base of the Torridonian sandstone is highly irregular. A pre-Torridonian buried hill formed of Lewisian gneiss rises on the right-hand side of the present-day mountain. The ancient screee formed on its lower flank is consolidated to form breccia.

seas did already exist. Several of the events have been dated radiometrically, and results show that the earliest episode of metamorphism took place almost 2700 million years ago. The rocks it affected must have been formed even earlier.

In contrast to the highly altered Lewisian gneiss, the Torridonian sandstone is composed of unaltered sedimentary rocks. Although these range in age from 950 to 750 Ma, most of them are near horizontal and free from tectonic disturbance other than simple faulting. The North-west Highland region, which suffered intense tectonic activity in early Precambrian times when the Lewisian gneiss was formed, has been *stable* since then. The commonest Torridonian rock type is a red feldspathic sandstone (arkose), often with cross bedding or other sedimentary structures. It forms mountains almost 1 km high and its total thickness in some localities, before erosion, may have been as much as 4 km. In the northern part of its outcrop, the predominant red colour of the cement indicates that the Torridonian sandstone was exposed to the air at some stage during deposition and cementation. It was deposited in temporary lakes or on land, and is not a marine deposit. Other evidence, such as wind-rounded sand grains, infilled stream channels and ripple marks on some slabs, can be knitted together to show that these rocks were formed in mountain-girt desert basins, involving deposition, sometimes by seasonal floods, sometimes by wind. The ancient geography and its modification can be interpreted from the distribution of rocks. Conversely, it is possible to infer something about the distribution of rocks from this general reconstruction. For example, the surface between the Torridonian sandstone and the Lewisian gneiss is highly irregular, and it rises and falls in altitude by up to 600 m, although the Torridonian sandstone layer is near horizontal. The Torridonian sandstone was deposited on the ancient land surface formed of hills and valleys cut into the Lewisian gneiss in pre-Torridonian times. It was buried eventually as sediment and was banked progressively against these hillsides as the region subsided. Breccia consisting of angular fragments of gneiss in a sandy matrix occurs on the lower slopes of these buried hills, and in the context of the stratigraphic model they are interpreted as ancient scree slopes. Their distribution may be understood and tentatively predicted from this relationship if enough is known about the ancient geography.

The inliers of Precambrian rocks in Wales and western England contain a variety of metamorphic rocks (for example, in Anglesey), great thicknesses of sandstones and other terrigenous strata (for example, in the Long Mynd of Shropshire), and also thick piles of acid volcanic rocks (most notably the Uriconian rocks of Shropshire). The Precambrian rocks of Shropshire are overlain by several hundred metres of coarse sandstones and other sedimentary rocks of Cambrian age. Like the Cambrian strata of the North-west Highlands, the Shropshire rocks are comparatively free of tectonic disturbance. In contrast, the Cambrian strata cropping out in the regions between, such as North Wales and the southern Scottish Highlands (Fig. 4.35), are much thicker and have been deformed by compressive stress. The Cambrian strata are a few kilometres thick in North Wales.

The common rock types in these regions are **turbidites** (including greywacke), which consist of poorly sorted mixtures of clay, silt, sand and rock fragments. They are formed where unstable sediment is disturbed and flows down a slope on the sea floor as a thick mixture of sediment and water (slurry) before settling in deeper water. Similar changes of thickness occur in other strata of Lower Palaeozoic age within a broad tract trending NE–SW across Britain between the stable areas of the North-west Highlands and Welsh Borders that form its margins. The traditional term used to describe such a major belt of thick sediment is **geosyncline**. The sediments of the geosyncline have been strongly compressed subsequently, as is shown by the intense crumpling and fracturing that they have suffered. It appears that the two stable margins have moved together to squeeze the intervening sedimentary pile. During the early depositional period and the later tectonic activity, there are characteristic types of volcanicity. The emplacement of granite batholiths and the metamorphism of some of the sediments at a late stage in the history of the belt are indicative of the rise of temperature within it. Active belts of this type, which have been or are being affected by tectonic activity and volcanicity, are referred to as mobile belts in contrast to stable regions (see Fig. 4.30b), such as the North-west Highlands since Torridonian times. A mobile belt often coincides with an earlier geosyncline, and in turn is often the site of a later mountain chain. The eroded roots of high mountains which rose on the site of the Lower Palaeozoic mobile belt in Britain are exposed in Wales, the Lake District, the Southern Uplands and the Scottish Highlands. The sequence of events – geosyncline, compression, granite emplacement, and finally the formation of a mountain chain – used to be described as the **mountain building** (or **orogenic**) **cycle**. A fuller understanding of it has come with the theory of plate tectonics (see Section 4.5). Ocean floor is destroyed as plates move towards each other. The thick 'geosynclinal' deposits are turbidites accumulated on the flanks of the continental masses, beyond the shallow seas of the continental shelf, and in deep marginal troughs like the oceanic trenches of the Pacific (Figs 4.30a & 35b). Melting of the underthrust plate gives rise to volcanic activity. Eventually the movement and the processes halt as the continental masses grind together and lock, leaving the thinner sediments on the continental shelves unaltered. As the compression is relaxed, the thickened plate of the mobile belt is free to rise buoyantly in the denser asthenosphere, in which it 'floats', and the tract of high ground becomes a mountain range.

(a)

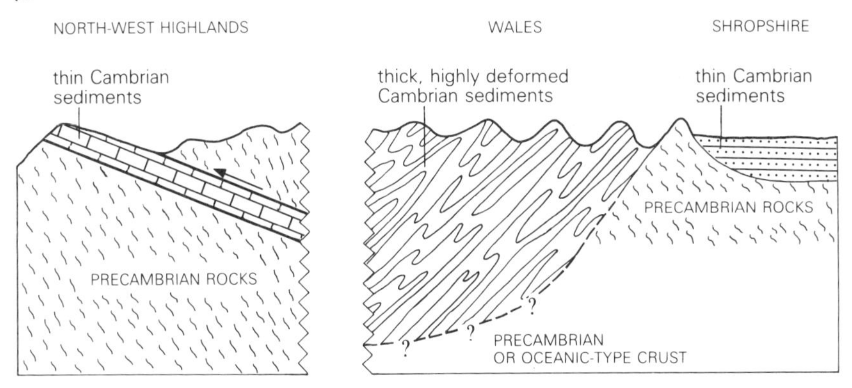

(b)

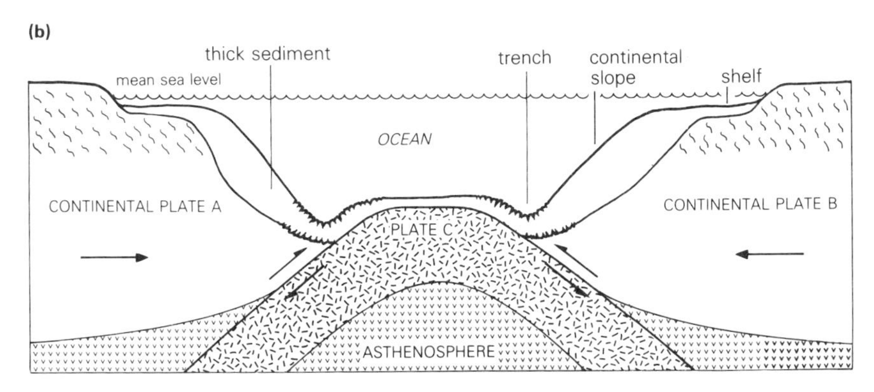

Figure 4.35 The Lower Palaeozoic mobile belt of Britain. (a) The diagram illustrates how the thickness and degree of deformation of the Lower Palaeozoic sedimentary rocks change between the stable margins of the North-west Highlands and Shropshire. Thick, highly deformed sediments of Lower Palaeozoic age are present in Wales and also (not shown) in the Southern Highlands, Southern Uplands and the Lake District. (b) The diagram offers a simplified explanation of the origin of a mobile belt in terms of plate tectonics. Plates A and B, both capped with continental crust, are converging on plate C. The upper part of C consists of oceanic crust, and its leading edges are being forced downwards, by compression, into the asthenosphere. The thicker sediments of the continental slopes and the oceanic trenches are deformed progressively as this subduction proceeds, and eventually they are squashed and highly folded between plates A and B as they collide. The process produces volcanic activity near the margins of A and B. Once A and B are locked together, further convergence ceases, and the thinner sediments of the continental shelves are left comparatively undeformed. See also Section 4.5.1.

In the Devonian period (Table 4.2) following the mountain building that had taken place on the site of present-day Britain in Lower Palaeozoic times, most of the British area was raised above sea level. Continental conditions of sedimentary deposition existed. The environments were similar to those in which the Torridonian sandstone was laid down, and the Devonian rocks have many features in common with it. The most obvious of these features is the prevalent red colour of the rocks, which is a result of the iron in them weathering in the presence of air to produce oxygen-rich oxides and hydroxides. These form the cement of the sandstones, which are the commonest rock type. The sandstones, together with interbedded conglomerates and shales also of Devonian age, are known collectively as the Old Red Sandstone.

CARBONIFEROUS ROCKS OF BRITAIN

Nearly all the coal mined in Britain comes from Carboniferous strata. They are also an important source of other economic rocks, such as limestone and fireclay. The extraction of these rocks presents special problems for the engineer (Section 4.5.5). Such difficulties are often encountered, since most of the large industrial cities of northern Britain are sited on Carboniferous outcrops.

At the beginning of the Carboniferous period, much of the British region was covered by shallow seas, which tended to be deeper towards the south. Some areas, such as the Highlands, remained as dry land. Thick limestones were deposited over large areas of England, but in the north the Lower Carboniferous succession also contains much sandstone, shale and basaltic lavas. As the Carboniferous period progressed, there was an overall shallowing of the area of deposition in Britain, until in Late Carboniferous times nearly all the strata were non-marine. The waste from forests growing in coastal swamps was later altered to coal, and the **seat earth** on which the trees grew became **fireclay** or **ganister**. Long ribbons of sandstone that wind their way through the coal seams and interrupt their continuity mark the courses of streams that existed, and were infilled, at that time.

A feature of Carboniferous rocks the world over is their frequent display of **cyclic sedimentation**, in which rocks of different types tend to occur repeatedly in a particular sequence throughout hundreds or even thousands of metres of succession. Knowledge of the sequence within one sedimentary cycle may be very useful in the early stages of site exploration as a basis for prediction from fragmentary evidence. In coal-bearing formations, a typical complete **cyclothem** consists, from the bottom upwards, of (a) shale (occasionally with marine or brackish-water fossils), (b) sandstone, (c) shale, (d) seat earth, and (e) coal. This is overlain by another cyclothem of shale, etc. The sandstone and coal are the units most likely to be missing, but no matter which unit is absent, the relative order is maintained. On average, cyclothems are between 6 and 10 m thick. There is considerable variation, depending mainly on the presence or absence of the sandstone unit, which may be up to 30 m thick. The coals are not always of economic thickness and they, like all units, may show considerable lateral variation. In some regions, marine limestones may be present on top of the coal as a unit additional to those already described.

More than one theory has been put forward to explain cyclic sedimentation. Coal-bearing cyclothems are thought to represent a normal sequence of events in a subsiding delta. Trees growing on the muds and sands of the delta died, and the debris from them was covered by water and sediment. A change in the pattern of sedimentation brought a new influx of mud and sand to make the water sufficiently shallow for another, younger forest to establish itself. This in turn was destroyed, eventually to become a coal seam, as the subsidence continued.

MESOZOIC ROCKS OF BRITAIN

The Mesozoic era consists of three periods, the Triassic, the Jurassic and the Cretaceous. The rocks deposited in the British region during the first of these were laid down in environments broadly similar to those of the Torridonian sandstone, and they are predominantly red in colour. Together with similar Permian rocks they form the New Red Sandstone. The Jurassic rocks consist largely of fossil-bearing clays, sandstones and limestones, which were deposited in the shallow seas of the continental shelf. The common rock types of Lower Cretaceous age are clays and sands, and the Upper Cretaceous rocks consist largely of chalk, which forms the best known rock layer in Britain.

Mesozoic rocks crop out over most of eastern and southeastern England, but it is the presence of thick Jurassic and Cretaceous strata at depth, under the northern North Sea and other areas of the continental shelf around Britain, that came to be of wider public interest in the 1970s. The decay of organic material (particularly dead marine plankton) buried within these rocks has produced petroleum. This mixture of oil and gas has migrated through pores in the rock and accumulated in **oil traps** where the geological structure prevented further movement (see Section 5.3.5). Natural gas may also be formed by itself (that is, as nearly pure methane), as in the southern North Sea, by the degasification of coal. The effects of deep burial transform peat into bituminous coal and then into anthracite, and result in the loss of natural gas from the seam. This may escape to the surface, sometimes cratering the sea floor with large pock marks, or it may be trapped in underground reservoirs of permeable rock.

References and selected reading

Ambraseys, N. N. 1978. Middle East — a reappraisal of the seismicity. *Q. J. Engng Geol.* **11**, 19–32.

Ambraseys, N. N. and S. K. Sarma 1967. The response of earth dams to strong earthquakes. *Geotechnique* **17**, 181–213.

Attewell, P. B. and I. W. Farmer 1976. *Principles of engineering geology*. London: Chapman and Hall.

Billings, M. P. 1942. *Structural geology*. Englewood Cliffs, NJ: Prentice-Hall.

Duvall, W. I. and D. E. Fogelson 1962. Review of criteria for estimating drainage to residences from blasting vibrations. *US Bur. Mines Rep. Invest.* **5968**.

Geological Society of London 1972. The preparation of maps and plans in terms of engineering geology – Engineering Group Working Party Report. *Q. J. Engng Geol.* **5**, 293–382.

Hills, E. S. 1972. *Elements of structural geology*, 2nd edn. London: Chapman and Hall.

Kennett, J. P. 1982. *Marine geology*. Englewood Cliffs, NJ: Prentice-Hall.

Lomnitz, C. and A. Rosenblueth (eds) 1976. *Seismic risk and engineering decisions*. Amsterdam: Elsevier.

Long, R. E. 1974. Seismicity investigations at dam sites. *Engng Geol.* **8**, 199–212.

Oborn, L. E. 1974. Seismic phenomena and engineering geology. *Proc. 2nd Int. Congr. Int. Assoc. Engng Geol.* **1**, Th II, GR1–41.

Oxburgh, E. R. 1974. The plain man's guide to plate tectonics. *Proc. Geol. Assoc.* **85**, 299–357.

Price, N. J. 1966. *Fault and joint development in brittle and semi-brittle rock*. Oxford: Pergamon.

Roberts, A. 1958. *Geological structures and maps*. London: Macmillan.

Seed, H. B. and I. N. Idriss 1967. Analysis of soil liquefaction – Niigata earthquake. *J. Soil Mech. Found. Div. Am. Soc. Civ. Engrs* **93**, SM3, 83–108.

5 *Subsurface (ground) water*

5.1 Behaviour of water in rocks and soils

5.1.1 Porosity, hydraulic conductivity and permeability

The two properties of a rock or soil which are most important in controlling the behaviour of subsurface water are (a) how much water the rock or soil can hold in empty spaces within it, and (b) how easily and rapidly the water can flow through and out of it. The first is defined in hydrogeology by the **porosity** of the rock or soil. This expresses the ratio of voids in it to its total volume. (In soil mechanics, **void ratio**, the ratio of the volume of void space to that of the solid component, is more commonly used.) The relationship between porosity and rock and soil textures is discussed elsewhere (Section 7.2.2). In general, crystalline igneous and metamorphic rocks have low porosities unless **secondary voids** such as joints are produced by fracturing, for example in a fault zone or at a fold axis, or are produced by chemical erosion to give solution cavities. Porosities of terrigenous sedimentary rocks and soils may also be affected locally by fracturing, but these rocks also have **primary voids** called **pores**. These are spaces left between the solid grains, and they are distributed fairly evenly throughout the body of the rock or soil when it is first formed. This intrinsic primary porosity tends to be greatest in young, poorly compacted sediments which have not been deeply buried and compressed at any stage of their geological history, and tends to be least in old, well compacted rocks buried under a thick overburden.

Flow of water through a soil or rock is described empirically by **Darcy's Law**, and the relative ease of flow by the **hydraulic conductivity** (K) (referred to as the coefficient of permeability in some older texts). When water, oil or gas is flowing through the voids in a rock, its hydraulic conductivity is dependent on its viscosity and on other factors as well as on the properties of the rock. The properties of the rock *alone* that affect ease of flow are defined by its **intrinsic permeability** (k), (usually shortened to **permeability**). This index property is used for definition in the petroleum industry, where more than one fluid is of interest, but is seldom used in hydrogeology. The relationships between hydraulic conductivity, permeability and water-yielding capacity are discussed later (Section 7.2.2).

Rocks must have voids (that is, they must be porous), in order to have permeability and hydraulic conductivity, but the relationship between porosity and permeability depends mainly on the size of the voids rather than on their frequency. Clay has high porosity combined with low permeability. In an idealised terrigenous rock or soil in which the grains were spherical, of one size, and uncemented, the permeability would increase with the $(\text{diameter})^2$ of the grain size. In real terrigenous soils there is a corresponding exponential increase in permeability as grain and void size increase linearly.

5.1.2 Darcy's Law

Slow, laminar flow of water through a porous medium within which the stream lines are smooth parallel paths, and where there is no turbulence or eddies in the wake of grains, is described by an empirical formula, Darcy's Law. This law is valid under most conditions found in nature, where the critical value of the Reynold's Number (at which turbulent flow develops) is seldom approached. The rare exceptions of non-Darcian behaviour in the subsurface include flow through leached limestones, especially near springs, and in the immediate vicinity of pumped wells.

In its simplest expression Darcy's Law is $v = -Ki$, where v, the **specific discharge**, is the volume of water (Q) discharged through a unit area (A) of a porous medium in which the pores are saturated, and inflow rate to the medium equals outflow rate, that is, $v = Q/A$. (It should be noted that though v has the dimensions (L/T) of a velocity, it is *not* the average linear velocity of flow through the medium, $\bar{v}$, which is controlled by the porosity in cross section, nA; $\bar{v} = \bar{v}nA$.). K is the hydraulic conductivity, i is the **hydraulic gradient**, that is, the rate, dh/dl, at which the hydraulic or pressure head of water, h, measured from a horizontal datum, changes laterally with horizontal distance, l.

The pressure head at a given point within a body of rock or soil which is saturated with water is the level to which water would rise if an open-ended tube were inserted to tap the hydraulic pressure at that point. A tube that serves this physical purpose in a laboratory experiment is called a **piezometer**. In geological conditions where the subsurface water in a permeable rock is sealed and confined below an impermeable layer (see confined water, Section 5.2.4), the level to which water rises in a cased well penetrating the permeable rock is usually referred to as the **piezometric level** rather than the pressure head. Piezometric levels are points on a **piezometric surface**, the maximum tilt (i) of which from the horizontal defines the hydraulic gradient dh/dl in the water saturating the confined permeable rock.

The physical meaning of these terms and Darcy's Law may be demonstrated by the apparatus shown in Figure 5.1. This consists of a reservoir (A) joined to a horizontal conduit (B) through which water can flow from A to a tap (C). Six open-ended tubes (manometers) rise vertically from B. The pipe B is filled with soil of uniform permeability k_1. The tap is closed and the reservoir A is filled until water reaches the level

h_0, and remains there after water has percolated through the soil to C and risen in each of the tubes to the **static water level**, h_0.

When the tap is opened and the reservoir is replenished continously to maintain the level at h_0 within it, a new equilibrium is attained eventually for steady flow through the soil, and for steady discharge v at C. Under these steady-state conditions the levels in the tubes fall to h_1, h_2, and so on. Each tube measures the pressure head at the point where it joins the conduit B. There is a progressive regular drop in head (dh) with horizontal distance (dl) from A to C, giving a steady inclination $i = dh/dl$ of the piezometric 'surface'. (The experiment is limited to a one-dimensional traverse of a permeable medium.)

If the soil k_1 between A and the midpoint (M) of the conduit were replaced by a less permeable k_2 soil, then, under steady-state conditions of flow with the reservoir level at h_0, there would be a reduction in the bulk hydraulic conductivity of the soils in B, in the rate of flow through them, and in the amount of water discharged at C. The section of the piezometric surface between A and C would have a deflection point at M, with the inclination i_2 of one segment corresponding to the hydraulic gradient through k_2, and the lesser gradient i_1 to k_1.

The relevance of this experiment to natural phenomena, particularly the flow of groundwater (Section 5.2.3) and the pressure of pore water (Section 5.2.4), is discussed as they are described.

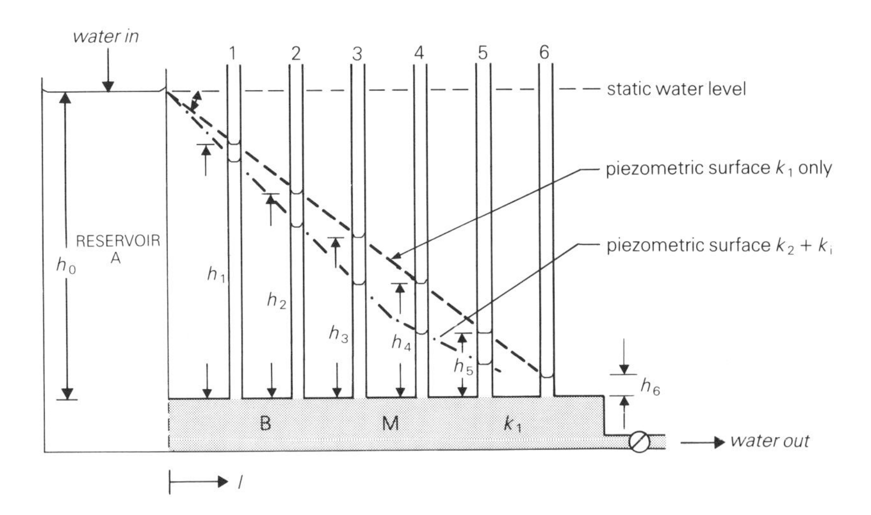

Figure 5.1 An experiment to illustrate the relationships between permeability, change of head with distance and inclination (i) of the piezometric surface. A description and explanation are given in the text.

5.2 Natural circulation of subsurface water

5.2.1 Precipitation and its dispersal

The water in soils and rocks below land areas comes from three sources if we disregard local incursions of brine at some coasts. It may be a product of recent volcanic activity and have separated from magma as **juvenile water**, newly involved in the great cycle of natural circulation. This juvenile water is often heavily contaminated with dissolved minerals and gases. Alternatively, it may have been isolated from the cycle for millions of years as **connate water**, sealed in porous sediment by surrounding impermeable rocks, or lying near-stagnant deep within a groundwater basin. If the sediments are marine, the connate water would be saline. Juvenile and connate waters form only a minor part of the water present in the uppermost kilometre of the solid Earth. Most of this water is **meteoric** and has its source in precipitation of rain, snow and dew. The last of these is important to many desert areas where survival of an animal, or of a human community, may depend on the ability to trap and use dew.

Precipitation ranges from an average of about 50 mm per year in desert regions like Death Valley to over 12 m per year in the Himalayas. The average British rainfall of just over 1 m per year is spread irregularly throughout the year. Such an annual fall concentrated into a wet season would, however, produce seasonal floods in a semi-arid climate. Measurements and statistics of precipitation for about 6000 stations in Britain are available from the Meteorological Office. A list of its publications is given in *Government publications sectional list no. 37* (HMSO). Enquiries should be addressed to The Director General, Meteorological Office, London Road, Bracknell, Berkshire RG12 2SZ. Other sources of rainfall data are the local Meteorological Stations and the Water Authorities. It should be borne in mind that, whereas surface water supplies are vulnerable to summer droughts, underground water is adversely affected by winter droughts when replenishment normally takes place – especially after a year when the summer rainfall was low, and a large soil moisture deficiency has to be made good.

Precipitation is dispersed in three ways: (a) by evaporation and by transpiration from plants, (b) by direct runoff in streams and rivers, and (c) by infiltration into the ground.

In the course of a year, between one-third and one-half of the total precipitation in Britain evaporates shortly after falling. Nearly all the light rainfall on a warm summer day is likely to be lost this way. The factors affecting evaporation are the intensity of the rainfall, the ground temperature, the humidity and the wind strength. It is difficult to measure it directly with accuracy. Estimates of potential annual evaporation are given in Penman (1950). The Meteorological Office publishes

Table 5.1 (a) Estimates of the Earth's water supply (in 10^6 km^3). (b) Estimates of the daily circulation of part of this water in the planet's hydrologic cycle of evaporation and return to the oceans (in km^3).

(a) Supply

total world water supply	1357
total water in oceans	1312
total water in icecaps	29
total water on land	8
total water at surface on land	0.2
total water in subsurface on land	7.8

(b) Circulation

evaporation from oceans	943
precipitated on oceanic areas	861
evaporation from land	205
precipitated on land	287
runoff from land (including discharged subsurface water)	82

information about potential evapotranspiration over Britain, and data on soil moisture and evaporation in its *Soil and moisture deficit bulletin*. The actual evaporation may be lower than the potential, when the soil is deficient in moisture, and is also affected by the type of vegetation cover.

Surface water runoff can be measured by gauging streams and rivers. Data are collected by the Water Authorities and by some other bodies. The results from selected stations are published annually by the Water Data Unit, DoE, in *The surface water yearbook of Great Britain*. Table 5.1 gives some details on the Earth's water supply and circulation.

The division of unevaporated rainfall between runoff and infiltration is controlled by the relative ease of flow in either direction (Fig. 5.2). The amount of runoff from a given area is related directly to its steepness of slope, and indirectly to the vegetation on it (which restrains flow and may act as a sponge) and the permeability of the soils and rocks at the surface. Water infiltrates easily into permeable soils such as gravel. The amount of runoff relative to infiltration after a fall of rain depends also, and significantly, on the amount and concentration of the precipitation. Most of a heavy rainstorm will be dispersed as runoff.

5.2.2 *Groundwater and the water table*

The first rain to infiltrate below the ground surface wets the grains of soil and adheres to them as **pellicular water**. The forces holding the water to the grain boundaries are so strong that it is not moved further and can be detached only by evaporation or plant roots. Percolation to greater

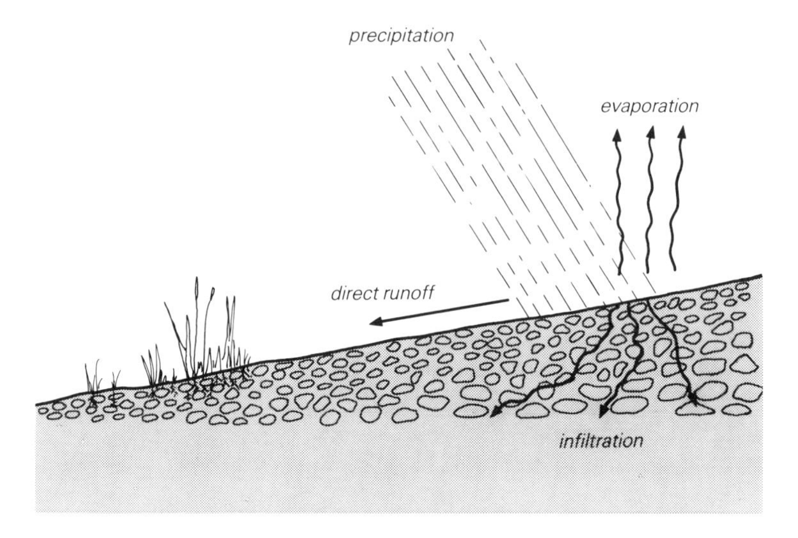

Figure 5.2 A globule of water precipitated as rain is dispersed by evaporation, by direct runoff and by infiltration into the ground. The division between runoff and infiltration is determined by the relative resistance to flow along either path. Vegetation and steepness of slope influence the flow into streams, and the near-surface permeability controls infiltration.

depths as rain continues to fall proceeds only after the soil reaches its **field capacity**, at which it cannot hold any more water against the downward pull of gravity. Then, the next water to infiltrate moves over these first films of water, but does not fill the voids in the rock completely. There is still air present in the centres of the larger voids. The pellicular and gravity water in this **unsaturated zone** (or **zone of aeration**) is called **vadose water**.

Eventually the gravity water percolates to a **zone of saturation**, where all the effective void space in the rock is filled with water. The journey from the surface may take up to a few weeks. The water in this saturated zone is referred to as **groundwater**, and its upper surface is often referred to (not in strict usage of the term) as the **water table** (Fig. 5.3). The saturated zone persists downwards until the compaction of the rock under the pressure of overburden reduces porosity to zero. This depth varies with local geological conditions, and may be as much as 10 km in regions of thick sediments.

The water table is shaped like a subdued replica of the topography above it. It is not static, as groundwater in a permeable rock is continually in motion. Highs in the water become flatter, and gradients are reduced, at a rate controlled mainly by the permeability of the rock. The gradients are usually less than 1 in 100, but may be as much as 1 in 10 in hilly country. If an uncased well were drilled into the saturated zone, water would flow into it and, given time, fill it to a level which is a point on the

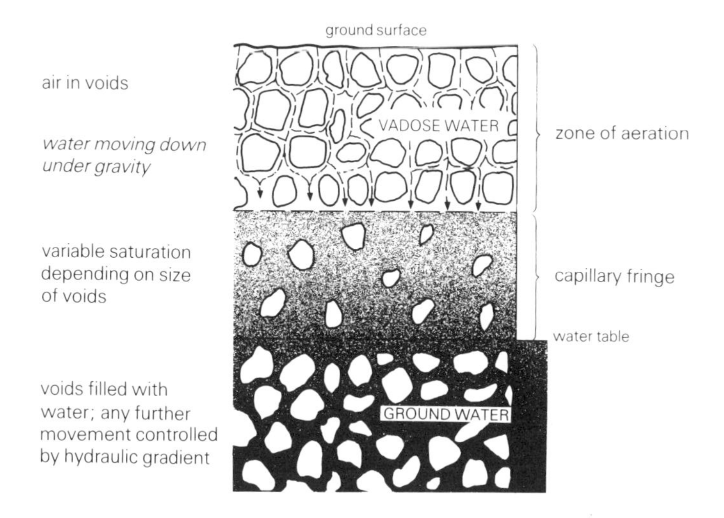

Figure 5.3 The distribution and behaviour of water in rocks and soils in the subsurface.

water table, as precisely defined. The water table lies below the top of the saturated zone in the rocks surrounding the well, separated from it by the **capillary fringe** (sometimes referred to as the **capillary zone**). This is usually between a few centimetres and a few metres in thickness, but may be over 10 m thick in very fine-grained rocks in which capillary pressures are high. The smaller the voids in the rock, the higher the fringe. In the open well, the water surface is at atmospheric pressure; and at the top of the saturated zone in the rocks, water is at atmospheric minus capillary pressure.

Under ideal conditions in a rock or soil of uniform porosity, there would be no physical distinction between the water in the capillary fringe and the water below it. Both would be affected by the same hydraulic pressure gradients inducing flow, and by the same capillary forces resisting flow. There would be no boundary to distinguish between them until a well was drilled. In most real conditions, however, the near-surface soils have variable grain sizes and permeabilities, and in practice the degree of saturation usually decreases upwards from the true water table. Only close to the water table are all the voids in the soil completely filled with water and, for this reason, most of the capillary fringe often lies within the zone of aeration rather than forming part of the zone of saturation in the strict sense.

If fine soil were dumped on swampy ground, then there would be a slow upward pull of the water to create a capillary fringe. In the case of an embankment built on swampy ground, this must be allowed for. The rise may take as long as three years before it is completed.

Groundwater normally contains impurities such as carbonates and sulphates, which were dissolved from the rocks with which it came in contact. The concentration depends partly on the time in contact, partly on the rate of flow, and partly on how reactive the minerals are. The water from bodies in gravel within a buried channel is usually purer than from other parts, because flow is greater and the gravel is inert compared with clay or limestone. The quality of the groundwater, and its suitability for particular purposes such as drinking or brewing, is thus related to the **aquifer** (Section 5.2.4) containing it. It may also affect the choice of engineering materials for foundations. Groundwater with a high sulphate content is present in some rocks, particularly certain shales and clays. The sulphate may come from gypsum in the rock group, from sulphides, from hydrogen sulphide in the groundwater reacting with other minerals, or, if the groundwater is connate, the sulphate may have been concentrated in the sea water. Sulphates speed the corrosion of iron, and react adversely with Portland and some other types of cement. It is recommended (BS 4027) that the use of sulphate-resisting cement should be considered once the concentration passes a specified figure.

Evaporation is not restricted to surface water, but also takes place from the top of the water table. The phenomenon is particularly important in semi-arid regions. As a result, a layer of strongly cemented material forms in the otherwise unconsolidated sediment of the soil (see Section 3.3.2). This may be mistaken for bedrock. It may consist of lime (calcareous), chalcedony (siliceous) or iron hydroxides (ferruginous). It is commonly referred to as **hard pan** and occasionally as **duricrust**. Other terms are used for particular types (for example, **iron pan**) and the terms are further multiplied by regional usage. For example, calcareous hard pan is called **caliche** in Spanish American regions but **kankar** in India. Hard pan is relatively impermeable, and its presence affects percolation through the soil. In some areas it may be thick enough to interfere with seismic refraction surveys (Section 6.3.3) and give spurious values to rock head by acting as the refractor.

Definition of the water table and of its variation seasonally and over longer periods is important for groundwater supply and other practical purposes. It can be located and monitored by wells, and less accurately by geophysical surveys (Section 6.3). In large investigations both would be used. The results are usually presented as contoured maps of the water table of a particular area. The statutory responsibility for systematic regional studies of this type is usually vested in a national or state Geological Survey. In Britain, the Water Division of the BGS carries out hydrogeological work. Data are also collected by the regional Water Authorities, and selected records have been published by the Water Resources Board.

5.2.3 *Flow of groundwater*

Groundwater is in continuous fluid contact such that a change of pressure at one point affects pressure at all other points. Its flow is controlled by the mechanical energy of the fluid per unit of mass, that is, by the change of head within it. The head at any point is a measure of the potential energy of the fluid relative to a specified state. Points of equal fluid potential define **equipotential surfaces** within the body of groundwater, and these in turn may be used to construct the **flow lines** of the groundwater. If the soil or rock is uniform and has the same permeability in all directions of flow, then the flow lines are orthogonal (that is, at right angles) to the equipotential surfaces. The pattern of equipotential surfaces and flow lines for a given model of permeability is called a **flow net**. An example is shown in Figure 5.4.

In this model of simple topography and uniform permeability, precipitation is everywhere added to the water table, and this source in the system is balanced under steady-state conditions by discharge into streams, which act as sinks.

Note that flow lines radiate at the sink and are most concentrated there. Accordingly, the most intense flow takes place in that part of the system. All the water in the model is moving, and there is no stagnant water at depth. In most real geological conditions, however, there is a decrease of permeability with depth because of the compaction of soil and rock under overburden, and there is a corresponding decrease in groundwater movement. In some cases, significant flow is restricted to the uppermost few metres of soil, but in others, high hydraulic gradients can produce measurable flow at depths of a few kilometres in permeable rock layers. In general, the direction of flow is normal to the contours of the water table. Flow usually follows the general features of the topography, and is usually greatest near the surface. Along buried valleys, the underflow may be as great as the surface flow in the present-day river.

Flow nets may be drawn using an empirical, graphical construction, if the conditions are simple, for example if permeability is uniform and the ground surface is even. This approach is adequate for many problems which involve seepage of impounded water. Mathematical solutions using formulae also exist for models where the topography is regular and definable in simple terms (see, for example, Fig. 5.4), but tend to be of academic interest rather than practical use. The effects of irregular topography and varied permeability, such as are found in most areas of the real Earth, on groundwater flow and the shape of the flow net can be tackled effectively only by using an analogue or digital computer.

Darcy's Law, which describes the slow flow of water in a granular medium, is analogous to Ohm's Law ($I = E/R$), which describes the flow of electricity. Each physical quantity in one equation has a matching (analogous) quantity in the other. The flow of water v is analogous to the

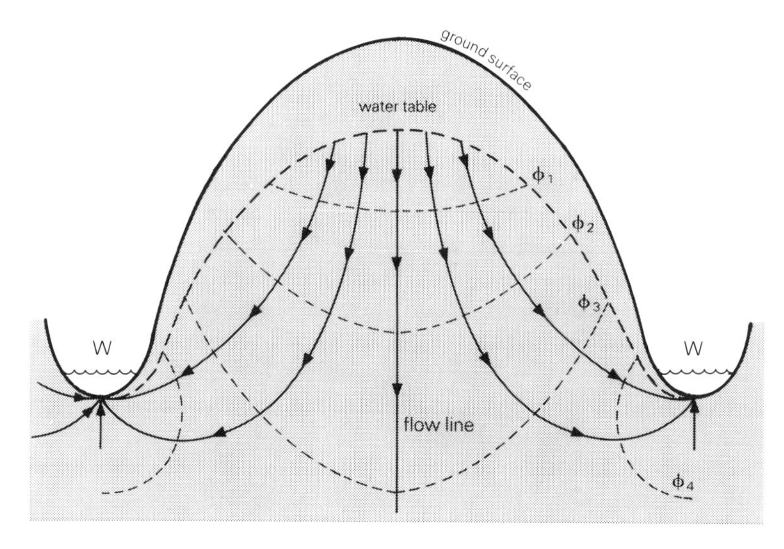

Figure 5.4 The diagram illustrates the flow of ground water under a symmetric hill flanked by valleys, and formed of rocks of uniform permeability. The flow net is correspondingly simple and regular and can be derived analytically. It consists of a set of equipotential lines (broken curves, of value ϕ_1, ϕ_2, etc.) and a set of flow lines orthogonal to them. Flow takes place from the water table to the points of discharge at W (which are sinks in the potential field and springs or seepages in physical reality).

flow of electricity I, the change of head i to the voltage drop E, and the permeability k to the inverse of the resistance, that is $1/R$. These relationships may be used to construct an **electrical analogue model**, which can compute the behaviour (flow, change of head, and so on) within an underground reservoir of groundwater, if the distribution of permeabilities and the heads in test wells are known. It can cope with a permeability model within which the topography is more irregular and the geological structure more complex than can be handled easily by simple equations and analysis; once built it can be used continuously to monitor conditions in an underground reservoir and manage it like a surface supply. Other analogous pairs of equations are sometimes used for this purpose.

5.2.4 Geological controls on movement of subsurface water

The behaviour of groundwater as it circulates through layers of different permeability underground and is eventually discharged on the surface or at the coast produces a range of phenomena. For simplicity in describing these, rocks are grouped into two categories. Those that have relatively high permeabilities because the pores of the rock are large, or because the rock is well jointed, are referred to for brevity as pervious. (This term is conventionally used to describe high fluid conductivity of a rock in which the voids are mainly secondary, that is, where there is appreciable flow along joints, fissures and bedding planes, in contrast to 'permeable', which is then restricted to a description of primary **pore permeability**.) A body of pervious (and permeable) rocks capable of yielding groundwater is called an **aquifer**. Rocks with low permeability and no fissures are referred to in the following account as impervious: they form **aquicludes**.

If gravity water percolating through the unsaturated zone meets a layer or lens of impervious rock, for example a lens of clay in gravel, then further flow downwards is hindered, and a local zone of saturation is formed above the main water table. It is called a **perched water table**. This drains slowly by flow over the edges of the lens. There may be more than one perched water table above the main water table. Each contains only a limited supply of water, which is vulnerable to contamination from the surface and is not usually suitable for development as a source of water supply.

Circulation of groundwater in the saturated zone takes place within aquifers, but the zone of movement is effectively limited by any aquicludes bordering the aquifer. Consequently, the subsurface is divided by geological structure into more or less self-contained **hydrological units**, or **groundwater basins**. Within each of these there is an approximate balance of supply and discharge, and flow is independent of other units. The units vary considerably in structure and in size. A hydrological unit may be a thick layer of sandstone several hundred kilometres square in area, a body of gravel in a buried channel, or an alluvial fan (Fig. 5.5) only 100 m across.

When an aquifer is covered by an aquiclude, retarding or effectively stopping any flow of water to the surface, the aquifer is said to contain **confined water**. Figure 5.6 shows an aquifer which crops out on high ground and dips under an aquiclude. The outcrop serves as an intake area for percolation, which recharges the confined water elsewhere in the

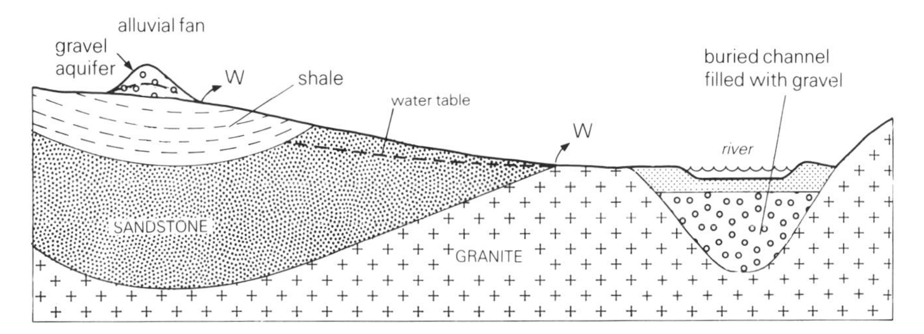

Figure 5.5 The diagram shows three examples of a hydrological unit, (a) an alluvial fan with a gravel aquifer and discharge at W, (b) a thick sandstone layer, and (c) the gravel infill of a buried channel.

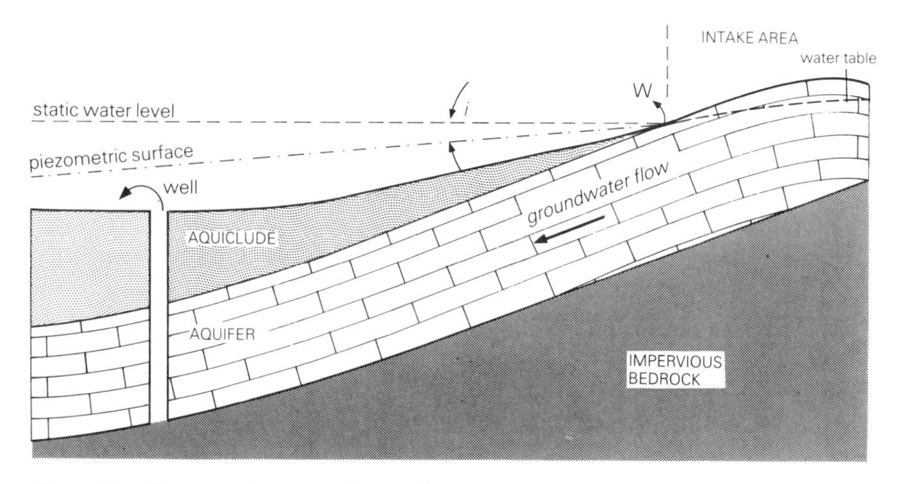

Figure 5.6 The ground water in the aquifer is confined by the aquiclude above it, except at the intake area where it crops out, and can be recharged by infiltration. The free water table within the intake area, and the spring line (W) at the contact of aquifer and aquiclude, are shown. Depending on the permeability of the aquifer and the amount of groundwater flow, the piezometric surface slopes at some angle *i* from the intake area. If the surface lies above ground level at the well, as shown, there is artesian pressure enough to make water flow from the well.

aquifer. The water table of the **intake area** is shown. Beyond its lower boundary (S), fluid potential in the aquifer is shown by a static water level, corresponding to stagnant groundwater conditions, or a piezometric surface, corresponding to flow away from the intake area. The flow may be taking place naturally within the hydrological unit, or induced artificially by flow to the surface through wells. The inclination of the piezometric surface is controlled by the rate of flow through, and the permeability of, the aquifer, as described by Darcy's Law, and as demonstrated by the experiment illustrating it (Section 5.1.2). If the piezometric surface is higher than the water table in a porous aquiclude, then groundwater conditions are **artesian** rather than normal or subnormal. If the piezometric surface lies above ground level within this **area of artesian flow**, then water will rise in an open tube to give a **flowing artesian well**.

The basic geological conditions for artesian pressure, apart from rare cases where there is a major source of supply to the aquifer other than at the outcrop, are that groundwater be confined and that strata be inclined. These are not uncommon, and neither are artesian conditions. They may occur very locally in drift where sand or gravel lenses are covered by clays, or they may affect hundreds of square kilometres underlain by a permeable rock group. The best known example in Britain is the synclinal structure known as the London Basin (Fig. 5.7). Pervious chalk crops out on the hills to the north and south, but in the Thames region it is covered

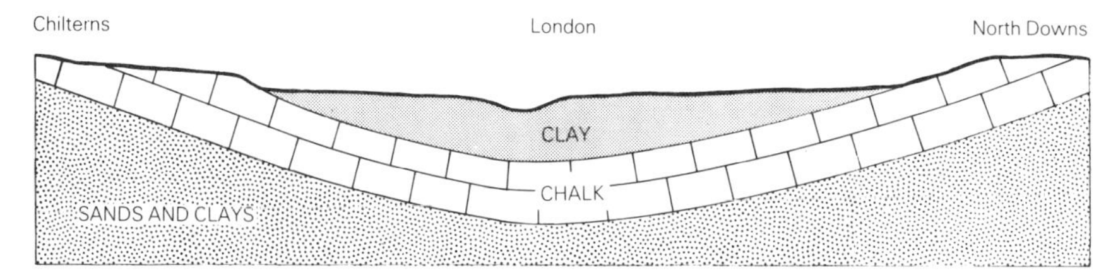

Figure 5.7 A simplified geological section across the London Basin, which shows the chalk aquifer confined by the impermeable London Clay.

by the London Clay. A supply of groundwater from the chalk is an attractive source of water to many small businesses, not only on economic grounds but also because of its qualities compared with processed river water. Early drilling in the London Basin produced flowing artesian wells on several sites. For example, the fountains in Trafalgar Square were originally fed by artesian flow. This success led to overexploitation of the groundwater resources and a consequent loss of head in the older wells, which now have to be pumped. The piezometric surface is now 125 m below its original level. Further damage to the resource is restricted by legislation.

The aquifer itself is affected by these changes of pore pressure in the water it holds, and the overlying beds are supported by this pressure. An increase in it expands the aquifer by a small amount and tends to lift the aquiclude. Artesian flow produces a relief of pressure near the bottom of the well, and results in settlement and compression of the aquifer. The initial yield from a well is augmented by a **flow from storage**, so that yield is not reflected directly, and simply, by a drop of the water table at the intake area. At a later stage of production, there is a decrease in the pressure of the confined water, and discharge into the well becomes less. The related subsidence is most marked near to the well point, where pressure gradients are highest, and may be sufficient to affect adjacent foundations if the aquifer is shallow. Prolonged withdrawal of groundwater from the aquifer affects wide areas eventually, and *subsidence may be serious*, especially in low-lying coastal areas. Venice has been damaged, and is still threatened, in this way. Further withdrawal from the aquifer below the city has been curtailed and some of the city's fountains, which were flowing artesian wells, are no longer allowed to play. A proposed scheme to restore the ground level of the city would involve sealing the aquifer under Venice with a continuous screen of impermeable material at depth, around the city limits, then pumping water down **dumb wells** to repressurise and inflate that part of the aquifer within the dam.

Flow to the surface may also come from unconfined aquifers, where a well is sited near the bottom of a steep valley. Figure 5.8 shows an example of this, with the water table, flow lines and equipotential lines near the well drawn in. If the well were uncased, it would fill with

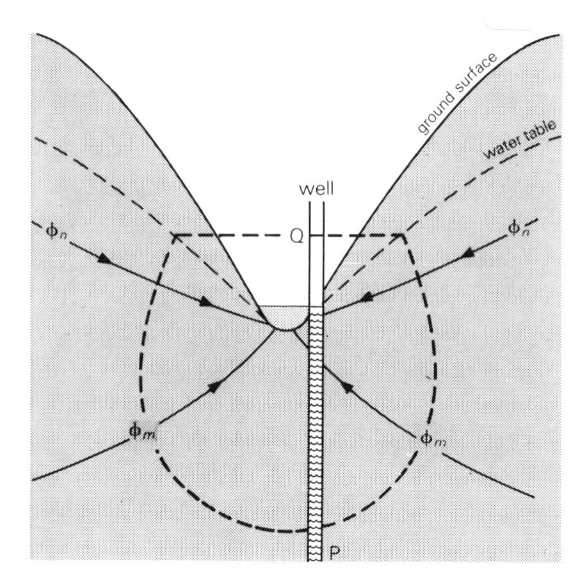

Figure 5.8 The well sunk into the side of a steep valley is perforated at point P only, which lies on the equipotential line ϕ_m. The head of water in the well rises to the corresponding level Q, to give a flowing well without a confining aquiclude being present. In an uncased well, water would rise only to the level of the water table.

groundwater to the level of the water table where it cuts it. If, however, the well were cased, and the tube is perforated at P only, then the potential at that point raises water in the well to the level Q and gives a flowing well. The level Q corresponds to the height at which the equipotential line through P intersects the water table.

5.2.5 *Natural discharge of groundwater*

Natural discharge of groundwater takes place where the ground surface intersects the water table. If the flow from the hydrologic unit is spread diffusely over an area of marshy ground, it is usually referred to as a **seepage**. If it is concentrated, say by a fissure acting as a channel, it is called a **spring**.

A **valley spring** (Fig. 5.9a) occurs where the water table intersects the bottom of a valley that is cut into pervious rocks. If the water table rises and falls seasonally, flow is intermittent, and the spring is called a **bourne**. These are particularly common in the chalk lands of southern England.

If the intersection of surface and water table is controlled by geological structure, and discharge takes place where impervious rock bounds the hydrologic unit (Fig. 5.9b), the overflow is called a **contact spring**, or sometimes a **stratum spring** if the rock is layered. The impervious barrier may be a layer of clay or shale, or an igneous intrusion. By converse argument, a **spring line** known to be related to a particular geological contact may be used to trace its outcrop laterally when mapping.

Fault springs occur where pervious rocks are faulted against impervious rocks (Fig. 5.9c), and may be thought of as a type of contact spring. If the rock in the fault zone is a pervious breccia, a natural artesian flow may bring confined water to the surface from a concealed aquifer. In desert areas this flow may produce an oasis.

In limestone areas, groundwater usually follows channels along bed-

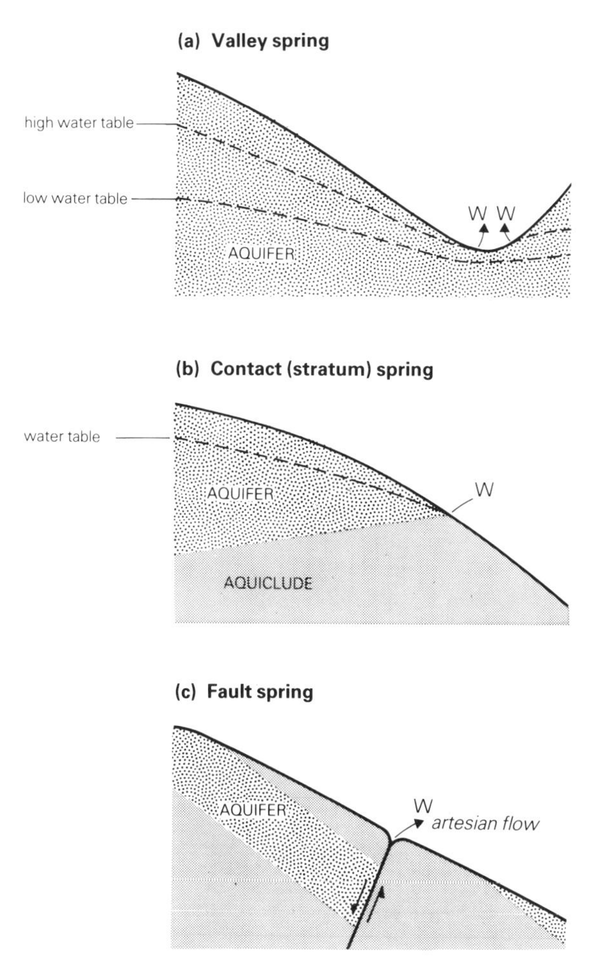

Figure 5.9 Springs and seepages. (a) The valley is cut into uniformly permeable rock (for example, chalk) and valley springs occur where the water table intersects the ground surface. A fall in the water table after drought lowers and flattens it, and may result in the springs drying up. (b) A stratum spring sited at the contact of an aquifer and an aquiclude. (c) Flow from a confined aquifer is taking place through the breccia in the fault zone, and produces a fault spring.

ding planes or fissures, which are progressively widened by solution, and discharge from limestone is often from **solution channel springs**.

The major discharge of fresh groundwater takes place at the sea coast, where it mingles slowly with saline groundwater. The fresh water and the brine behave, over short periods, like immiscible liquids, and rock is 'wet' with either one or the other. Mixing takes place within a relatively narrow subsurface zone of contact. The shape and position of this contact surface are determined by the way in which the body of fresh water 'floats on' the brine. The fresh groundwater is less dense than the brine and tends to override and displace it. Where the relief of the ground surface is low, with correspondingly low hydraulic gradients and rates of flow, the body of fresh groundwater is in approximate hydrostatic equilibrium with the brine below it. This relationship has been observed in many low, sandy islands, including the Frisian Islands and Long Island, and is usually referred to as the **Ghyben–Hertzberg Balance**. The brine of the North Sea surrounding the Frisian Islands has a specific gravity of 1.027, and hence the fresh groundwater (sp. gr. 1.000) in **hydrostatic** balance with it has the form shown in Figure 5.10.

For each unit of height of the water table above mean sea level, the fresh brine contact surface lies 37 times that number of units vertically below it. This provides a useful rule of thumb in simple cases, but becomes an increasingly inaccurate approximation as groundwater flow increases and conditions become **hydrodynamic**. For example, if the effects of an engineering project in a hilly coastal region on the groundwater regimes (say the construction of a canal through valuable farmland) are being

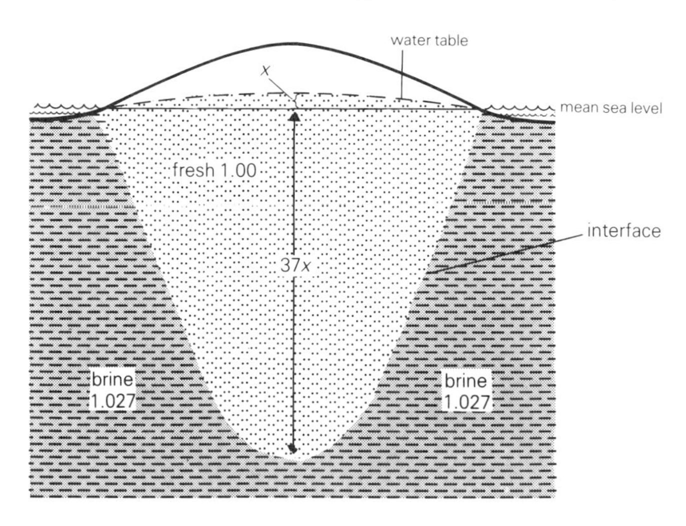

Figure 5.10 The distribution of fresh and salt ground water under one of the Frisian Islands is shown. The fresh water is in hydrostatic equilibrium with the denser salt ground water and floats on it.

assessed carefully, then a more complex mathematical solution than the Ghyben–Hertzberg Balance should be sought.

Tidal changes in sea level usually produce only a small sympathetic rise and fall in water wells near the coast, and seldom a measurable effect at distances of more than 1 km from the shoreline.

In limestone areas, discharge at the coast is often at solution channel springs, which may occur in the sea bed at distances of a kilometre or more from the shoreline, for example in the Gulf of Corinth, Greece. They are a potentially useful source of water in limestone islands prone to drought. The colder fresh groundwater can be recognised in subtropical areas by the use of infrared-sensitive air photographs.

5.3 Some practical engineering aspects of groundwater

5.3.1 *Groundwater and engineering*

The presence of water in rocks and its movement through them are of great importance to many human activities such as farming. However, this brief account is limited to some of the more important geological aspects of groundwater which are linked to engineering. These are: (a) water supply from subsurface sources, (b) drainage of marshes, and (c) disposal of toxic waste underground. The other aspects of equal or greater importance to most engineers, which are normally covered in a separate course of soil mechanics, are omitted or have been referred to briefly. They include pore fluid pressure in soils, the effect that the presence of water has on the strengths and other mechanical properties of soils, and the relationship between settlement of buildings and the loss of pore water from clay.

5.3.2 *Groundwater inventory*

In arid regions, groundwater may be the only dependable source of water supply for much of the year, unless water is piped over long distances. It has the bonus quality of being filtered naturally. There are advantages also, which may outweigh any extra cost per litre, in having some supply from wells in areas like England that depend mainly on surface reservoirs. Problems such as creating a watertight reservoir in chalk lands, or acquiring its site, are avoided. The water is often more palatable than alternative surface water. Underground reservoirs are seldom drained entirely during droughts, and deep wells can draw on reserves temporarily, on the understanding that replenishment will be allowed to take place during the next winter. An important consideration in many schemes of supply is that groundwater resources can be *developed more rapidly* than a surface reservoir can be built. The lead time between the conception of a reservoir and its completion is conditioned not only by construction times and technical delays, but also by the statutory procedures of acquiring approval, permissions and land, and allowing appeals by those affected. Where a relatively quick solution must be found to a supply of water or to insure against drought, a groundwater scheme should be considered.

In assessing the feasibility of any scheme of groundwater supply and in planning it, geological and hydrological information is essential. A first step is to evaluate the permeabilities of the rocks and soils present in the area, and to determine their structure. This identifies aquifers and aquicludes and outlines the hydrologic units present. The common aquifers include: (a) sand and gravel deposits, occurring as drift in glaciated areas, or as alluvial cones (especially in semi-arid hilly regions); (b) sandstones, for example the Bunter Sandstone and Keuper Sandstones (Waterstones) in the English Midlands; (c) certain permeable limestones such as chalk; and (d) fissured or deeply weathered igneous rocks, especially in tropical areas. A small local supply may be obtained from a single well in an alluvial cone, but a large scheme usually draws on the groundwater in an extensive, thick, permeable layer.

Potential supply of groundwater from the hydrologic unit selected for investigation is then assessed by drawing up a **groundwater inventory**, by evaluating the equation

$$\text{total precipitation} = \text{evaporation} + \text{direct runoff} + \text{groundwater increment (by infiltration)}$$

The values of these quantities for the hydrologic unit can be assessed from records or measured. In the natural state there is equilibrium within the unit, and the groundwater increment is balanced over long periods by natural discharge. In a scheme of groundwater supply, it is artificially discharged through wells. The **safe yield** of the hydrologic unit is equal to its groundwater increment. If more water than this is withdrawn by pumping from deep wells, then finite reserves rather than current resources are being exploited. This may be justified during a summer drought, but as standard practice over years it will produce a serious drop in level of the water table or the piezometric surface of the aquifer. These can be monitored using wells. Recovery to normal after a major lowering of the water table is likely to require a span of time comparable to the period of overexploitation. Some semi-arid and desert areas in the USA, and in other countries, have been overexploited in this way. The water which makes the desert bloom is sometimes being irreversibly mined, and in these cases it will run out. In coastal areas, careless exploitation may lead to serious settlement (Section 5.2.4) and to invasion of irrigation wells by brine. Some areas of South-east England have been overexploited, but drilling is now controlled by legislation. **Recharging of aquifers** by artificial means is done by pumping surface water down dumb wells into them, or by guiding its flow across spreads of gravel or other pervious rock on the aquifer, and so encouraging infiltration.

The investigations that preceded a scheme of groundwater supply from Permian strata in south Durham have been described in an unpublished paper to a conference in 1969 by A. S. Burgess and T. Cairney. The decision to opt for wells rather than a reservoir was weighted in this project by the need for a source of water capable of rapid development. The potential of the aquifer, the Magnesian Limestone, to give substantial yields was already appreciated from limited exploitation of it. The limestone is 150 m thick, but less than 30 m of it behaves as a good aquifer. The best yield comes from an oolitic layer near the middle, with a significant contribution from the heavily jointed limestone above and below it into those boreholes that intersect major fractures. The limestone has a gentle dip to the east, which is complicated by very open folds about east–west axes, and by low pre-Permian hills that protrude through the aquifer. Recharge is limited to the western outcrop of the limestone, where drift is thin or absent, and where some streams vanish down sink holes. Wells in this part of the area are of the water-table type (see Section 5.3.3). To the east, water in the aquifer is confined by overlying impervious strata and by a thicker cover of glacial clays.

An area of 160 km^2 was investigated by a primary programme of 42 small-diameter boreholes used to prove the local geology, to provide samples for testing and to give information about water levels. A secondary programme of 15 larger-diameter holes were bored for test pumping, and from these the aquifer's transmissibility and storage (Appendix B) were calculated. Aquifer conditions proved to be more variable than expected. To investigate the safe yield and the reaction in the aquifer to various ways of withdrawing groundwater, an analogue computer was built to model the aquifer and its piezometric surface on a scale of 1:25000. Screw terminal nodes were fixed in a grid of 38 mm intervals on a perspex sheet. The electric potentials at each node were related to the fluid potential at the corresponding point in the aquifer, and an initial estimate of the comparative transmissibilities between nodes was made from the potential gradients. The corresponding resistor values between nodes were calculated using a working flow-current scaling factor. These were adjusted iteratively until the model potentials showed reasonable agreement with the piezometric contour map values. Pumping tests could then be simulated, and effects such as the significant recharge from two streams, which caused initial mismatching, could be allowed for. The model can be used not only for studying possible well distributions but also for monitoring and managing the underground reservoir.

5.3.3 *Siting and testing of individual wells*

Wells may be dug or bored, either to get fluids out of the ground or to put them into it. Those for water supply are best bored in such a way that the sides are supported without the use of a mud flush, which might clog the aquifer, and they should penetrate its full thickness. They vary from a few metres in depth to 600 m in some cases. An average yield is about 40 litres per minute, but may be as high as 4000 litres per minute. To complete the well after boring it, a gravel filter is placed near the bottom, and is held in place by a slotted inner lining tube. An excessive rate of flow through these slots (more than 6 cm s^{-1}) may lead to deposition and obstruction by ferrous and carbonate deposits. Improvement of a poor flow may be achieved, if the screen or strata is clogged, by surging a phosphate solution up and down the borehole. Weak acid is pumped down oil wells producing from limestone, to improve flow around the well by opening fractures. This acidisation treatment can add about 15% to the total costs of a water well in chalk or other limestone, and has not been common practice, though it is thought worthwhile by some experienced engineers. Since the flow lines converge on the well, the permeability immediately around the well, and the presence of any fractures, critically affect its yield. For this reason, the prime geological factor in choosing the precise site for a boring is the probability that the permeability of the aquifer is higher at that locality, or that flow is assisted by fracturing. In soils, a lens of gravel may expedite flow from a larger body of sand; in solid rock the yield from a well sited on a fold axis, where jointing is well developed, may be several times that from massive chalk less than 100 m away. Since rocks weakened by fracturing are often eroded preferentially into hollows, the water table is likely to be at shallower depth also. Electrical methods of geophysical surveying are used to locate such water-filled fractures (see Section 6.3.6).

In **water table wells**, water is pumped from an unconfined aquifer, whereas in **confined water wells**, the whole aquifer penetrated by the well is saturated and there is usually artesian pressure.

Once withdrawal of water from a well is under way, by pumping or artesian flow, an inverted **cone of depression** of the water table forms around a water table well. Its slope is related directly to the level below the water table at which pumping is taking place, and inversely to the permeability of the rocks surrounding the well (see Appendix B for details). The top of this unsaturated zone may extend for distances of several hundred metres and may interfere with the cone of depression of an adjacent well, so that the water table does not attain its normal level between the wells. The flow into both wells from this sector of their surroundings is reduced, and the yields are less than might be obtained, if each were clear of the other's influence (Fig. 5.11). Spacing of wells must be planned to allow for this effect. The phenomenon has also been

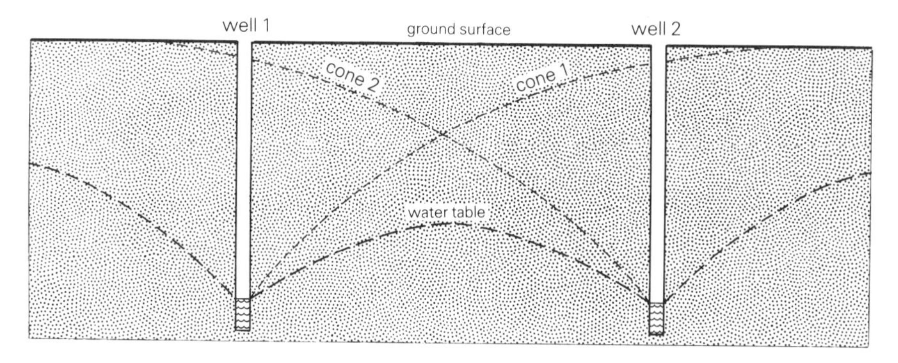

Figure 5.11 Wells 1 and 2 are sited so close to each other that their cones of depression interfere with one another to produce a lowering of the water table between the two wells. This reduces the hydraulic gradients in that sector around the wells and lowers the flow into them and their yields. The same phenomenon can, however, be put to use in draining loose soils prior to excavation, as the ground between the two wells can be de-watered in the same way.

exploited to drain soils where excavation would otherwise be difficult, by pumping groundwater from a line of deep, closely spaced wells.

After drilling is completed, pumping tests are often carried out to determine the yield characteristics of the well, and to learn more about the hydraulic properties of the aquifer, particularly its **coefficient of storage** and its **transmissibility** close to the well. These terms are defined and discussed briefly in Appendix B.

5.3.4 *Drainage of groundwater*

The removal of groundwater from soil is the main aim in reclamation of marshes, and it may also be important in other projects where, for example, the stability of a slope is affected by saturation. This excess water may be replenished by infiltration or by subsurface flow into the saturated zone, of which it is part. If marshy ground is an area of seepage, then a system of drainage should be devised not only to drain the excess water stored in the soil but also to cut off the source of supply by lowering the water table. For example, Figure 5.12 shows a stretch of persistently marshy ground on a gentle concave slope. It may be inferred that the water table meets the ground surface there and seepage is taking place. Its shape is represented by water table 1. A spring line (W) is present at the upper limit of the area. The flow of groundwater is from the higher ground on the right of the diagram towards the marshy area. The most effective siting of a deep, main drain would be to place it just below the spring line. Flow from the uphill side is diverted into the drain, and initially there is some reversed flow from the ground under the marsh, as an asymmetric 'cone' of depression (exhaustion)—actually wedge-shaped—forms in the water table (water table 2 on the diagram). This is

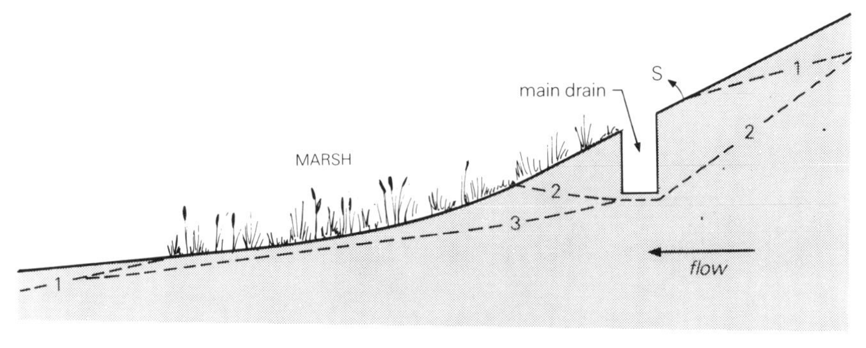

Figure 5.12 The marsh on the concave part of the slope is maintained by seepage of ground water flowing to it from the uphill direction. A main drain dug near the spring line (W) would produce successive changes, 1, 2 and 3, in the water table (see text) so that in its final position (3) it would no longer intersect the ground surface, nor replenish the marsh.

followed by a slow adjustment of the water table to its equilibrium position (water table 3) as stored water is drained downhill. This sequence may be complicated temporarily after heavy precipitation.

Foundations may be de-watered using the same general principle of letting subsurface water move to low spots (sinks), and by pumping it out at these spots, so lowering the water table. In small excavations, a deep sump dug at a corner, filled with gravel or other coarse filter material and pumped, may be adequate. Alternatively, in large open excavations, lines of well points may be used. Wells are set in one or more lines at the edge of the intended excavation, with a spacing which ranges from 1 m to more than 6 m, depending on the conditions and the depth of the wells.

De-watering can be a slow process in clay soils, and drainage may be speeded by **electro-osmosis**. If two electrodes are placed in a saturated soil that has a colloidal content, then water molecules tend to be carried by movement of positive ions towards the negative terminal (cathode). The flow is greater than would otherwise be achieved by normal gradients of the water table, as the electro-osmotic permeability is much higher than k. Using a perforated tube for the cathode, water can be pumped from it. A potential of the order of 100 V is generally suitable. Clays can be drained this way, but not sands, however fine.

5.3.5 *Disposal of toxic waste, and storage of fluids, underground*

Disposal of toxic waste and noxious fluids underground, by tipping them down disused mine shafts or pumping them down dumb wells, is practised more often in the USA than in Britain. Much of the brine pumped out with oil is disposed of in this way, and at present more than 200 wells are used in the USA to inject acid and alkali wastes, pumping liquors and uranium mill wastes into deep rocks. The technique has become of increased interest in Britain as controls on dumping at sea and on the surface become more strict and are fully implemented.

Underground disposal is an acceptable practice where the local geology and hydrology are favourable, and where there are rigorous engineering and administrative controls. The most important aspect, geologically, is that there must be a negligible rate of groundwater movement near the storage area. For example, salinity of the groundwater at the storage area would indicate that it is connate, and would be a useful criterion of limited flow through the rock.

Disposal wells for toxic waste basically require a pervious reservoir layer of sandstone for injection, overlain by an impervious layer of shale or salt which seals it. Groundwater flow should be *very slow*, since any movement carries a risk of contaminating neighbouring water wells or springs. If, however, the fluid added to the groundwater were immiscible with it, and also less dense so that it floats on top, its movement laterally would be restricted by certain types of structure affecting the reservoir layer. Where the reservoir layer is concave downwards under the impervious seal, for example in an anticline (Fig. 5.13), the lighter fluid rises and is trapped there. Accumulations of petroleum (that is, of oil and natural gas) occur naturally in structures of this type, which are called **oil traps**. (There is not always petroleum present to fill them.)

Oil traps can be used to store imported natural gas, and a few schemes

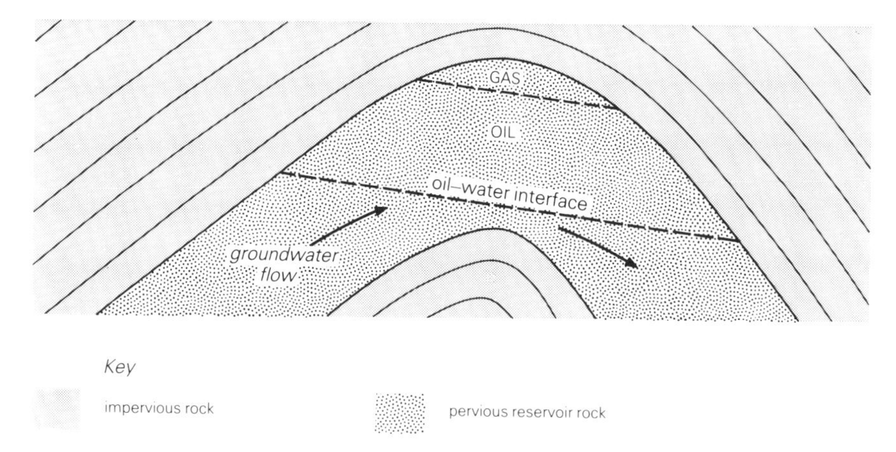

Figure 5.13 A thick pervious layer (the reservoir rock) and an overlying impervious layer (the roof rock) has been folded to form an anticline. This has trapped petroleum which has moved through the reservoir rock under the action of buoyancy and the hydraulic gradient, but cannot move further under the same forces. The less dense gas fraction has risen to the top, and the oil is concentrated below it. The voids in the rock below the oil/water interface are filled with ground water, which is flowing from left to right. The hydraulic gradient which produces this flow has also caused a tilt of the gas/oil and the oil/water interfaces in the direction of flow. If there were no flow the interfaces would be horizontal.

of this type exist. A feasibility study of some structures in England was made before the development of North Sea resources. Theoretically they could be used to store strategic oil reserves, but only at an unacceptable cost, with present technology, of not being able to recover a large percentage of the oil in the reservoir. So far, underground storage of oil and of liquefied frozen gas has been in large artificial caverns. For oil, these are excavated in impervious rocks; for liquefied gas they are excavated in soils or soft rocks, and the groundwater frozen by refrigeration makes the walls of the cavern impermeable. In Germany, salt domes provide excellent sites for stores, as the salt is easily excavated and impermeable. In Scandinavia and America, unlined caverns in crystalline rocks are used. They are not completely sealed as, unlike the salt chambers, they are not free of fractures, and groundwater tends to flow into them. The buoyancy and immiscibility of the oil prevent any leakage, however, by outward flow against the hydraulic gradient of the groundwater.

Faulting that affects any underground storage area is a hazard that should be avoided, particularly in areas of known seismicity or where the extraction of oil, gas or coal is causing subsidence. There is the possibility of leakage of fluid along the fault zone, and the changes of pore pressure in the fault plane may trigger off local movement along the fault. Disposal of waste down a 3600 m deep well in Colorado triggered off such a series of small earthquakes.

The disposal of radioactive waste is an increasingly important problem that demands careful study because of its scale and because of the consequences of any misjudgment. The conventional solutions, depending on the level of radioactivity, have been put succinctly as: (a) dilute and disperse, (b) delay and decay, and (c) concentrate and contain. Small quantities of highly radioactive waste can only be disposed of safely in the last way. Methods used or proposed for long-term storage include the following.

(a) Excavating a storage chamber in a layer of bedded salt at a depth of 300 m in Kansas: the salt is dry and impervious, and its flow under pressure would close any fracture that develops. It is a good heat insulator, and has enough strength to support shafts and room during mining.
(b) Placing the liquid waste in a cement mix and injecting it under high fluid pressure into deep shale layers: the high pressure produces local fracturing and makes possible the injection into shale.

At present, the UK is participating in a European programme of assessing how best to dispose of small quantities of highly radioactive waste from nuclear power plants. The UK Atomic Energy Authority has commissioned studies at Harwell (Robins *et al.* 1981) and by the BGS (Gray *et al.* 1976, Chapman 1979). These studies deal with relevant geological criteria concerning waste disposal (Gray *et al.* 1976), geochemical considerations of granites, argillaceous rocks and evaporite formations (Chapman 1979) and an investigation into the characteristics of certain argillaceous strata (Robins *et al.* 1981). Subsequent follow-up work has examined, in more detail, southern Scottish granites and evaporite (gypsum) formations in northern England (Northumberland) as potential radioactive waste disposal sites, but no firm conclusions have been reached as yet, and investigations are proceeding.

General criteria by which sites will be graded include population density, present land usage, accessibility and local amenities. The main geological criteria being assessed are the nature of the rocks, their physical and chemical properties, the seismic stability of the site from the geological history and the site's proximity to seismic zones, the permeability of the rocks and their nearness to pervious strata (or pervious zones), their susceptibility to the effects of major climatic changes including possible glaciation, and the possible effects on the potential site of nearby mining or drilling for minerals or dam construction.

An issue of *Geoscience Canada* (in reference list under Fyfe *et al.* 1980) was devoted to nuclear waste disposal, and the papers included the use of seismic techniques in identifying potential disposal sites, seismic risk, and geochemistry of radioactive waste disposal. The papers in this journal also provide an excellent bibliography of other relevant papers in this important and complex subject.

References and selected reading

Bell, F. G. 1978. Petrographical factors relating to the porosity and permeability in the Fell Sandstone. *Q. J. Engng Geol.* **11**, 113–26.

British Standards Institution 1972. *Sulphate-resisting Portland cement.* Br. Stand. Inst. Rep. BS 4027: Part 2.

Cedergren, H. R. 1977. *Seepage, drainage and flow nets,* 2nd edn. Chichester: Wiley.

Chapman, N. A. 1979. *Geochemical considerations in the choice of a host rock for the disposal of high level radioactive waste.* Rep. Inst. Geol. Sci. no. 79/14.

Clark, L. 1977. The analysis and planning of step drawdown tests. *Q. J. Engng Geol.* **10**, 125–43.

Domenico, P. 1972. *Concepts and models in groundwater.* New York: McGraw-Hill.

Fyfe, W. S. and many other authors 1980. Nuclear waste disposal. *Geosci. Canada* **6** (4).

Galley, J. E. (ed.) 1968. Subsurface disposal in geologic basins – a study of reservoir strata. *Mem. Am. Assoc. Petrol. Geol.* **10**.

Geological Society, London, and Institution of Water Engineers 1963. Joint symposium. Ground water: geological and engineering aspects. *J. Inst. Water Engrs* **17** (3), 153–291.

Gray, D. A., J. D. Mather and I. B. Harrison 1974. Review of ground water pollution from waste sites in England and Wales with provisional guidelines for future site selection. *Q. J. Engng Geol.* **7**, 181–98.

Gray, D. A. *et al.* 1976. *Disposal of highly active, solid radioactive wastes into geological formations – relevant geological criteria for the United Kingdom*. Rep. Inst. Geol. Sci. no. 76/12.

Hodgson, A. V. and M. D. Gardiner 1971. An investigation of the aquifer potential of the Fell Sandstone of Northumberland. *Q. J. Engng Geol.* **4**, 91–110.

Hubbert, M. K. 1940. The theory of groundwater motion. *J. Geol.* **48**, 785–944.

Jones, G. P. 1971. Management of underground water resources. *Q. J. Engng Geol.* **4**, 317–54.

Krynine, D. P. and W. R. Judd 1957. *Principles of engineering geology and geotechnics*. New York: McGraw-Hill.

Penman, H. L. 1950. Evaporation over the British Isles. *Q. J. R. Meteorol. Soc.* **76**, 372.

Robins, N. S., B. A. Martin and M. A. Brightman 1981. *Borehole drilling and completion: details for the Harwell research site*. Inst. Geol. Sci. Rep. no. ENPU 81–9: Harwell.

Water Data Unit (current). *The surface water yearbook of Great Britain*. London: HMSO.

6 *Geological exploration of an engineering site*

6.1 General considerations

6.1.1 Introduction

Good professional practice in the exploration of an engineering site is described in BS 5930: 1981, Code of Practice for Site Investigations, which should be referred to by professional engineers as a source of definitions, and as an authoritative statement of the objects, scope, procedures and methods of site investigation. It covers testing of mechanical and other properties of soils and rocks, which conventionally forms part of a civil engineering course taught by engineers. This chapter deals, however, only with the geological and geophysical surveys used to investigate what rocks and soils are present, and how they are distributed under the site.

The exploration of a site to assess the feasibility of a project, to plan and design appropriate foundations, and to draw up bills of quantity for excavation normally requires that most of the following information be obtained:

(a) what rocks and soils are present, including the sequence of strata, the nature and thicknesses of superficial deposits and the presence of igneous intrusions;
(b) how these rocks are distributed over, and under, the site (that is, their structure);
(c) the frequency and orientation of joints in the different bodies of rock and the location of any faults;
(d) the presence and extent of any weathering of the rocks, and particularly of any soluble rocks such as limestone;
(e) the groundwater conditions, including the position of the water table, and whether the groundwater contains noxious material in solution, such as sulphates, which may affect cement with which it comes in contact;
(f) the presence of economic deposits which may have been extracted by mining or quarrying, to leave concealed voids or disturbed ground; and
(g) the suitability of local rocks and soils, especially those to be excavated, as construction materials.

Special information such as the seismicity of the region or the pattern of sediment movement in an estuary may also be required.

Much of this exploration, particularly the making of geological maps, is normally carried out in large projects by a professional engineering geologist. The role of the engineer, and his need to understand geological terms, concepts and methodology even while employing a consultant, has been discussed earlier (Section 1.1). In limited sites the engineer may have to collect his own geological data, and make elementary, but crucial, geological decisions on, for example, whether or not a boring has reached bedrock, or has struck a boulder in the overlying till.

6.1.2 *Inductive reasoning*

Most conclusions reached by mathematical arguments are achieved by **deductive reasoning**. In using this method, a relationship between physical quantities is accepted as being generally true, and expressed as a scientific principle. Then, in the simplest case, the answer to a specific problem is found by substituting numerical values into the general equation that describes the relationship. In other uses of deduction, more subtle general relationships between the physical quantities may be *deduced* by mathematical argument. This type of reasoning is objective, and a logical person or a logical machine, suitably skilled, arrives at the same conclusion. The answer is usually unique, as distinct from 'most probable', and its error is governed by the precision with which the physical quantities are determined. The method is tidy and intellectually satisfying, and there is understandably a temptation to think of it as the only respectable way to arrive at a scientific conclusion.

Its limitation is that it depends on the existence of established principles and on their suitability to take *full* account of the physical conditions in which the problem is set. There has to be a logical method, other than deduction, of deriving principles; problems can only be solved if these principles assist prediction in the complex circumstances that occur naturally, as distinct from in laboratories and other environments controlled and refined by man.

This other method of reaching a conclusion is called **inductive reasoning** or **inference**. A general principle is arrived at by recognising the pattern in a set of observations, and predictions are then made. That is to say, the general principle is *inferred* from the specific cases. It is comparable to a game where card players each play a card, then a bystander indicates who has won each round of play. The point of the game is that each player tries to discover what the rules are.

The uses and nature of inference, intertwined with deduction, are shown by the following adaptation of a game described by Gardner (1969). Any number can play. Each player draws a block of 7 × 7 squares.

Each square is a discrete location in a small two-dimensional, timeless universe, and may be occupied by any one of four types of phenomenon, characterised by the symbols +, ○, #, * (Fig. 6.1). The distribution is known only to the designer, who has planned the universe before the game starts. The aim of each competitor is to discover what symbol the designer has put in each square, that is, the complete distribution of symbols in the block of squares. Discovery can be made by asking the designer what is present in any square, at any stage of the game, before the final scoring. Enquiry may be made as often as is preferred. Alternatively the player may infer what symbol is present; but whether rightly or wrongly is not confirmed till the final scoring. A correct inference scores +1 and a mistake scores −1. The squares filled by asking the designer neither add to, nor detract from, the final score. This can theoretically range from −49 to +49. A cautious player can score zero, and possibly win, by asking the designer for his entire plan.

The general points that emerge are as follows:

(a) Prediction cannot be made by deductive reasoning alone, even after a certain amount of random data is obtained from the designer, to sample the distribution of symbols. Understanding comes from inference, which is a subjective approach requiring imagination.

(b) It is only possible to predict (infer) the distribution of symbols if they are systematically arranged, that is, if they form a pattern. The distribution must be capable of definition in a simpler way than by saying what is in each square.

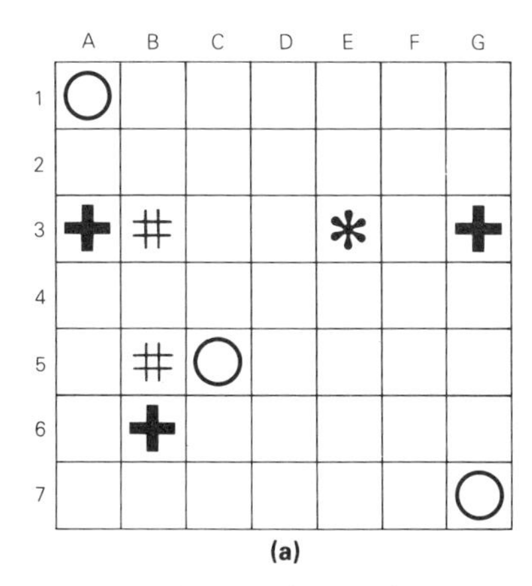

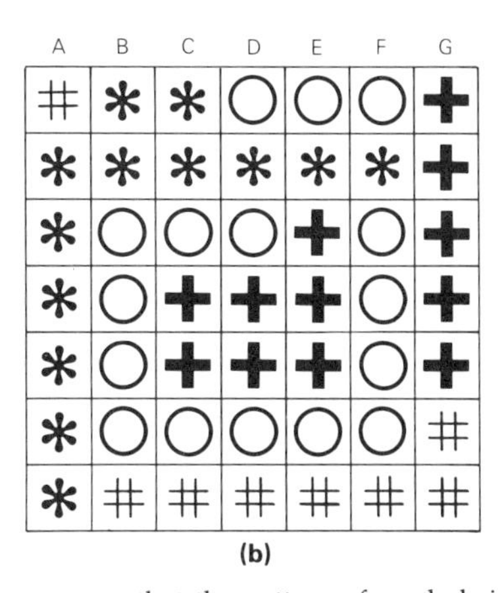

Figure 6.1 (a) This first random scatter of data suggests that the pattern of symbols is symmetric, and further testing to prove this hypothesis is made. (b) The pattern of symbols in this block of squares may be puzzling at first inspection. It is a sequence of prime numbers starting in the top left-hand corner which spirals towards the centre. It would remain puzzling if the inspection were done by someone unfamiliar with prime numbers.

(c) The pattern must be one known to the player. For example, a representation of successive prime numbers remains a meaningless jumble to someone unfamiliar with the concept of primes. An uncommon pattern will be correspondingly difficult to recognise. (For example, what is the pattern in Fig. 6.1b?)

(d) The systematic approach to a solution is as follows (i) to get a random sample of data from the designer, such as is shown in Figure 6.1a; (ii) a number of possible patterns are suggested, especially by the coincidences (squares A1 and G7; B3 and B5; A3 and G3) that are consistent with symmetry; (iii) the simplest pattern (a subjective judgment) consistent with the observations is then tested by choosing further information from the designer to prove its existence, and allow discrimination between it and other hypothetical patterns; (iv) feeling moderately certain because the new data are similar to what we predicted, the postulated pattern now becomes the guiding principle that allows us first to see what additional information from the designer is needed to define one symmetric segment of the pattern adequately, and secondly to predict the content of the remaining squares.

6.1.3 *Systematic exploration of a site*

The methodology of site exploration hangs on a framework of inductive reasoning similar to that used in the game described earlier. Geological information available at the start of exploration is assessed against known patterns of rock distribution (that is, against stratigraphical and structural models). The working hypothesis obtained is then checked by results acquired at later stages, and there is a progressive defining, and refining, of the geological model. Choices, similar to those of the game, are offered to the player. He can insure himself against a disastrous negative score by intensive checking. This may require a profusion of drill holes, or even excavation of the entire site, and he may be defeated by a more skilled competitor. Successful exploration not only acquires all necessary information about the subsurface, but does so as economically as possible. Information should not be bought if existing data are available on file. A well-sited drill hole which tests a geological model will be as informative as two or three randomly sited. Local detail that is of little service in planning or costing an extended site should not be bought at excessive cost.

The most elaborate exploration is needed for a large area where no systematic exploration or construction has been carried out, that is, an extended greenfield site. Systematic exploration *may* involve up to five procedural stages:

(1) preliminary investigation using published information and other existing data;
(2) a detailed geological survey of the site, possibly with a photogeology study;
(3) applied geophysical surveys as a reconnaissance of the subsurface geology;
(4) boring, drilling and excavation to provide confirmation of the previous results, and quantitative detail at critical points on the site; and
(5) testing of soils and rocks to assess their suitability, particularly their mechanical properties (soil mechanics and rock mechanics), either *in situ* or from samples.

In major projects all of these investigations may be carried out. In minor projects, or where the site is small and has previously been built on, only stages (4) and (5) may be necessary, together with (1), which is cheap and always worth doing. A report by Dumbleton and West (1972a) is a useful and comprehensive guide to site investigation.

6.2 Preliminary investigation

6.2.1 Sources of information for UK site investigation

The first stage of investigation is to search for and study existing information. This may be adequate for an *initial* estimate of feasibility, planning and costing of the project, and to plan the full exploration of the site. It provides a geological model which is progressively refined by data from detailed surveys and drilling. It is referred to as a desk study.

The main sources of information are the following:

(a) *The British Geological Survey* (see Appendix C). The publications of most use to engineers are the 6 in : 1 mile and 1 : 10 000 maps, and the District Memoirs. The Memoir describes the geology of the area covered by the 1 in : 1 mile (or 1 : 50 000) map which encompasses the ground of the larger-scale map. Information about the rock types in the strata at any locality is obtained by referring to the section of the Memoir dealing with rocks of that age, and then to the specific area. Any lateral changes of sequence across an extended site may be inferred from comparing sequences of the same age at different localities. Attention is drawn to rocks of economic interest in the Generalised Section on the right-hand side of the map, and this may be supplemented by a brief account of mineral veins in the Memoir, or a fuller one in a special Economic Memoir dealing, for example, with a coalfield. This information provides warning that there might be old mineral workings or other disturbed ground (such as fill at the site of an old quarry) where coals, ironstones and fireclays come to outcrop. The other statutory duties of the BGS include acquisition and curating of information, such as borehole logs and measurements of the engineering properties of rocks.

It is important that, before any actual exploratory work is carried out on the site, the engineer assesses, by studying the BGS information available, any problems that are likely to be encountered. Since many 1:10000 geological sheets are not published, a preliminary study of a site should involve a visit to the nearest BGS offices, where unpublished maps can be made available for examination. The librarian at a BGS office will usually help in this, and Appendix C gives the addresses of the main BGS offices throughout the UK.

(b) *National Coal Board Area Offices.* Old coal workings present one of the most frequent problems affecting sites in industrial Britain. The problems increase as land previously rejected for building is reassessed by developers as part of larger schemes of rejuvenation. Coal has been mined in the UK from formations other than Upper Carboniferous Coal Measures. If coal-bearing strata crop out at the site, information about past and present extraction should be sought from the NCB. The Board holds mining plans on behalf of HM Inspectorate of Mines, and these can be inspected at the Area Offices. *The catalogue of abandoned mines and plans,* with supplements up to 1938, may be available there. Copies are rare, but are available in a few national libraries. The Estates Manager, NCB, will provide a copy of plans and a report, which offers no firm opinion on the site but is useful preliminary information, for a small fee.

(c) *Library of Health and Safety Division, Department of Energy, London.* Information about abandoned mineral workings other than coal must usually be sought from the Department of Energy. Their service is, however, only that of a reference library. The searcher must visit London and make what notes and tracings he or she requires.

Information about disturbed ground and the stability of areas, where the cause of possible trouble precedes the 19th century laws about mining records, *may* be available from local archives, libraries, private firms or individuals. A comprehensive study (see Appendix D) might include the following:

(d) *The Ordnance Survey:* to compare past and present site topography, including changes in benchmark levels.

(e) *The British Museum:* to study old maps and records (these may be available locally and should be sought there first).

(f) *Local Authority Offices and local museums:* for information about records of ground collapses.
(g) *Local mining and engineering organisations.*

Before committing himself to extensive research, the engineer should, however, consider whether the information from these minor sources might not be obtained more economically and quickly from investigation of the site itself.

6.2.2 *Geology of the site*

After the desk study has been completed, an assessment of the on-site geology must be made before any geophysical work, drilling or rock testing is carried out.

First, the engineer or geological surveyor should be able to recognise and evaluate any *landforms* (that is, characteristic shapes of the ground surface) that occur on the site. This allows him to trace boundaries across areas in which there are few outcrops, and to draw inferences about the distribution of solid rocks or superficial deposits concealed under a cover of soil. Section 3.6 gives full details on landforms. In general, landforms produced by *erosion* of the land surface convey information about the distribution of the *solid rocks*, whereas landforms resulting from *deposition* are related to *superficial deposits*. Identification of landforms may be greatly assisted by a **photogeology study**. Air photographs of the area (see Fig. 6.2) are obtained from an air survey, which is a means of rapid reconnaissance of an extended site, such as a new road, and also provides a bird's-eye view that allows the geological interpreter to see each landform in its entirety, so making appreciation of its complete shape easier than it would be on the ground. He or she is further assisted by the accentuation of relief produced by vertical exaggeration when he views photographs stereoscopically. The vegetation patterns and the tone and texture (that is, the pattern of varied tone) often provide additional information about the near-surface rocks, and in some cases about their water content. A survey done on colour film produces more data, but is more expensive. The rarer surveys employing infra-red photography usually serve aims other than geological, but are a way of tackling such esoteric problems as recognising the presence of cool freshwater springs discharging into the warmer sea off arid limestone coasts (Section 5.2.5).

Surveys can be commissioned from specialist contractors, who may also offer geological interpretation of the air photographs as an additional service. Photographs taken by the RAF during practice flights are available in certain specialist libraries to approved users, but are not suitable for accurate survey work.

A *site inspection* should be made as part of the preliminary geological

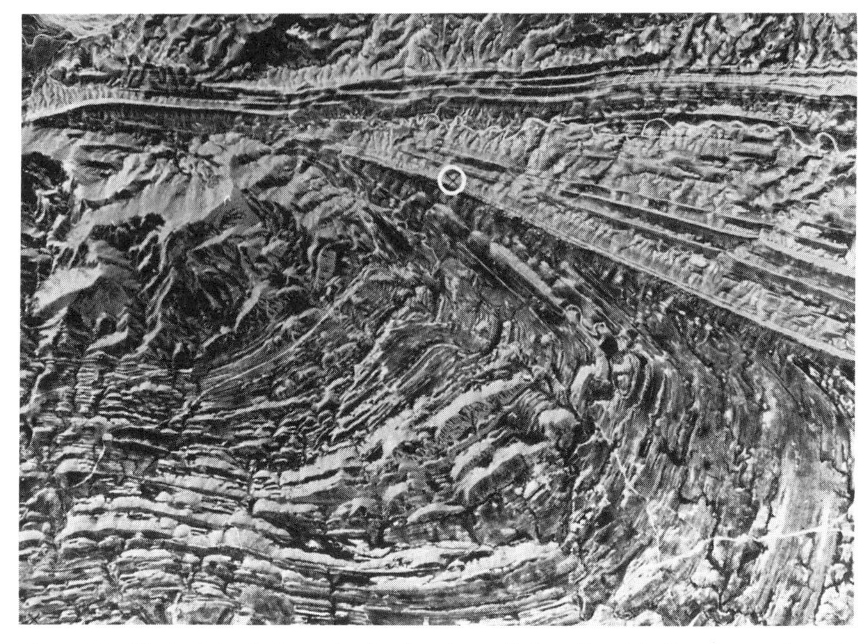

Figure 6.2 Vertical air photograph of a plunging anticline, thrust fault and unconformity in Iran. The geological structure can be surveyed from this, and other, overlapping photographs which provide stereoscopic coverage. (Photograph by Huntings Surveys, reproduced by kind permission of British Petroleum Co.)

exploration. The investigator should ideally have a copy of a BGS or other geological map to which he can refer. He should compare his observations and his inferences about boundaries with those of the map. A systematic check will show whether any additional exposures produced by natural or manmade developments allow a fuller understanding of the geology and its relation to the site. All landforms should be studied, and particular attention paid to ground probably disturbed by dumping of waste, by quarrying, by mining or by other features that might indicate earlier use of the site. Landslip scars or minor terracing may indicate past or present instability. Some impression of groundwater conditions, including the position of the water table or of a perched water table, may be gained from seepages, springs and vegetation.

If a preliminary geological map of the site is not available, the investigator should attempt to make one. He should examine all exposures of rocks and superficial deposits in river banks, cuttings and excavations, and plot their position on a map or plan of the site, together with any landforms that might serve to determine boundaries. In general, his first aim should be to determine the rock types present, the sequence of strata, the lateral changes in sequence and the other age relationships between the rocks and soils present, before using the vertical sequence as a measuring rod to determine structure.

An account of how to inspect and describe rock masses and the discontinuities that affect their properties systematically is given in Appendix H.

6.2.3 *Old mineral workings*

SUBSIDENCE FROM UNKNOWN WORKINGS

Several types of rock and mineral have been extracted in Britain by quarrying at the surface and by mining underground. Either type of disturbance may produce complex and hazardous conditions of the subsurface. Old quarries within urban areas may be backfilled with loose waste, sometimes consisting mainly of rock fragments similar to those previously *in situ* but very different in bulk engineering properties. A more common hazard arises from shallow mining carried out before the middle of the 19th century, as pre-1873 plans of coal mines are rare and may be inaccurate.

The Carboniferous strata, especially the Coal Measures, provided the main targets for the early miners. Mines for coal, fireclay, ganister and ironstone are common, and underground workings for sandstone (as building stone) and limestone have been encountered. Small mines for lead ore (galena) are common in the major outcrops of Carboniferous limestone; copper, tin and silver ores give rise to rashes of small abandoned mines in certain localities. In general, however, the metalliferous ores in Britain occur away from urban centres, and, though worrying in certain ways, they do not produce the same steady crop of site problems that old coal workings do. Among the obscure causes of ground disturbance are shallow pits dug in chalk by Stone Age men in search of flints.

The oldest type of mine workings is the medieval **bell pit** for coal or ironstone. The general form is a shaft about 1.25 m in diameter, going down vertically to a depth that rarely exceeds 12 m. Coal is extracted radially from the bottom of the shaft for several metres, until the technical difficulty of supporting the roof leads to abandonment of this bell pit and to a new one being dug nearby. The pits may be partially backfilled with waste but the compaction may not be good.

Room and pillar (**pillar and stall**; in Scotland **stoop and room**) workings are much more common, and provide frequent problems in areas of early mining (see Fig. 6.3). Coal is extracted from the rooms, and pillars are left to support the roof. Both vary in size, as does the percentage of the coal seam extracted. Complete subsidence may not take place until more than a century after abandonment, in some cases by deterioration and crushing of the pillars, in others by creep of seat earth, but in most by failure of the roof strata and collapse into the rooms (see Fig. 6.4). The roof void migrates slowly upwards, and usually migration either ceases below a bed of sufficiently high tensile strength, such as a massive

Figure 6.3 Abandoned room and pillar workings for coal, which have been exposed during excavations at a locality in Central Scotland. Notice how close to the soil cover the workings extend. (Reproduced by kind permission of Mr George Archibald, Subsidence Engineer, National Coal Board.)

Figure 6.4 A quarry face showing abandoned room and pillar workings which are partially collapsed. The voids (rooms) have migrated upwards as the roofs have collapsed, and in one case (extreme left of photograph) only a thin layer of rock separates the void from the overlying soil. (Reproduced by kind permission of Messrs J. W. H. Ross & Co., Mining Engineers, Glasgow.)

sandstone that spans the void, or reaches the surface. The surface feature corresponding to each collapsed room, a shallow roughly circular depression, is called a **crown hole** or a **plump hole**. The factors governing collapse are described in the *Subsidence engineers' handbook* (National Coal Board 1963). In shallow workings, voids have been known to persist under a cover of stiff boulder clay only.

SUBSIDENCE FROM KNOWN LONGWALL WORKINGS

The near-universal method of mining coal at the present day is by **longwall workings**, in which the seam is extracted along a continuous coalface. The roof is supported temporarily and later allowed to subside, with only roadways left open. Access may be by an inclined roadway (adit), or more commonly by a vertical shaft. From about 1864 it became compulsory in Britain to have two shafts for each mine, and these are normally together. Mining could only be developed efficiently up to a certain distance from the shafts and usually only to the rise of the seams. After the mine was abandoned, the shafts were normally rafted over at ground level or at a depth of a few metres, and backfilled with local waste such as ashes from the spoil heap. In Scotland, shafts were normally rectangular with a wooden lining, and may have been roofed over with wrought-iron rails. In England and Wales, they were nearly always circular with a brick lining. How serious a hazard the shafts represent depends partly on the thickness and cohesion of the superficial deposits around the top of the shaft. Any crater produced by failure of the raft widens out from the shaft at rock head and, as in a recent collapse near Glasgow, an area of as much as 30 m in diameter may be affected.

Collapse behind the working face underground produces differential vertical subsidence within the **area of draw** (Fig. 6.5) at the surface, and a general lowering of the ground surface above those older longwall mines where settlement is complete. In industrial regions, where production engineering firms are active, even slight inclinations of the surface can upset the accuracy of some production tools. The ground surface of the outer area of draw, where differential subsidence is initiated, is convex, and buildings in this zone are liable to develop tension cracks. As the coalface advances to bring the same buildings within the inner area of draw, the ground surface continues to subside, but is now slightly concave. The buildings are now subject to compression, which closes the older cracks and produces new ones associated with crushing, for example upheaval of pavements and similar effects. The subsidence may be concentrated along a pre-existing fault, and give rise to a step at the surface. The risk of serious damage is greatest where coal is worked below the overhang of an inclined fault. A standoff of 15 m on either side of the surface position of a known fault is recommended in one 1959 publication on mining subsidence but, as a general rule, 15 m is dangerously too short where a seam is worked under an overhang.

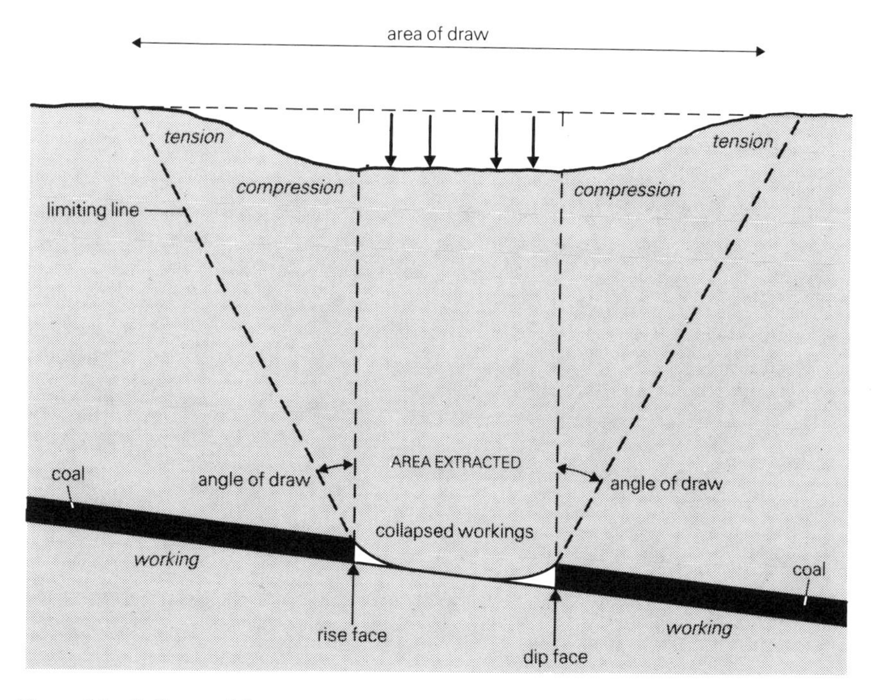

Figure 6.5 Collapse of the area where coal has been extracted produces subsidence at the surface. Within the areas of draw, subsidence is at different stages, and as a result part of the surface is affected by tension and part by compressive stresses.

The greater the depth of the coal workings, the wider the effects of differential subsidence are spread at the surface, and the lesser is the damage to property. Surface subsidence is less than the thickness of the extracted seam, and the ratio is affected mainly by depth to the collapsed workings. A rough guide is given by the formula for the surface subsidence as the percentage thickness of extracted seam:

$$\frac{s}{t} = \frac{400}{\sqrt{D} + 4}$$

where s is the surface subsidence and t is the thickness of the seam (both in feet or metres), and D is the depth below surface in hundreds of feet or metres.

Other factors include dip of the seam, method of stowage of waste, and nature of overlying strata. The last will normally have their engineering properties (such as permeability) affected by the process of subsidence. Subsidence starts within 24 hours of extraction, but its *full* effects are transmitted slowly upwards, and it may be as long as 12 years before the surface is completely stable. If this lag in the transmission of stresses, which depends on depth, is large compared with the rate of advance of the coalface, then the surface may suffer essentially vertical subsidence only. On the other hand, if a face is abandoned, the effects of draw are marked.

The problem of locating old mineral workings, particularly shafts and room and pillar workings, is a common one in regions of Britain which were industrialised early. It is usually difficult, since detection of voids of room or shaft size is uncertain and laborious by present geophysical methods, and may be prohibitively expensive by drilling. Nevertheless, sterilisation of land by blocking off approximate locations is costly, and may be dangerous if the approximations are poor. An outline of procedure and methods is given in Appendix D.

6.3 Applied geophysical surveys

6.3.1 Introduction

Geophysics is the study of the physical processes active within the Earth, and of the physical properties of the rocks forming it. From it stemmed **applied geophysics**, a set of exploration methods used to infer the distribution of rocks underground from physical measurements made at the surface. For example, measurements of the pull of gravity at different places (stations) along a traverse can be used to predict how rocks of different density occur below the traverse. If we consider a simple density model, say the left-hand side of Figure 6.6, with only two materials of density 2.0 and 2.5 g cm^{-3} present and separated by a horizontal interface, then the attractive mass distributed under any point at the surface is the same, and the pull of gravity at the surface is constant along the model. Using this as a standard, let us redistribute the materials. If the deeper, denser layer is stepped up towards the surface by a fault (F on Fig. 6.6), part of the 2.0 g cm^{-3} layer in the standard model is now substituted by 2.5 g cm^{-3} material. The new density model may be described as differing from the previous, standard model by the addition of a positive **anomalous mass** of density +0.5 g cm^{-3}, and of cross-sectional area abcd. This extra mass increases the pull of gravity relative to the 'standard' value. There is a **gravity anomaly** with respect to standard gravity that corresponds to the anomalous mass and indicates its presence.

Note that *at most* our measurements tell us about the lateral distribution of densities underground. Any inference about the type of rock corresponding to a particular density must depend on another argument. How one physical property of a rock differs from those of its surroundings is seldom diagnostic of the rock. Another limitation is that the precision with which a boundary can be fixed depends on the magnitude of the

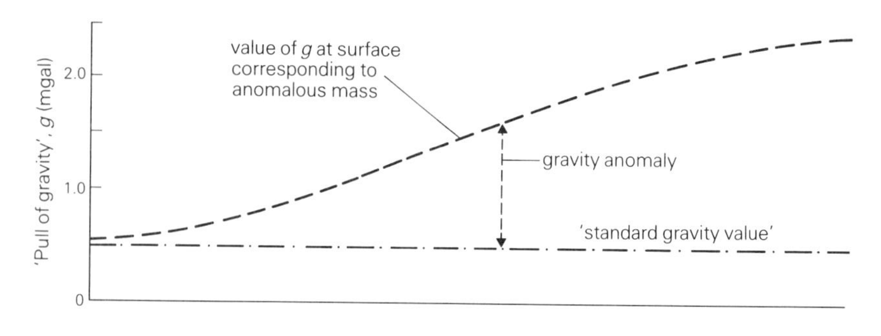

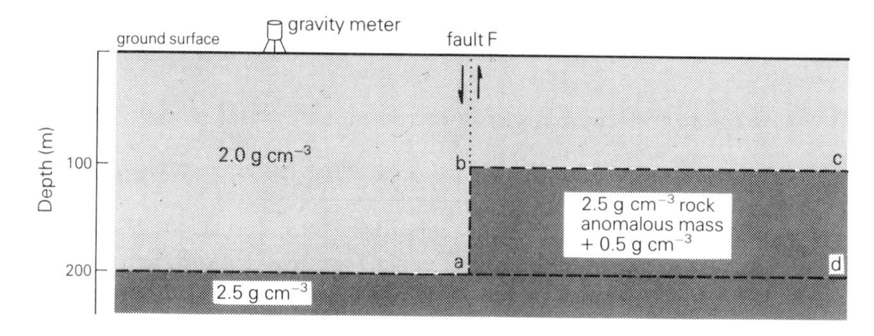

Figure 6.6 The 'pull of gravity', g, which results from the simple horizontal density layers, is plotted as 'standard gravity value'. The raising of a block of the denser 2.5 g cm^{-3} layer by upthrow along the fault F increases the value of g along the traverse. The greatest increase is on the upthrown side, and the most rapid change of g along the traverse (that is, the greatest horizontal gradient of g) occurs immediately above the fault. The difference between the second set of values, which include the effects of the anomalous mass, and the 'standard value' at any surface point is called the 'gravity anomaly' at that station. The units used for g are milligals (mgal); see Section 6.3.8 for definition.

contrast of density across it. In the extreme case where there is no contrast (for example, greywacke and weathered basalt have the same density), there is no anomalous mass arising from the fault step, no gravity anomaly, and no way of detecting the fault by a gravity survey. The boundary is invisible unless some other contrast in a physical property exists and can be employed in another geophysical method.

A wide range of physical properties of rocks are used as the bases of particular geophysical methods. The most common in engineering surveys are: (a) elastic properties, used in the seismic methods; (b) magnetic properties, used in the magnetic methods; (c) electrical properties, used in the electrical and electromagnetic methods; and (d) density, in the gravity method. The choice of a method depends, first, on whether an appropriate physical contrast occurs across the boundary which is being investigated. Criteria, other than geophysical, are the resolution sought, and the relative cost of coverage and time taken, by a particular type of survey; and the nature of the site, for example whether it is water covered, whether it is permissible to explode small charges on it, and so on.

6.3.2 *Seismic waves*

The behaviour of wave motions in rocks is dependent on the elastic properties of the rocks. The applied seismic methods are used to determine the distribution of elastic properties under a site, as a step to inferring what rocks and structures are present.

Elasticity is the reaction of a given substance to stress, where the stress is not sufficiently large to deform it permanently. It is the **strain** produced by small stresses acting over a short period of time, which is *reversible* when the stress is removed. Two parameters are required to describe the elasticity of an isotropic material. Young's Modulus and the Poisson Ratio are the pair most familiar to engineers. The two aspects of elasticity, *resistance to change of volume* and *resistance to change of shape*, are, however, most simply described by the **coefficient of incompressibility** (bulk modulus) k and the **coefficient of rigidity** (shear modulus) μ. If F/A is the force per unit area (that is, the stress), then k and μ are defined by the stress–strain equations:

$$F/A = k\Delta V/V$$

where a volume of the material V is reduced by ΔV under a confining pressure F/A, without change of shape (Fig. 6.7):

$$F/A = \mu\theta$$

where a rectangular block of material is deformed through an angle θ (Fig. 6.7) by a shearing stress F/A without change of volume.

If an elastic medium, such as a rock, is suddenly disturbed at a point 0, two types of elastic wave (**body waves**) spread out from 0 in all directions. Each individual particle of rock vibrates harmonically about a mean position as the wave motion is propagated through the rock. In **longitudinal waves**, the particle motion is parallel to the direction of propagation, and each part of the rock is periodically compressed and dilated by the wave motion. Their velocity of propagation, V_p, is given by

$$V_p = [(k + 4/3\mu)/\rho]^{\frac{1}{2}}$$

where k and μ are the elastic constants, and ρ is the density of the rocks, or

$$\rho V_p^2 = E(1 - \sigma)/[(1 + \sigma)(1 - 2\sigma)]$$

where E is Young's Modulus and σ is the Poisson Ratio.

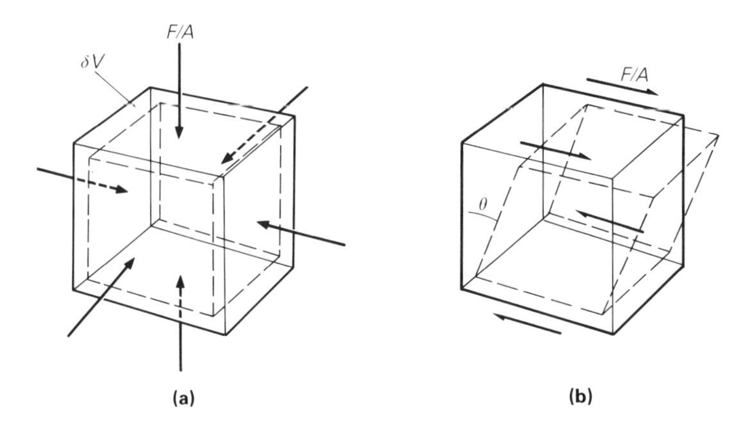

Figure 6.7 (a) Uniform confining pressure F/A produces a decrease in volume δV, which is related to the coefficient of incompressibility of the rock. (b) A shearing stress F/A produces deformation through an angle θ, without accompanying change of volume.

In **transverse waves**, the particle motion is normal to the direction of propagation. Their velocity, V_s, is

$$V_s = (\mu/\rho)^{\frac{1}{2}}$$

or

$$\rho V_s^2 = E/[2(1+\sigma)]$$

(see above for explanation of symbols).

From the equations, (a) both velocities are independent of amplitude or period of the waves, and (b) V_p is always greater than V_s, hence the longitudinal waves are the first to arrive at any detector.

Other types of wave motion can exist only close to a free boundary, such as the surface of the Earth, and they decrease exponentially in amplitude with depth. Their velocities are lower than those of body waves. They are generated as a by-product of the shot, but they do not provide useful information for the seismic methods commonly used in site investigation.

The passage of waves through rocks can be represented by **rays** (trajectories) which show the paths taken by the waves, and by **wave-fronts**, normal to the rays in isotropic material, which show how far the wave motion has travelled since it left the **shot point**, the source of disturbance. Each wavefront is a surface of equal time that has elapsed since the moment when the waves were generated. For most purposes, the propagation of elastic waves is described adequately by the laws of geometrical optics.

When waves meet an interface separating rocks of different seismic velocity V_1 and V_2 (Fig. 6.8), such that the incident ray makes an angle i_1 with the normal to the interface, then part of the energy is reflected along a ray making the same angle i_1 to the normal, and part penetrates into the lower layer, V_2. In doing so, it is refracted such that the ray makes an angle i_2 with the normal. Snell's Law describes this refraction and states that $\sin i_1/\sin i_2 = V_1/V_2$.

If the incident ray meets the interface at an angle i_c, such that $i_2 = 90°$, then i_c is called the **critical angle**. Rays meeting the interface at angles greater than i_c are *totally reflected* and waves are not transmitted into layer V_2. For the critical ray, where $\sin i_2 = 1$, Snell's Law simplifies to $\sin i_c = V_1/V_2$.

Seismic velocity in rocks is influenced by depth of burial, degree of compaction and cementation, fluid saturation of voids, texture and chemical composition. In general, the lowest velocities are found in non-coherent, dry, superficial deposits, and higher velocities are found in older sediments which have been, or are, deeply buried, which have a low porosity and which are saturated with ground water. The highest velocities occur in crystalline rocks: in certain limestones and evaporites

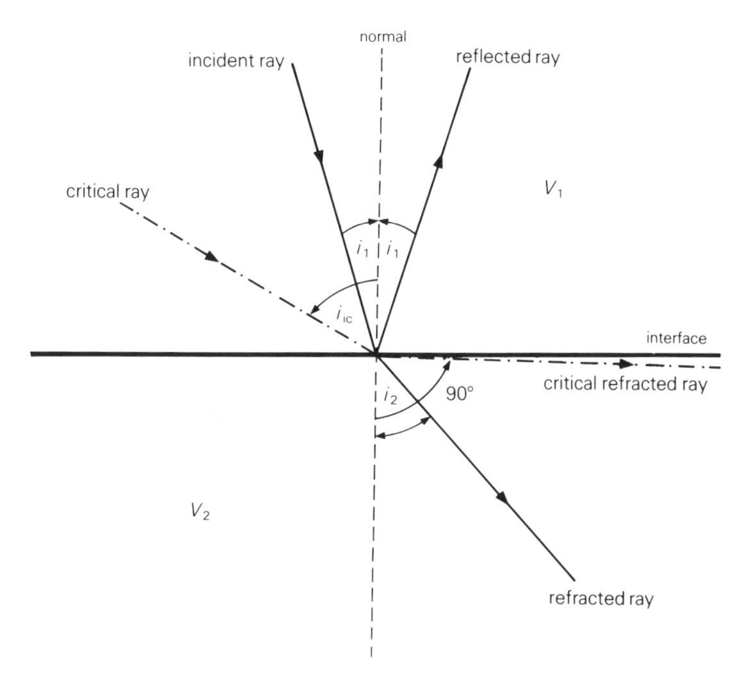

Figure 6.8 The interface separates two materials of different (propagation) velocity, V_1 and V_2. The lower layer has the higher velocity ($V_2 > V_1$). Part of the energy of the incident longitudinal ray is reflected, and part is refracted (away from the normal since $V_2 > V_1$). The critical longitudinal ray is refracted to 90° from the normal, and is propagated along the top of layer V_2. Transverse waves are not shown, but also suffer reflection and refraction.

(salt and anhydrite), in metamorphic rocks and in igneous rocks. In the last, velocity is greatest in rocks of basic composition. A range of velocity is found within any rock formation, and there is ambiguity in interpreting lithology purely from seismic evidence. The typical values of longitudinal (p) wave velocity given in Table 6.1 should provide a sense of quantity, but should not be considered as diagnostic of the rocks mentioned.

Table 6.1 Typical values of longitudinal wave velocity V_p

	V_p (km s^{-1})		V_p (km s^{-1})
air	0.33	Carboniferous sediments	3.30
fresh water	1.45	Lower Palaeozoic greywackes	5.50
dry soil	0.60	granite	5.00
saturated sand	1.50	gabbro	6.50
boulder clay	2.00	Lewisian gneiss	6.00

6.3.3 *Seismic-refraction method*

In the applied seismic methods, waves are generated at a shot point usually by firing off a small explosive charge electrically or, in adaptations of the method, by dropping a weight or by hitting a metal plate with a sledgehammer. The arrival of the waves at various stations on a line of traverse is detected by **seismometers** planted at the surface, or by **hydrophones** in water-covered areas. These convert ground or water motions into a varying electrical current. This is transmitted along wires linking each detector to a **seismograph**, where the output from each channel is amplified and recorded. On the seismic record (**seismogram**), the ground motion at each detector is represented by a line of wiggles on one trace of the record. The time taken by the waves as they follow different paths between shot point and detector can be measured as the moment of explosion, and time marks (usually at 0.01 s) are recorded also.

In the simplest and most commonly used seismic-refraction method (Fig. 6.9), only the times of the first pulse of energy to arrive at each detector are used. For this reason, seismic-refraction apparatus is simpler and cheaper than that needed for seismic-reflection surveys, in which more complex electronic equipment is essential to give a useful record of later, reflected arrivals. The **first arrivals** at detectors close to the shot point travel directly along the ground surface. A plot of arrival times (T) against distance (X) of each corresponding detector from the shot point gives a straight line. Its gradient is the reciprocal of the velocity (V_1) of the rock at the surface. If the rock layer below layer V_1 has a higher velocity V_2, there is a critically refracted ray which travels along the top of V_2. If layers V_1 and V_2 are coupled mechanically, such that the top of V_2 cannot vibrate without exciting waves at the base of V_1 (and this is the natural state of subsurface rocks), then waves pass through the interface from layer V_2 to V_1, and in the process are refracted to continue towards the surface at an angle i_c to the normal. Paths of this type (where the ray first

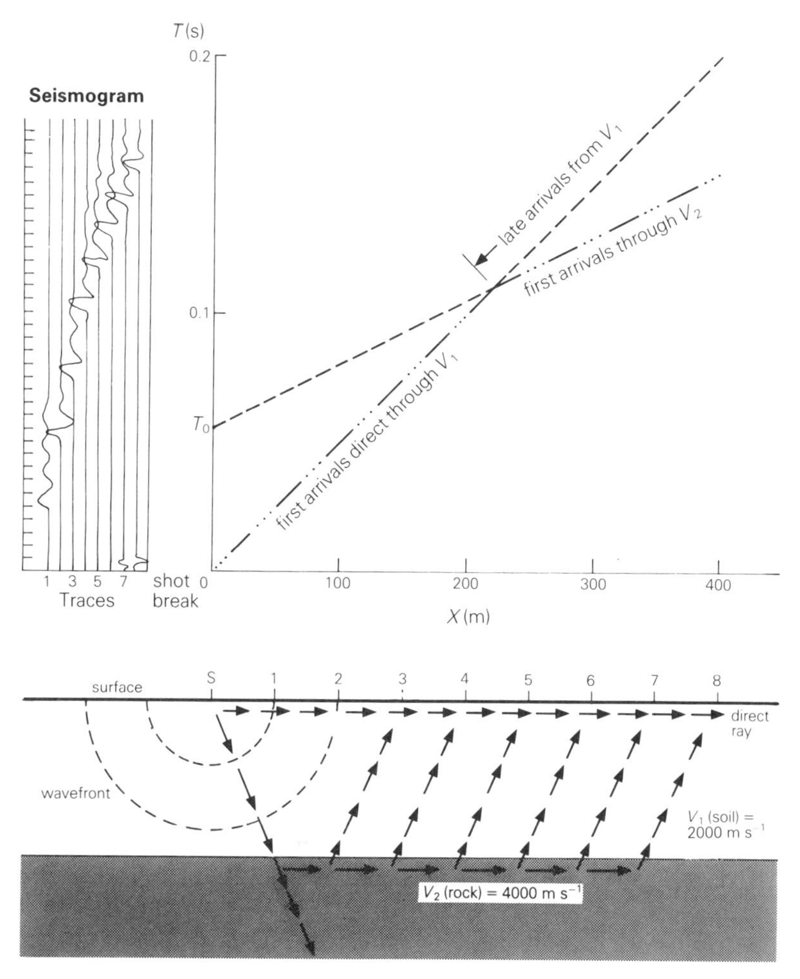

Figure 6.9 An explosive charge, fired at the shot point S, generates body and surface waves which propagate in all directions from S. Only the rays and wavefronts of the longitudinal (P) waves are represented, since with their higher velocity relative to the other types of wave they arrive first at the detectors laid out along the spread (traverse) being surveyed. The times of first arrivals (see text) at detectors 1 to 8 can be read from the corresponding trace on the seismogram. The first seismic pulse at each detector produces a sudden sharp deflection of the trace. The time can be read from timer lines on the seismogram, numbered relative to zero time at the shot break. The corresponding time–distance (T–X) graph is a plot of these first-arrival times against distance X from S. The graph consists of two straight-line segments, each of which is related to one of the two velocity layers V_1 (soil) and V_2 (bedrock). The interpretation of the T–X graph is discussed in the text.

plunges through layer V_1 at the critical angle, then the wave travels rapidly along the top of refractor V_2, emerging through layer V_1 at the critical angle) gradually overtake the direct ray. The delay in going down to and up from V_2 is compensated as X increases by faster travel through the refractor. Critically refracted compressional waves eventually become the first to arrive at the detectors beyond a particular value of X. The gradient of the T–X values of these arrivals bears a simple relation to the velocity of a horizontal refractor. The depth d_1 from surface to interface may be obtained from the T–X graph, using the equation

$$D_1 = V_1 V_2 T_0 / [2(V_2^2 - V_1^2)^{\frac{1}{2}}]$$

T_0 is the intercept on the time axis ($X = 0$) of the projection of the velocity segment on the graph which corresponds to V_2 arrivals. The velocities V_1 and V_2 are obtained from the gradients of the corresponding velocity segments. The derivation of the equation is given in Appendix E.

This theory can be extended to derive the depth of each of n horizontal interfaces separating velocity layers where the velocity increases with depth and every layer is recorded on the T–X graph as a straight-line segment.

Arrival times are measured usually with a *precision* of about 0.001 s. The extrapolation from these values to obtain T_0 usually results in a larger standard error in that value. The minimal error in depth d_1 is related to this, and to the contrast in velocity between V_1 and V_2. For $V_1 = 600$ m s^{-1} and $V_2 = 3000$ m s^{-1}, which gives a strong contrast comparable to that between dry soils and Carboniferous sediments, an error of 0.001 s in T_0 produces an error of approximately 0.3 m in the value of d_1. The refinements of timing in certain seismic instruments for civil engineering purposes, with claimed precision of 0.0002 s, offer, at least theoretically, corresonding accuracy in determining depths. In practice, *accuracy* is controlled mainly by whether there are significant differences between the simple elastic model and the geometry of its refracted rays and the real geological structure to be investigated. That is to say, whether the assumption on which the above equation is based holds in specific, real circumstances.

If the interface dips at an angle α (Fig. 6.10), the gradient of the second velocity segment in the T–X graph is not related simply to the **true velocity** of the refractor, V_2, but is an **apparent velocity** which is a function of both V_2 and α. In the down-dip direction, the apparent velocity U_d is lower than V_2 because of the steadily increasing delay as the emerging ray traverses a greater and greater distance through V_1. In the up-dip direction, the apparent velocity U_u is greater than V_2. Since the dipping interface model is more complex, with an additional parameter α needed to define the distribution of velocities, more information must

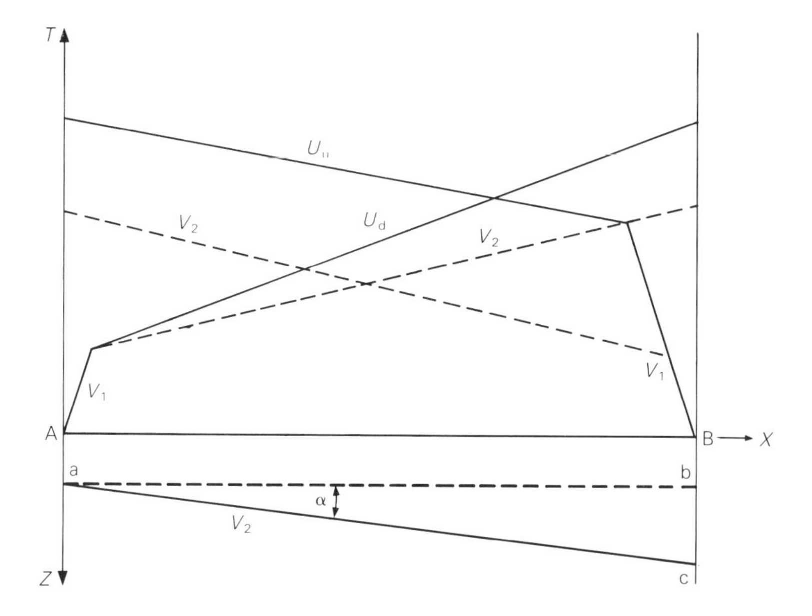

Figure 6.10 A simple velocity model, where two layers V_1 and V_2 are separated by a horizontal interface ab, is superimposed for comparison on a model where the interface ac dips at an angle α. The corresponding T–X graphs of first arrivals, shooting from shot points A and B into traverse AB, are shown. The direct rays are unaffected by the interface, and segments (with a slope corresponding to V_1) are common to both sets of graphs. The refracted arrivals from A shooting down-dip plot as a straight-line segment U_d of velocity lower than V_2. The up-dip apparent velocity is higher than V_2. Note that the travel time from A to B equals that from B to A.

be obtained. This is done by '**shooting both ways**' into the same surface spread of detectors, and getting two different T–X graphs for the same subsurface coverage. Shot holes are made at both ends of the spread. This double coverage should be standard practice in seismic-refraction surveys, as apparent velocities are an unresolved mixture of true velocity and dip and they are potentially misleading, especially about the character of the refractor.

A basic assumption in the refraction method is that the critically refracted ray follows the top of the refractor. This will not be the case where the spread straddles an acute anticline or a sharp rise in rock head. The results are accordingly in error. If the rock head surface is irregular, a commonplace state of affairs in glaciated areas, with glacial gouging and *roches moutonnées* producing rapid rises and falls of a few metres amplitude, the seismic results will give a blurred, generalised picture of bedrock along the spread, and there may be a poor match between this and depth values from individual boreholes. For certain purposes, such as estimating quantities of soils, the averaged values from seismic measurements may, however, be of more use than those derived from a few, unluckily sited boreholes.

Another basic assumption of that method is that each and all of the velocity layers present in the model are recorded as a segment on the *T–X* graph, and hence that the distribution of velocity with depth above the principal refractor (usually bedrock in site exploration) can be determined from the graph. This is not always the case. The conversion of measured times to depths to the refractor is dependent on this velocity distribution. If, however, there is a decrease of velocity with depth across any interface (that is, a **low-velocity layer** is present in the subsurface), then critical refraction does not take place, and there are no refracted arrivals through that layer. The velocity distribution inferred from the *T–X* graph is in error. If the layer is of uniform thickness, then the refractor is plotted systematically too deep. Its shape is almost correct, but effectively the vertical scale of the cross section is wrong. Some independent check on velocity distribution, or of scale, is essential. This is most simply, and most commonly, done by integrating seismic-refraction traverses and borings into one exploration programme. A boring provides not only greater accuracy at a key point, but also a measure of the 'scaling error' of the seismic profile around it. Geological conditions likely to give rise to a low-velocity layer above rock head include a raised beach where estuarine clays rest on peat, or a sub-arctic region of thick soils where the ground water near the surface is frozen, or where boulder clay rests on sand and gravel.

The seismic-refraction method is commonly used in site exploration to determine depth to rock head in a reconnaissance survey, or in an area where boring is difficult or forbidden.

Some engineering properties of the bedrock may be inferred from its velocity. This is used empirically as a crude measure ('rippability', Section 7.2.2) of suitability of the rock for extraction by a caterpillar tractor 'ripper'. If density and V_p and V_s are known, the bulk elastic properties *in situ* can be derived, for example below a dam site. Field determinations of velocity differ in value from those made in the laboratory using small coherent specimens of the same rock. The ratio of the *in situ* value to the laboratory value is called the **fracture index** (F), and is a measure of fracturing of the rock mass. F is a criterion used to recognise areas of rock of poor quality for foundations so that the intake of grout into broken rock can be assessed, and it can be correlated with core recovery. If a **borehole geophone** (seismometer) is used to make a **velocity log** of a drilled hole, the time delay across a fracture may be observed and the fracture pinpointed. This technique has been used to locate major joints which affect the stability of a rock face, and which need to be rock bolted.

6.3.4 *Seismic-reflection method*

The physical property that controls the amplitude (and hence clarity on a seismogram) of a reflection is **acoustic impedance**, ρV, the product of density and seismic velocity of the rock. The ratio of the amplitude of the reflected wave to that of the incident wave is called the **reflectivity**, R, at an interface. It is related to the contrast of acoustic impedance and to the angle of incidence. For incidence normal to the interface, we have

$$R = \frac{\rho_2 V_2 - \rho_1 V_1}{\rho_2 V_2 + \rho_1 V_1}$$

where $\rho_1 V_1$ is the acoustic impedance of the upper layer and $\rho_2 V_2$ that of the lower layer. If the lower layer 2 has a greater acoustic impedance, there is no phase change on reflection. If, however, it has a lesser acoustic impedance, the reflected wave suffers a phase inversion. That is, if the reflection from an interface where $\rho_2 V_2$ is greater than $\rho_1 V_1$ appears as a peak on a seismogram, the reflection from an interface where $\rho_1 V_1$ is greater would have the same amplitude, but record as a trough. As a result reflections from the top and bottom of a limestone interbedded in shales, which is thin compared with the wavelength of the seismic waves being used, interfere destructively with one another. The layer could only be located by higher-frequency waves of shorter wavelength. In effect, small objects, of fine detail, cannot be detected by waves longer than their dimensions. The resolution in locating larger objects is *at best* about one-quarter of a wavelength and in some cases, especially sparker surveys (see below), is several times worse because of the slow decay of the source pulse oscillation. These arguments for using high-frequency seismic waves must be balanced, however, against their limited penetration. They are reflected and scattered as they meet minor structures which longer waves ignore, and are rapidly attenuated in comparison with lower-frequency waves.

The geometry of the reflection method is simple. There are only two elements of path from shot point to geophone – an incident ray to, and a reflected ray from, the interface. If the average velocity between surface and a horizontal interface is $\bar{V}$, and the depth is Z, then at a seismometer at distance X from the shot point we have

$$(\bar{V}\,T)^2 = X^2 + 4Z^2$$

For normal incidence, where $X = 0$, this simplifies to $Z = \bar{V}\,T/2$.

Reflected waves are the basis of the best systems of detection, for example when light waves are used in human sight, or when seismic waves are used in applied geophysics. The seismic-reflection methods yield the least ambiguous answers, and the highest resolution in locating interfaces, but only where structural complexity does not produce sudden and unpredictable lateral changes in the distribution of velocity ($\bar{V}$). They have been developed and refined by affluent patrons, the petroleum companies, to serve as their principal tool of exploration. Much of this work is done by large contractors who offer this one special service. Their techniques, as evolved to investigate depths of a few kilometres, and the large structures in which petroleum occurs, use frequencies in the range 10–100 Hz and are unsuitable for nearly all engineering purposes. The requirements of exploration at most sites are: (a) penetration of less than 100 m, (b) high resolution, (c) simple logistics, and (d) modest costs.

These have been met to a large extent for *surveys of water-covered areas* by **continuous acoustic profiling** but these techniques cannot be used on dry land, and no substitute comparable in ease and resolution has been developed. The recording and other electronic equipment can be accommodated on a large launch, and the seismic source from which the seismic pulses emanate, plus the hydrophone, are towed behind the vessel. Seismic waves are usually generated by the rapid discharge of a large condenser across a spark gap in the water, at intervals which range between about one every 4 seconds, to twelve per second, depending on the type of seismic source. Sources used are chosen according to the frequency range appropriate to the resolution and penetration required. The **sparker** has a broad frequency range from about 200 Hz to 2 kHz, giving a resolution in soft sediments of roughly 2–20 m. Higher resolution can be obtained by higher-frequency sources called **pingers**. Greater penetration can be obtained by an **air gun** which discharges compressed air as explosive bubbles. The reflected waves are detected by the hydrophone, amplified, processed and automatically plotted on a continuous paper strip. This record appears as a section of the sea floor along the traverse, with reflection times, not reflection depths, plotted vertically. A depth section may be made from this using the velocities of sea water, and of superficial deposits. In simple cases it is easy to interpret, and can give answers without ambiguity. A sparker record is shown in Figure 6.11.

6.3.5 *Seismic surveys: case history*

As part of the Lake Winnipeg Regulation Project (Prior & Mann 1972), an 8-mile channel and a 2-mile channel were needed to supplement natural outflow. It was intended to remove 30 million cubic metres of soil by dredging, and it was essential in selecting a route that rock (Precambrian granite; $V_p = 5400 \text{ m s}^{-1}$) excavation be avoided. The exploration of the 2-mile channel was attempted initially by drilling alone, until it was clear that the grid of holes needed to cover the large areas of the possible route would cost an estimated \$3 million and take two years. A reconnaissance by seismic-refraction surveys was the only way of carrying out the job in

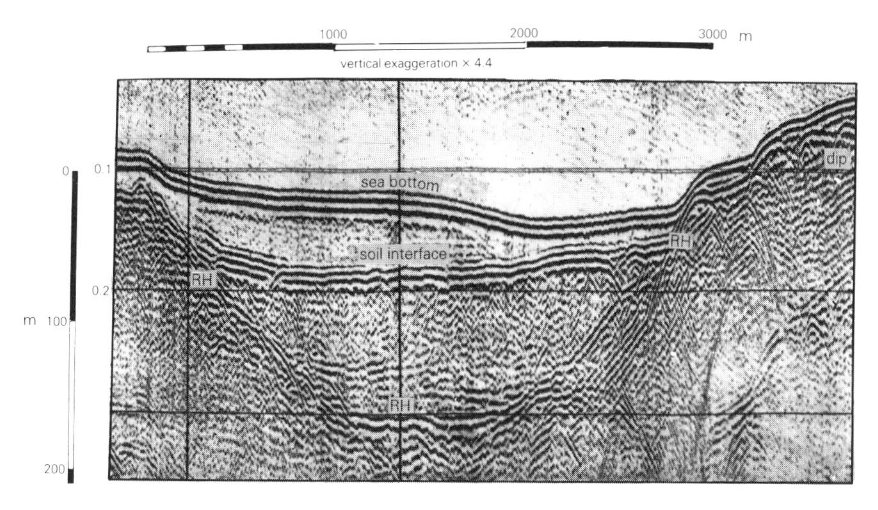

Figure 6.11 This continuous seismic profile, made with a sparker wave source, crosses a deep buried channel of glacial origin cut into bedrock in the Firth of Clyde. In addition to the normal time lines printed as a scale on the seismogram, a depth scale based on assumed velocities, plus a horizontal scale, are added to make comprehension easier. The sea bottom and the rock head (RH) are indicated. Since the reflected energy consists of trains of waves rather than a single pulse, each interface is underlain by two black lines on the record. Stratification with low dips to the right can be distinguished near rock head at the right-hand margin. The apparent layering elsewhere in the bedrock is spurious, and results from reverberations in the water. The sedimentary infill of the channel is of two types, separated by a regular, distinct bedding plane. The nature of the infill may be *surmised* from a knowledge of regional geology, but can be *proved* only by drilling and sampling.

the time available. Since rock head is critically close to the channel invert in several areas, an accuracy of ±1.5 m (5 ft) in defining it was desired. This would have been impossible by the seismic method alone, as the velocity distribution above bedrock is complex and is not defined fully by the *T–X* graphs of first arrivals. Part of the area was covered by shallow lakes, where ice (V_p = 3300 m s^{-1}) covered lake water (V_p = 1450 m s^{-1}). On land, frozen soils (V_p = 1830–2440 m s^{-1}) overlay a variable thickness of unfrozen peat and muskeg (as low as V_p = 340 m s^{-1}); between these deposits and bedrock were glacial clays (V_p = 1570 m s^{-1}) with patches of till (V_p = 2130–2750 m s^{-1}). That is, there was both a velocity reversal with depth (a low-velocity layer), and a lateral change in the velocity distribution. Severe rockhead relief was another source of error, as computed seismic depth is the minimal radial depth from a seismometer location, whereas drill depth is obtained from rock head directly under the seismometer.

For these reasons, the seismic results had to be checked and calibrated by a relatively large number of drill holes. These were usually sited at culminations in rock head, and profiles were 'draped' between the control points. Over the ice, overall accuracy of the seismic results, compared with drill depths, was better than 10%, that is, within 1 m. At a third of the check points, the error was less than 0.3 m, but at three places where bedrock relief was severe, seismic depths were out by more than 3 m. Over land, predicted depths were within 2 m of the drill depths in the nine cases where a check was possible.

To prove 1.5 km of channel through an area of shallow bedrock required 20 km of seismic traverse and 70 drill holes, each about 10 m long. The combination of seismic surveys and drilling was the cheapest and most efficient way of exploration, including the examination of nearby alternative routes. The total cost was $1.2 million.

6.3.6 *Electrical methods*

Several methods of prospecting for minerals and of investigating geological conditions have been developed using the electrical properties of rocks. The **resistivity method** is the one that has been most used in engineering. It depends on the property of resistivity, which is the electrical resistance between opposite faces of a unit cube of a given rock. The SI unit is the ohm metre (ohm m). The values of resistivity of rocks and minerals range from 10^{-6} to over 10^{12} ohm m, with no simple relationship of the spread of values to a genetic classification of rocks. Some minerals, like clays, conduct electricity well, but the predominant factor controlling the resistivity of most rocks is the presence or absence of ground water in them, and the relative concentration of electrolyte (dissolved salts). A sand saturated with brackish water will have a very low resistivity. The same sand, if it lay above the water table and had air, not water, in its pores, would have a high resistivity. Crystalline rocks normally have high resistivities, except locally where they are broken by faults or joints and ground water fills the voids.

Resistivity is measured along traverses in the field by inserting two metal stakes (the current electrodes C_1 and C_2 in Fig. 6.12) and applying a current. The circuit is completed by flow through the subsurface. There is a matching drop of electrical potential at the surface, which is measured over a distance a by inserting two similar stakes (the potential electrodes, P_1 and P_2) into the ground. In the commonly used **Wenner arrangement**, one has $C_1P_1 = P_1P_2 = P_2C_2 = a$. If the traverse is underlain by an isotropic rock of uniform resistivity, the flow lines of the electricity are circular arcs passing through C_1 and C_2, and the corresponding equipotential surfaces are orthogonal to them. For the Wenner arrangement of electrodes, the resistivity ρ of the rock is given by

$$\rho = 2\pi(V/I)a$$

where V is the voltage between the potential electrodes, I is the current

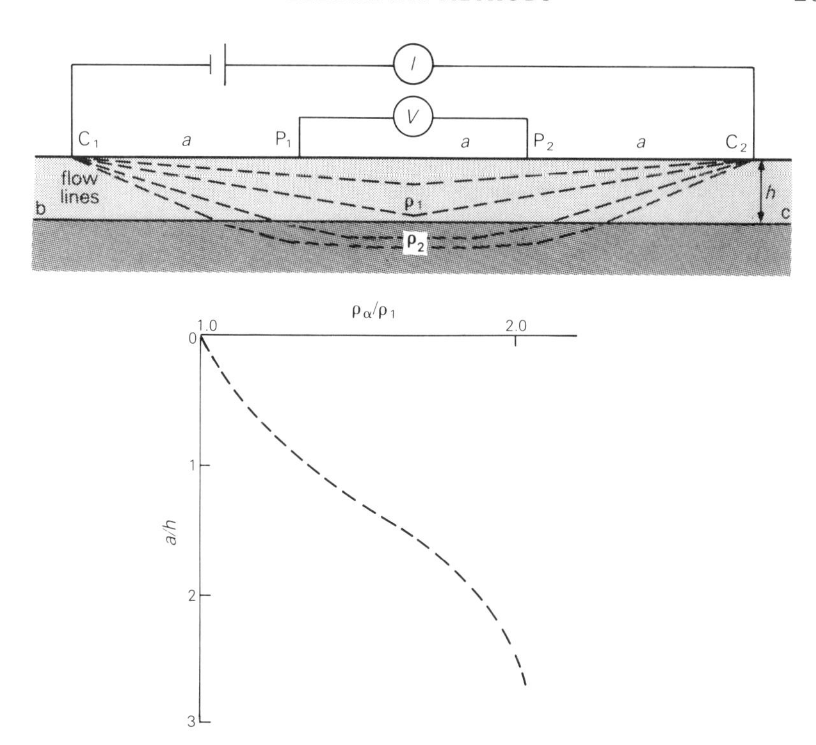

Figure 6.12 The upper part of the diagram shows the layout of electrodes on the ground surface. In the subsurface two layers of different resistivity ρ_1 and ρ_2 ($\rho_1 > \rho_2$) are separated by a horizontal interface bc, and the flux is denser in layer ρ_2 because of its lower resistivity. The graph shows how apparent resistivity ρ_α changes with electrode spacing a in a given two-layer case with the interface at depth h, and with a particular contrast between ρ_1 and ρ_2. Observed results may be interpreted by comparing them with a set of master curves, each of this type and corresponding to a particular contrast of resistivities. Master curves are usually presented on logarithmic scales, and observations are conventionally plotted on log graph paper. If direct current is applied, as shown in this diagram, non-polarisable electrodes must be used. This awkward procedure is obviated by the use of alternating current in most resistivity apparatus.

flowing between the current electrodes, and a is the separation of the electrodes.

Where a layer ρ_1 of lower resistivity overlies one of high resistivity ρ_2, the flow lines are distorted from circular arcs, with flow preferentially concentrated in the good conductor ρ_1. If resistivity is measured and calculated from the above equation, it would have a value intermediate between ρ_1 and ρ_2. It is referred to as a value of **apparent resistivity** ρ_α. The percentage of the total current flowing through the lower rock layer ρ_2 depends on its depth relative to a, and on the contrast in resistivity with ρ_1. As a is increased, the flow of current through layer ρ_2 increases and ρ_α tends towards the value ρ_2. That is, for small values of a, ρ_α approximates to ρ_1, and for values large compared with the depth to ρ_2, to ρ_2.

In the commonly used procedure of a **depth probe**, the centre of the electrode arrangement is kept fixed, but the spacing a is increased by progressive steps to give deeper and deeper penetration by the current. A typical two-layer curve, where ρ_α/ρ_1 is plotted against a/h, is shown in Figure 6.12. The inflection point is related to the depth to the interface separating the two layers. A rule of thumb (valid for a range of resistivities and thicknesses only) is that the depth to the interface is equal to two-thirds of the value of a at which the corresponding inflection point occurs. Accurate values may be obtained by comparing the observed curve with families of standard theoretical curves for a range of models. More accurate resistivity models may be obtained using a computer programme. The accuracy of the depth value in simple geological conditions, for example a cover of boulder clay on dolerite, is about 0.5 m (20%), but as the number of layers of different resistivity increases, each additional layer steps up the error. Indeed, with more than three layers present, the results obtained have little value in most real problems of site exploration.

In **horizontal investigations**, the electrode spacing is kept constant, and the spread is moved laterally in successive steps. This technique is used to trace an interface of marked resistivity contrast, and the electrode spacing is chosen after a depth probe has shown what value of a corresponds to the inflection point. It is also used to locate steep zones of low or high resistivity, such as water-bearing fractures or dykes.

The resistivity method has been, and to a limited extent still is, used in surveys to determine depths to rock head over a wide area. Since a depth probe, with maximum values of a about 15 m, can be completed by a team of three in about 10 min, extended sites can be covered quickly and relatively cheaply. The geological interpretation of the resistivity model is, however, more prone to error than the interpretation of seismic results, and the results must be treated with greater caution. More borings must be made as a control. In one survey where the thicknesses of a boulder clay overburden obtained by resistivity and by boring were compared, a difference of 5 m in depth to rock head between resistivity value and bore was found at one point, though the average difference was only 0.5 m.

The method is best suited to problems where ground water is involved, such as surveying the water table or locating water-bearing fissures. It is a useful guide in siting wells for water supply, or in detecting where leakage is taking place. The more ingenious and unusual applications include following the flow of grout underground into an ill-defined cavity after lowering its resistivity with some added salt.

Two other electrical methods have some engineering applications, though their main use is to prospect for sulphide ore bodies. The **self-potential method** (often abbreviated to SP) measures natural changes of potential at the ground surface by connecting a millivoltmeter to two electrodes. These must be non-polarisable and are normally copper rods immersed in a copper sulphate solution held in porous pots. In permeable rocks, filtration of ground water through pores in the rock produces an increase in the electrical potential in the direction of water flow, and the potential gradient is proportional to the intensity of seepage. The method is used to detect leakage paths in earth dams. Periodic measurement at such localities shows whether leakage is increasing with time, and whether fines are being washed out of the fill.

The **induced polarisation method** (IP) requires that a current be passed through the ground for a short period, then interrupted. The difference of potential between the electrodes does not disappear immediately, but decays at a rate determined by the type of rock and its degree of saturation by ground water. The applications to engineering are broadly similar to those of the resistivity method, but in some cases afford greater accuracy.

6.3.7 Magnetic method

The Earth's magnetic field (that is, the **geomagnetic field**) changes slowly with time, and from place to place. The variation with geographical position includes a gradual change of **magnetic intensity** from about 0.3 oersted (Oe) at the Equator to 0.7Oe at the Poles. The SI units of magnetic flux density are the weber per square metre (Wb m^{-2}) and the tesla (T). The unit commonly used in applied magnetic surveys is the **gamma** (1 gamma = 10^{-5} Oe = 10^{-9} Wb m^{-2} = 10^{-9} T). Acute local differences of intensity from the normal background value of a region are called **magnetic anomalies**, and are caused by the presence of magnetised rocks in the ground. Rocks may be *magnetised permanently* at the time of their formation. As magma solidifies to form igneous rocks, it only becomes magnetised in the direction of the prevailing (ambient) field as it cools below a certain temperature, the **Curie Point** of the individual minerals. This temperature is between 600 and 700°C, when the rock is still glowing a dull red. Sedimentary rocks, such as red sandstones, may acquire weaker permanent magnetisation during deposition, as the more magnetic grains rotate on settling with a preferred alignment in the then geomagnetic field, or as haematite is deposited chemically to form a cement. Part or all of the magnetisation of a rock may be *induced* temporarily by the present field; that is, it would disappear if the rock specimen could be screened magnetically from it. The degree of magnetisation in a given intensity of field is related to the **magnetic susceptibility** of each mineral in the rock. In contrast to density, which is related to *all* the minerals in a rock, only a *few* accessory minerals, particularly magnetite but to a lesser extent magnetic pyrite, ilmenite and haematite, contribute significantly to the bulk magnetic susceptibility of a given rock. Magnetic properties may be variable in an otherwise uniform rock. How magnetic a rock is depends to a large extent on the percentage of magnetite it contains and to a lesser extent on the size of the magnetite crystals. Accordingly, basic igneous rocks have the highest magnetic susceptibilities, acid igneous and most metamorphic rocks are much further down the scale, and most sedimentary rocks are effectively non-magnetic.

There are two important exceptions to these generalisations. Baked clay or shale and black topsoil have relatively high magnetic susceptibilities, and in certain situations give small but detectable magnetic anomalies which are of use in site exploration. Baked clay is most familiar as manmade bricks or as burnt colliery waste ('red ash'), the presence of which may be associated with a derelict, concealed mineshaft (see Appendix D). The position of ancient excavations, such as a Roman trench, may be detectable from magnetic anomalies, if the infill is partly topsoil or organic waste.

The more general application of the magnetic method is to locate boundaries where igneous rocks lie on one side. It is, for example, the quickest, cheapest and most certain way of locating a dyke close to the surface. Since a magnetised body has positive and negative magnetic poles, both positive and negative anomalies are produced by the same mass of rock. The shape of the anomaly curve is controlled not only by the shape of the magnetised body but also by the direction in which it is magnetised by the geomagnetic field.

A variety of instruments have been designed to measure different parameters of the geomagnetic field. The most commonly used in site exploration is the **proton magnetometer**, which gives the value of intensity at a station to within 1 gamma. A reading can be made in a few seconds and, once the stations are pegged out, a survey can be done rapidly. There is usually no need to level the stations or make systematic corrections to the readings if the survey is over a small area only and is completed in a day. If a close grid of stations is needed, surveying may be simplified and speeded by the use of a nylon cord net, of say $10 \times 10\ m^2$, knotted to locate stations at 1-m intervals. The outer corners of the net are fitted to survey pegs.

The contoured results (an isogram map) of a survey carried out by Dr George Maxwell over an abandoned mineshaft are shown in Figure 6.13. The position of the shaft's rectangular outline is approximate as there is no confirmatory surface feature. It was sunk after publication of the First Edition of the Ordnance Survey 1:2500 plans, and abandoned and landscaped before publication of the Second Edition. The nature of the capping of the shaft, which produces the positive magnetic anomaly, is unknown, but it seems unlikely that iron railings are present. Other

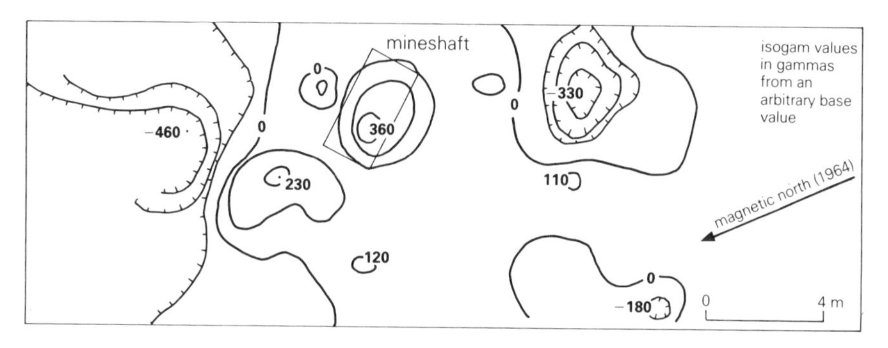

Figure 6.13 The position of a concealed mineshaft is shown by the rectangle on the map. The magnetic anomalies related to it and to the colliery yard are derived by contouring the readings taken by a proton magnetometer at 100 gamma interval between isogams. Further information is given in the text. (The results are reproduced by courtesy of Dr George Maxwell.)

anomalies in the geomagnetic field are typical of a colliery yard. The locating of such abandoned shafts is discussed in Appendix D.

6.3.8 *Gravity method*

The gravitational attraction between two masses, m_1 and m_2, which are a distance r apart, is defined by Newton's Law:

$$F = G(m_1 m_2 / r^2)$$

where G is the universal gravitational constant. It has a value of only 6.67 $\times 10^{-11}$ m^3 kg^{-1} s^{-2} (6.67×10^{-8} cgs units), which means that gravity is a weak force compared, for example, with magnetic attraction.

If the Earth is one of the two masses, then its gravitational attraction is more generally described in terms of the acceleration g it evokes in the other mass. Since $F = m_1 g$, then we have

$$g = G(m_2 / r^2)$$

Gravitational acceleration g has traditionally been expressed in cgs units of cm s^{-2}, called for this purpose a gal (after Galileo). Its value ranges from just under 978 gal at high altitude at the Equator, to over 983 gal at sea level at the Poles. A smaller unit, the **milligal** (equal to 0.001 gal) has conventionally been used in exploration surveys.

Gravity meters, for example the commonly used Worden Meter, can be read to a precision of 0.01 mgal, though slight irregularities in instrumental drift make the accuracy less. Errors in correcting for elevation, irregular terrain around the meter station, and tidal effects normally reduce the accuracy of a corrected reading to about 0.03 mgal under favourable circumstances (a small flat site), but the error may be closer to 0.1 mgal where topography is irregular. Although this is a small fraction of the total Earth's gravity field, it is an insensitive measure of changing subsurface conditions, and for this reason the gravity method has had limited application to engineering problems.

If the Earth were composed of concentric shells of rock, each of different density, then the corrected values of g would change systematically, and predictably, from the equator to the poles. A **theoretical value of gravity** corresponding to each value of latitude can be computed for the simple Earth model. Comparison of observed values of g corrected for elevation, etc. with the theoretical value for the same station normally shows a difference called the **Bouguer anomaly** at the station. Bouguer anomalies exist because the density layering in the Earth is not simply concentric. Folds, faults and igneous intrusions bring denser rocks close to the surface in some areas, resulting in anomalous masses with respect to a horizontally layered Earth (see Section 6.3.1 and Fig. 6.6).

The magnitude of a Bouguer anomaly is related to the size of structure and to the density contrast across the interface being investigated. Since structures of interest in site exploration are relatively small, the gravity method is usable only if the density contrast is relatively large, and even then only as a reconnaissance method because of poor resolution. This condition is met in two types of problem. The first is searching for *large* underground caverns in thick limestone, where the contrast may be as high as 2.7 g cm^{-3} between limestone and air. The second is tracing buried channels where an infill of unconsolidated drift deposits may differ in density by about 0.5 g cm^{-3} from the solid rocks, for example Carboniferous sediments. The cavities produced by mining do not give rise to gravity anomalies large enough to be recognised with certainty in a survey with a standard gravity meter. A new generation of more sensitive instruments with a sensitivity of 0.001 mgal may be of use for this purpose (Appendix D).

6.4 Drilling, boring, trenching and pitting

6.4.1 *Drilling and boring*

Quantitative information about a site at various points is usually required and can be obtained in several ways. Making a hole in soils is referred to conveniently as 'boring' and in solid rock as 'drilling'. Drilling is usually carried out using one of two techniques.

ROTARY CORE DRILLING

This is often called diamond drilling, and consists of a crown (either diamond-studded or made of tungsten carbide) rotated at the end of a

Table 6.2 Factors influencing core drilling.

Term	Rock types	Compressive strength ($MN\,m^{-2}$)	Percentage recovery (minimum figure)
very strong rocks	quartzite, granites, most igneous rocks, gneisses, hornfelses	>100	85
moderately strong rocks	sandstones,* limestone,* schistose metamorphic rocks, weathered igneous types	12.5–100	60
weak rocks	shales, mudstones, poorly compacted sedimentary rocks, highly weathered igneous and metamorphic types	<12.5	50

*Sandstones and limestones have a wide range of properties. The well cemented, well compacted types will be the strongest sedimentary rocks, with silica-cemented sandstones strongest of all.

hollow barrel which holds the cylinder of rock produced by the drill. High core recovery is important using this method, and this is influenced by core size and type of rock being drilled (Table 6.2). Recovery in the vicinity of faults is extremely poor, usually less than 30%, because of the brecciation of the rock.

Normally drilling is vertical, but in some instances, particularly if ore bodies are present, if the rocks are steeply dipping or if a sloping tunnel is required, the drill holes may have to be angled. This should not present any problems to the professional driller. Drill holes may also be angled so that a fault plane can be clearly intersected and an examination made of it (by measuring the electrical resistivity in the hole, Section 6.3.6) so that rock quality in the vicinity of the fault can be determined, and also to check whether any water flow occurs along the plane.

Drill bit sizes are chosen to suit the type of rock and depth of hole being drilled. The common bit sizes are listed in Table 6.3. For igneous or other very hard rocks, EX, AX or BX core drill sizes suffice, but for softer rocks much larger core drills should be used (HX and greater). In drilling holes in excavations preparatory to blasting, 105 mm ($4\frac{1}{8}$ in) holes are commonly used, although bigger-diameter holes are tried occasionally.

In most engineering site investigation work, 75 mm diameter cored holes or larger are recommended. An NM core barrel may be used in soft rocks to give a high core recovery. It consists of a double-tube core barrel, the inner tube being fixed while the outer one, in contact with the bit,

Table 6.3 Sizes of coring bits and solid bits for percussive drilling.

	Coring bits (hollow)		Solid bits for percussive drilling	
Name	mm	(in)	mm	(in)
XRT	17.5	($\frac{11}{16}$)	80	($3\frac{1}{8}$)
EX	22	($\frac{7}{8}$)	85	($3\frac{11}{32}$)
AX	28	($1\frac{1}{8}$)	89	($3\frac{1}{2}$)
BX	41	($1\frac{5}{8}$)	102	(4)
NX	54	($2\frac{1}{8}$)	105	($4\frac{1}{8}$)
HX	76	(3)	110	($4\frac{5}{16}$)
PX	92	($3\frac{5}{8}$)	115	($4\frac{1}{2}$)
SX	113	($4\frac{7}{16}$)	140	($5\frac{1}{2}$)
UX	140	($5\frac{1}{2}$)	152	(6)
ZX	165	($6\frac{1}{2}$)	165	($6\frac{1}{2}$)
			191	($7\frac{1}{2}$)
			216	($8\frac{1}{2}$)

rotates and water circulates between them. The hole size is about 75 mm diameter in total, and the rock core obtained is about 50 mm in diameter.

Recommended methods of logging cores for engineering purposes are given in a report written by a Working Party of the Engineering Group and published by the Geological Society of London (1970, 1977). The Working Party recommended that each borehole log should contain the following information:

(a) borehole identification, grid reference, contract details (client, contractor, site and reference number);
(b) drilling method, inclination and equipment used;
(c) drilling progress, rock resistance to drilling, *in situ* tests;
(d) groundwater levels and changes in standing water level;
(e) detailed geological description including rock types, discontinuities encountered (see Appendix H), core recovery, rock quality designation (RQD), degree of weathering and any simple field tests carried out.

An example of a borehole log from a civil engineering site is given in Figure 6.14.

Rock Quality Designation is obtained by examining a length of rock core and summing only those lengths that are greater than 100 mm long and are competent. The fractures splitting the core must be natural and *in situ*, for example foliations, natural joints, bedding planes, cleavages and so on. Fractures caused by drilling or handling should be discounted. The total sum of lengths greater than 100 mm is divided by the total length of rock core examined and expressed as a percentage. Table 6.4 gives descriptive terms for RQD.

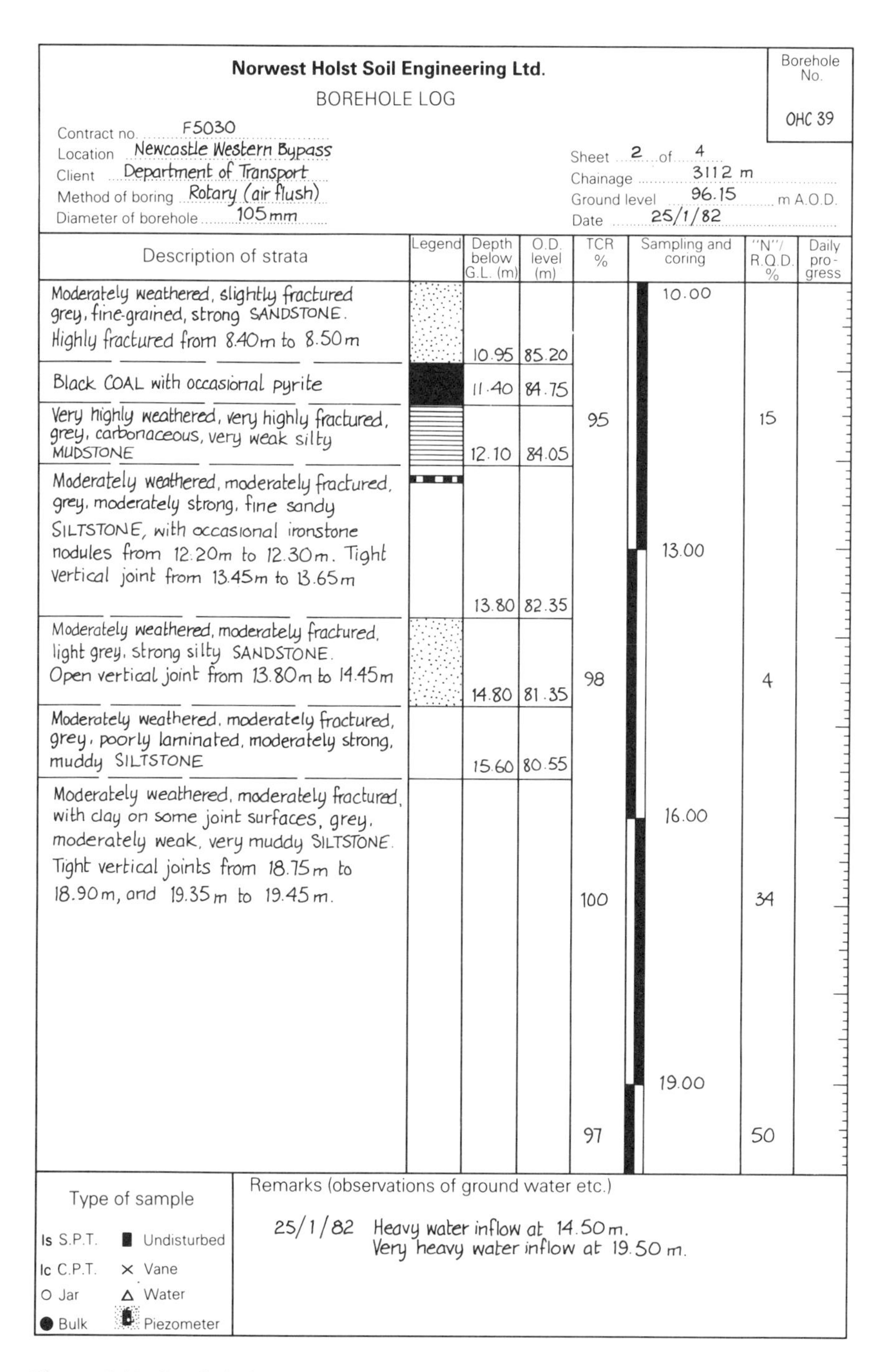

Norwest Holst Soil Engineering Ltd.

BOREHOLE LOG

Borehole No. OHC 39

Contract no. F5030
Location Newcastle Western Bypass
Client Department of Transport
Method of boring Rotary (air flush)
Diameter of borehole 105 mm

Sheet 2 of 4
Chainage 3112 m
Ground level 96.15 m A.O.D.
Date 25/1/82

Description of strata	Legend	Depth below G.L. (m)	O.D. level (m)	TCR %	Sampling and coring	"N"/ R.Q.D. %	Daily progress
Moderately weathered, slightly fractured grey, fine-grained, strong SANDSTONE. Highly fractured from 8.40 m to 8.50 m		10.95	85.20		10.00		
Black COAL with occasional pyrite		11.40	84.75				
Very highly weathered, very highly fractured, grey, carbonaceous, very weak silty MUDSTONE		12.10	84.05	95		15	
Moderately weathered, moderately fractured, grey, moderately strong, fine sandy SILTSTONE, with occasional ironstone nodules from 12.20 m to 12.30 m. Tight vertical joint from 13.45 m to 13.65 m		13.80	82.35		13.00		
Moderately weathered, moderately fractured, light grey, strong silty SANDSTONE. Open vertical joint from 13.80 m to 14.45 m		14.80	81.35	98		4	
Moderately weathered, moderately fractured, grey, poorly laminated, moderately strong, muddy SILTSTONE		15.60	80.55				
Moderately weathered, moderately fractured, with clay on some joint surfaces, grey, moderately weak, very muddy SILTSTONE. Tight vertical joints from 18.75 m to 18.90 m, and 19.35 m to 19.45 m.				100	16.00	34	
				97	19.00	50	

Type of sample

Is S.P.T. ▮ Undisturbed
Ic C.P.T. × Vane
○ Jar △ Water
● Bulk Piezometer

Remarks (observations of ground water etc.)

25/1/82 Heavy water inflow at 14.50 m.
Very heavy water inflow at 19.50 m.

Figure 6.14 Borehole log. (Reproduced by kind permission of Bullen and Partners, Glasgow.)

Table 6.4 Rock quality designation.

RQD (% recovery of lengths > 100 mm)	Description
0–25	very poor
25–50	poor
50–75	fair
75–90	good
90–100	excellent

In some cases, a huge hole may have to be bored (1–2 m in diameter, using an auger or calyx drill). This produces a cylindrical pit into which an engineer or geologist can be lowered to inspect the rock *in situ*. Great care should be taken to ensure that the sides of this hole are adequately supported. Features that can be examined include frequency and size of boulders, amount of weathering, position of water table, and whether a hard pan (ironstone) is present in peat-covered areas. In many building sites in central city areas underlain by soil or weak rock, when boring holes for piling, percussive or vibratory techniques are not used because they may damage existing buildings. Instead a cylindrical hole is bored out by means of a large auger, and the concrete pile is constructed *in situ*. This technique is common in London and its surrounding areas because the underlying rock is London Clay, but is not common in certain other parts of the country where new buildings may be built on strong rock or boulder clay, since the auger cannot bore through the former and is rarely used in the latter because of the occurrence of large boulders.

PERCUSSIVE DRILLING

Drill holes may be sunk using chisel drill bits and percussive hammers inside the hole to break the rock. Holes drilled in this way vary in diameter between approximately 110 mm and 80 mm (or less in some cases); 105 mm is the most common. In the traditional technique, only small rock chips and dust were recovered for inspection so that information on the rock mass obtained by this method was very limited. Recently a Swedish company (Hagconsult) has introduced an improvement. Their technique involves lowering a miniature television camera into the hole and examining the rock quality and other properties of the rock mass, including fissures, joints, weathering and the location of any fault planes. (Note that this technique does not work if the hole is filled with water: the drill hole must be dry.)

Such small-diameter boreholes may be used for hydraulic pressure testing to provide information on permeability of the rock mass and indicate zones where grouting is necessary. Miniature-television inspection will confirm these data, as well as providing information on joints and faults including their dips and strikes.

This technique has been widely used in recent years for inspecting a rock mass *in situ*, and ascertaining whether it would be suitable as an underground storage site for oil or gas. For example, it was employed in examining the rock at Cromarty in northern Scotland where the construction of a large oil refinery was being considered.

Borehole deformation devices can give details on plate and jack loading tests and *in situ* stress and shear movements.

Drill cores provide material for laboratory testing of seismic velocity, permeability, saturation moisture content and compressive strength of the rock type(s) drilled as well as producing material for petrographic examination. A chemical analysis of the main ore minerals in a core sample may be carried out to detect any anomalous fraction, such as high sulphur content and high water content – both hydroxyl (OH) and absorbed water. From this the presence of minutely sized minerals such as clays and sulphides can be predicted.

6.4.2 *Trenching and pitting*

In superficial deposits, digging a trench or sinking a pit or shaft to bedrock is usually the best means of obtaining information. Trenches may be dug by hand, but a mechanical back-acting shovel is the neatest and quickest method to use. Once again it must be emphasised that all pits or trenches must be adequately side-supported before any inspection is made.

Trenches and pits have two purposes:

(a) They allow a visual inspection of a deep and wide section of the superficial deposits. Complete sections can be drawn and layers of strata delineated. Provided that enough trenches or pits have been sunk, an accurate three-dimensional picture of the superficial geology of any site can be obtained. In this, the different types of overburden and their thicknesses will be shown, and also data such as frequency and size of boulders, areas of good quality sand and gravel (see (b)) and location of buried channels.

(b) They allow large-scale sampling and investigation of the material of the superficial deposits to be made. The general map of the superficial deposits will reveal the main varieties of strata and their approximate volumes. Any sands and gravels found will have to be tested so that their mineralogy (percentage of quartz grains present) and particular size distribution (using sieves of varying sizes) can be determined. If, for example, the grading of the sand conforms to a 'zone 2' type, then it is highly probable that any concrete required in site construction work can be made using these sand deposits.

Any impurities found in sands or gravels must always be clearly defined. Such impurities include salts, pyrite or other sulphides, and organic matter – particularly coal. Coal is often found in British sands, particularly from North-east England and the Midland Valley of Scotland, because of the presence of Carboniferous rocks containing coal seams underlying many of the superficial deposits in these regions.

6.4.3 *Groundwater conditions*

All of the trenches and pits (and large-diameter drill holes) should be left open and not backfilled immediately so that groundwater conditions can be observed over as long a period as possible. The level at which water is encountered in each hole or pit must be recorded, and also which horizons yield water and which strata are under artesian pressure. Artesian conditions in strata below a road route or a heavy building must be assessed carefully in case there is a danger of hydraulic failure of the soil or rock mass. From such studies the direction of flow of water can be estimated (Section 5.2.3) and drainage measures decided, especially in cases where the water table lies close to the surface.

Where the strata consist of dipping beds of alternating low and high permeability, a simple water table does not exist. Each aquifer has its own **piezometric surface** (see Section 5.1.2) and each of these surfaces is a measure of the lateral change of pore fluid pressure in a particular layer. The study of pore pressure is an important part of a complete site investigation programme. To determine pore pressure or water level and to monitor changes in them, **piezometer tubes** or tips are installed in the ground so that pits and trenches need not be left open but can be backfilled after inspection.

References and selected reading

Attewell P. B. and I. W. Farmer 1976. *Principles of engineering geology* (Chapter 7). London: Chapman and Hall.

Bogolovsky, V. A. and A. A. Ogilvy 1970. Application of geophysical methods for studying the technical status of earth dams. *Geophys. Prospect.* **18**, 758–73.

British Standards Institution 1981. Code of Practice for Site Investigations. BS 5930: 1981 (formerly CP 2001).

Buist, D. S., A. D. Burnett, and M. K. Sanders 1978. Engineering geological case study of the site investigation and design of a major trunk road cutting in Carboniferous rocks. *Q. J. Engng Geol.* **11**, 161–76.

Clark, A. R. and D. K. Johnson 1975. Geotechnical mapping as an integral part of site investigation — two case histories. *Q. J. Engng Geol.* **8**, 211–24.

Cooke, R. U., A. S. Goudie, and J. C. Doornkamp 1978. Middle East – review and bibliography of geomorphological contributions. *Q. J. Engng Geol.* **11**, 9–18.

Cratchley, C. R., P. Grainger, D. M. McCann and D. I. Smith 1972. Some applications of geophysical techniques in engineering geology with special reference to the Foyers hydro-electric scheme. *Proc. 24th Int. Geol. Congr.* **13**, 163–75.
Dearman, W. R. and P. F. Fookes 1974. Engineering geological mapping for civil engineering practice in the United Kingdom. *Q. J. Engng Geol.* **7**, 223–56.
Dumbleton, M. J. and G. West 1972a. *Guidance on planning, directing and reporting site investigations.* Transp. Road Res. Lab. Rep. LR 625.
Dumbleton, M. J. and G. West 1972b. Preliminary sources of site information for roads in Britain. *Q. J. Engng Geol.* **5**, 15–18.
Eastaff, D. J., C. J. Beggs and M. D. McElhinney 1978. Middle-East – geotechnical data collection. *Q. J. Engng Geol.* **11**, 51–63.
Gardner, M. 1969. A new pencil-and-paper game based on inductive reasoning. *Scient. Am.* **221** (5), 140–4.
Geological Society Engineering Group Working Party 1970/1977. (1st and 2nd Rep.) The logging of rock cores for engineering purposes. *Q. J. Engng Geol.* **3**, 1–24; *Q. J. Engng Geol.* **10**, 45–52.
Geological Society Engineering Group Working Party 1972. The preparation of maps and plans in terms of engineering geology. *Q. J. Engng Geol.* **5**, 295–382.
Grainger, P. and D. M. McCann 1977. Interborehole acoustic measurements in site investigation. *Q. J. Engng Geol.* **10**, 241–55.
Griffiths, D. H. and R. F. King 1965. *Applied geophysics for engineers and geologists.* Oxford: Pergamon.
Knill, J. L. and D. G. Price 1972. Seismic evaluation of rock masses. *Proc. 24th Int. Geol. Congr.* **13**, 176–82.
Lee, A. J. 1966. The effect of faulting on mining subsidence. *Trans Inst. Mining Engrs* **125**, 735–43.
McDowell, P. W. 1970. The advantages and limitations of geoelectrical methods in the foundation investigation of the tracked hovercraft experimental site in Cambridgeshire. *Q. J. Engng Geol.* **3**, 119–26.
McQuillin, R. and D. A. Ardus 1977. *Exploring the geology of shelf seas.* London: Graham & Trotman.
Miller, V. 1961. *Photogeology*. New York: McGraw-Hill.
Moore, J. F. A. 1974. Mapping major joints in the Lower Oxford Clay using terrestrial photogrammetry. *Q. J. Engng Geol.* **7**, 57–68.
National Coal Board 1963. *Subsidence engineers' handbook.* London: National Coal Board.
Newbery, J. and A. Siva Subramaniam 1977. Geotechnical aspects of route location studies for M4 north of Cardiff. *Q. J. Engng Geol.* **10**, 423–41.
Norman, J. W., T. H. Leibowitz, and P. G. Fookes 1975. Factors affecting the detection of slope instability with air photographs in an area near Sevenoaks, Kent. *Q. J. Engng Geol.* **8**, 159–76.
Powell, M. (ed.) 1970. *Preventing foundation failures in new dwellings.* Regd. House Bldrs Fdn Man., Nat. Housebldg Council.
Prior, J. W. and W. K. Mann 1972. Location of by-pass channels in Manitoba using a combined seismic-drilling programme. *Proc. 24th Int. Geol. Congr.* **13**, 217–24.
Robbie, J. A. 1972. The Institute of Geological Sciences and the Scottish environment. *Proc. 24th Int. Geol. Congr.* **13**, 55–63.
Wakeling, T. R. M. 1972. The planning of site investigations for highways. *Q. J. Engng Geol.* **5**, 7–14.
Watson, I. 1971. A preliminary report on new photogeological studies to detect unstable natural slopes. *Q. J. Engng Geol.* **4**, 133–7.

7 *Rocks and civil engineering*

7.1 Classification of rocks for engineering purposes

7.1.1 Classification of rocks by simple criteria

The genetic classification of rocks (Section 2.2.1) commonly used by geologists groups rocks in a way that can be of practical use as well as offering theoretical insight. The relationships of rock types in the classification often explains those between different rock types within one large rock mass and provides a model to be tested by exploration. It may be difficult or impossible, however, to assign a rock to an appropriate category in the classification using the simple criteria available in the field, and a more simple alternative classification of rocks is sometimes employed. This uses certain textures, fabrics or structures, and hardness (as a simple indication of composition).

The texture of the rock can be described as follows:

(a) *Crystalline.* All fresh igneous rocks, however formed, including porphyritic types with large crystals set in a finer-grained matrix, and glassy types with crystals too small to be seen by the naked eye. Metamorphic quartzites and marbles fit into this category, as do coarse-grained metamorphic rocks (gneisses, granulites, migmatites), which may possess weak banding as well. Inner-zone thermally metamorphosed hornfelses are also included.

(b) *Foliated.* Metamorphic schists, and also slates and phyllites with well developed cleavages.

(c) *Clastic.* Identifying features are given for each type. All clastic rocks are sedimentary and may show bedding planes and other features:

sandstone, arkose, greywacke	sand-sized material <2 mm to >0.06 mm; arkose and greywackes contain 75% quartz grains
conglomerate, breccia	more than 30% grains >2 mm in size; conglomerate contains rounded grains and breccia angular grains
siltstone	fine grained, <0.06 to >0.002 mm

	shales, mudstones	rich in clay minerals; very fine grained, <0.002 mm
	tillite, boulder clay	rock fragments in clay matrix; wide range of grain sizes
(d)	*Clastic and chemical precipitates*	
	limestone, dolomite, chalk, oolitic limestones	>50% carbonate; very fine grained; crystalline appearance
	marl, calcitic mudstone (or calcilutite)	variable clay content plus carbonate; calcitic mudstone >50% carbonate; marls have between 35 and 65% clay (see Fig. 2.35)
	evaporites	'pure' chemical precipitates – anhydrite, halite, gypsum deposits; some limestones and dolomites ('cement stones')
(e)	*Bioclastic*	
	shelly limestone, reef limestones (coral, algae, etc.)	limestones with fossil remains
	diatomite	>50% diatoms (remains of siliceous plants); very fine grained, porous
	peat, lignite, coal	partly or fully carbonised plant remains

Note that some pyroclastic rocks formed from igneous ejectamenta during eruptions, and tuff and agglomerate resemble clastic sedimentary rocks more than igneous; thus tuff is a volcanic 'sandstone' and agglomerate a volcanic 'conglomerate' or 'breccia'.

The textures of all rocks may be drastically changed by weathering, and may vary within a rock mass. Many of the rocks described in the five categories listed here also display internal structures, namely bedding, foliation and cleavage, and the terms used to describe the spacing of these planes are defined in Table 7.1.

7.1.2 Engineering group classification of rocks

Quarrying and construction companies are interested primarily, and sometimes exclusively, in certain index properties of rocks (see Sec-

Table 7.1 Descriptive terms applied to the spacing of rock structures.

Description	Spacing
very thick	>2 m
thick	600 mm–2 m
medium	200–600 mm
thin	60–200 mm
very thin	20–60 mm
narrow (igneous and metamorphic) thickly laminated (sedimentary)	6–20 mm
very narrow (igneous and metamorphic) thinly laminated (sedimentary)	<6 mm

tion 7.2.2). Since there are often no significant differences among several rock types using the criteria of these index properties, no distinction is made in naming them for this usage. Several rock types are grouped together according to their most important engineering properties and the group is given the name of a commonly known rock within it. In many groups the rock types included have obvious affinities with each other, but three, the basalt, gabbro and granite groups, include rock types that seem out of place. In particular:

(a) in the *basalt group* epidiorite and hornblende-schists have been included even though these are both metamorphic rocks possessing a strongly developed foliation. Their grain size may also be considerably larger than those of basalt and other fine-grained igneous rocks of the group. The densities of all rocks in this group are similar, but their engineering properties may differ widely.

(b) in the *gabbro group* both basic and ultrabasic plutonic igneous rock types have been included, which means that variation in density will be large (from ~2.8 to ~3.3), and basic gneisses are also included even though such rocks are banded. The rocks in this group have a similar grain size.

(c) in the *granite group*, acid gneisses possessing banding are included along with acid and some intermediate plutonic igneous rocks. Granulite, a very high grade metamorphic rock, is included, which is reasonable since it has a poorly developed fabric and is usually coarse grained. Pegmatite, however, although an acid igneous rock, has a huge crystal size (often called a pegmatitic texture, Section 2.2.2) and its inclusion is difficult to justify since it really does not resemble any other rock type in this group.

However, apart from the inconsistencies just described, the engineering group classification is a reasonable one. The groups are listed in BS 3618 and BS 2847, and Table 7.2 is based on these codes.

Table 7.2 The engineering group classification of rocks (based on BS 3618 and BS 2847).

Basalt group	Flint group	Gabbro group
andesite	chert	basic diorite
basalt	flint	basic gneiss
basic porphyrite		gabbro
diabase (similar to dolerite)		hornblende-rich rocks
dolerites of all kinds		norite (gabbro-type)
epidiorite		peridotite
hornblende-schist		picrite
quartz dolerite		serpentinite
spilite (basalt-type)		
Note: All dolerites and some basalts are called whinstone.		

Granite group	Gritstone group	Hornfels group
gneiss	agglomerate	contact-altered rocks of all kinds except marble
granite	arkose	
granodiorite	breccia	
granulite	conglomerate	
pegmatite	greywacke	
quartz diorite	grit	
syenite	sandstone	
	tuff	

Limestone group	Porphyry group	Quartzite group
dolomite	aplite	ganister
limestone	dacite	quartzitic sandstones
marble	felsite	recrystallised quartzite
	granophyre	
	microgranite	
	porphyry	
	quartz porphyry	
	rhyolite	
	trachyte	

Schist group		Artificial group
phyllite		clinker (BS 1165)
schist		air-cooled slag (BS 1047)
slate		
all severely sheared rocks		

Traditional names are also used in some branches of industry to describe certain rock types. Thus **whinstone** is used in the quarrying industry for any fine-grained basic igneous rock, and **freestone** is a term used in the blockstone industry to describe any sedimentary rock with thick beds which can be worked in any direction. The blockstone trade also employs trade names, e.g. 'Baltic Brown' to describe a certain natural rock used for cladding panels. These names have no geological or engineering origin, being based often upon geographical location and colour, etc. 'Baltic Brown' is the trade name for an orbicular granite.

7.2 Engineering properties of rocks

7.2.1 Isotropism

The arrangement of mineral constituents and textural elements of a rock in three dimensions is described as its *fabric*. In metamorphic rocks parallel orientation of minerals occurs (e.g. in schists and well banded gneisses) and such rocks are said to be anisotropic. Rocks that have no preferred arrangement of their components (e.g. many plutonic and hypobassal igneous rocks, and metamorphic hornfelses) are said to be isotropic.

7.2.2 Rock index properties

Certain properties of rocks are of particular importance to the engineer as they may affect the planning and cost of a project. Prior knowledge of these **index properties** from tests will guide decisions such as whether a body of rock will be blasted by explosives or be removed more simply for ripping (see p. 228), and whether excavated rock can serve as suitable constructional material for a specific purpose or lacks the essential characteristics. These index properties are now defined.

SPECIFIC GRAVITY

(1) **Dry apparent specific gravity determination** (G_b). Two methods need to be employed, depending upon the porosity of the rock sample:

(a) Method for rocks with higher porosities (>10%); this laboratory test is used for many sedimentary rocks and highly weathered igneous and metamorphic rocks. First, the sample is oven dried at 105°C for 12 h (or until constant weight), and weighed (W_1). Then the rock sample is immediately coated with paraffin wax or some other material whose density (γ_w) is known. The wax layer is cooled and the sample weighed (W_2).

$$\text{Weight of wax } (W_w) = W_2 - W_1$$

The volume of water (V) displaced by the sample is measured.

$$\text{Volume of rock } (V_1) = V - \frac{W_w}{\gamma_w}$$

Then dry apparent specific gravity of rock $G_b = W_1/V_1$.

(b) Method for rocks with lower porosities (<10%); well compacted or cemented sedimentary rocks, all fresh igneous and metamorphic rocks. A Walker Balance is used. The sample is suspended in air and weighed (W_1). It is immersed in water and weighed again (W_2). Then $G_b = W_1/(W_1 - W_2)$. If a White Specific Gravity Balance is used, G_b can be read directly from the balance (accurate to 0.01).

(2) **Saturated apparent specific gravity determination** ($G_{b(sat)}$). If the saturation moisture content (i_s or S) and its dry apparent specific gravity (G_b) are known, then:

$$G_b = \frac{G_{b(sat)}}{1 + i_s} \quad \text{or} \quad G_{b(sat)} = G_b\,(1 + i_s)$$

(i_s is defined on p. 222.)

(3) **Solid mineral grain specific gravity determination** (G_s). If i_s and G_b are known, then G_s can be calculated since

$$G_s = \frac{G_b}{(1 - i_s)G_b}$$

G_s can be obtained directly using a density bottle:

(a) Dry the density bottle and stopper at 105°C and weigh (W_1).
(b) The rock should be crushed to a mesh size about equal to the grain size of the rock. A sample should be dried (105°C for 4 h) and put into the density bottle (about one-third full) and weighed (W_2).
(c) Distilled water is added to the density bottle until the sample of rock powder is covered, and the density bottle is then placed in a desiccator which is evacuated slowly of air.
(d) Release vacuum and vibrate bottle gently. Repeat (c) and (d) until no more air issues from sample.
(e) Fill density bottle with distilled water, put in stopper, and keep in a constant temperature bath for 1 h, adding water to the bottle if its volume decreases.

(f) Wipe dry the stoppered density bottle and weigh (W_3). Empty, clean and refill density bottle with distilled water, and keep at a constant temperature for 1 h. Then wipe clean and reweigh (W_4).

$$G_s = \frac{W_2 - W_1}{(W_4 - W_1) - (W_3 - W_2)}$$

Two determinations should be made for each sample and the results then averaged.

MOISTURE CONTENT

To determine the moisture content (w or m), the rock sample is weighed immediately on being removed from its airtight container, e.g. a plastic bag (W_1). The rock sample is dried to constant weight in an oven at 105°C for 12 h. Cool in a desiccator and reweigh (W_2). Then:

$$w = \frac{W_1 - W_2}{W_2}$$

SATURATION MOISTURE CONTENT

If the sample has already had its moisture content determined, it has therefore been oven dried and weighed (W_2). Place sample in wire basket and immerse in water for at least 12 h. Remove and weigh after surface drying (W_3). Then, saturation moisture content (i_s or S) is expressed by the equation:

$$i_s = \frac{W_3 - W_2}{W_2}$$

If the sample is very weak, immerse in water first (*before* oven drying) and obtain weight of saturated sample. Then oven dry and reweigh, before obtaining i_s.

POROSITY

Porosity, n, is dimensionless, and is the ratio of the volume of voids in a soil or rock, V_v, to the total volume, V_T, that is $n = V_v/V_T$ (cf. **void ratio** which is the ratio of the volume of voids, V_v, to the volume of solids, V_s; or V_v/V_s). It is expressed as a percentage or as a decimal fraction. In rocks, 10% is average, 5% is low, and 15% or less is high.

Relationships between the index properties G_b, G_s, i_s and n can be expressed in the form of a nomogram (Fig. 7.1) designed by Duncan (1969).

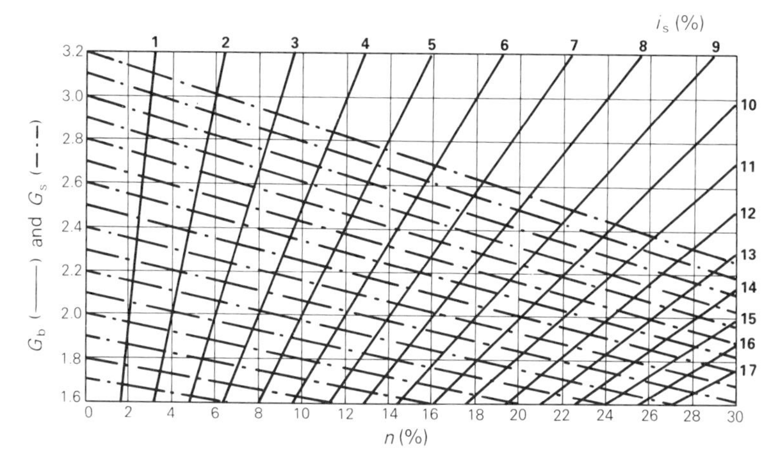

Figure 7.1 Relationships between the dry apparent specific gravity (G_b), the solid mineral grain specific gravity (G_s), the saturation moisture constant (i_s) and porosity (n). From Duncan (1969).

Not all voids in a rock are interconnected and accessible to penetrating fluids. Pumice is an extreme example of a very porous rock of this type. Many igneous and high-grade metamorphic rocks contain tiny cracks or microfractures, which are not interconnected.

The effective porosity is the ratio of the volume of interconnected voids to the total volume of the rock. Voids in rocks are most commonly of two types: primary voids (pores) between the fragments of clastic rocks, and secondary voids, produced by later fracturing or chemical weathering. The first is characteristic of the whole rock mass and is its 'porosity' by strict definition. The second depends on the rock's subsequent history and is highly variable within the body of rock. Representative values of (true) porosity are given in Table 7.3.

Table 7.3 Porosity values of some common rock types. These serve only as a guide and a porosity value must be determined for each specific rock.

Rock or soil type	Maximum porosity (%)	
soil	>50	
sand and gravel	20–47	soils
clay	>49	soils
cemented sand	5–25	soils
sandstone	10–15	soils
limestone (and marble)	5	
oolitic limestone	10	
chalk	up to 50	
igneous rocks	<1.5	
metamorphic rocks	generally very low	

The factors that control the porosity of sedimentary rocks and soils are as follows:

(a) *The degree of cementation* (that is, to what extent pore space is replaced by cement) and the extent of recrystallisation at points where grains touch. Both are influenced by the age and history of burial of the rock (see Section 2.2.4).
(b) *The grain size variation,* since small grains can fill the voids among larger grains. A sediment with a large variation in grain size (a well graded sediment, see p. 46) has a lower porosity than a poorly graded sediment.
(c) *The packing of the grains,* which, if the grains are spherical, can give a range of porosities from 26 to 47%. The looser packing is a less stable arrangement of grains (Fig. 7.2a), and a change from this to a more stable arrangement (Fig. 7.2b) will reduce porosity and may lead to the expulsion of water from a sediment.
(d) *The shape of the grains,* since angular laths such as occur in clay minerals often form bridges between other grains, hold them apart and increase porosity. Since porosity is dimensionless, the size of grains and voids does not affect the ratio of their volumes, and in a sediment formed of perfect uniform spheres the porosity would be independent of the size of the spheres. In practice, the different characteristics of clay minerals compared with quartz grains lead to an increase in porosity with *reduction* of grain size. This is because all clays adsorb water on to their outer surfaces. By this means, every clay **micelle** (tiny crystal) is surrounded by a thin film of water, which situation effectively increases the porosity of clay rocks.

In crystalline limestones, the void space is mainly secondary and is controlled by the presence of fossils and bedding planes, by leaching of carbonate and redeposition by acidic ground water, and by fracturing, on both a large and small scale. Because of progressive leaching, the void space in limestone usually increases with time, and caverns may develop.

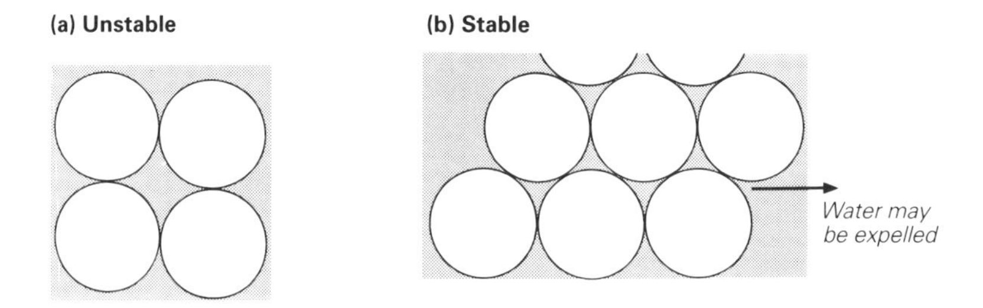

Figure 7.2 Packing arrangements of grain in sediments: (a) unstable, (b) stable.

WATER-YIELDING CAPACITY

Not all of the water in a rock can be removed from it by flow under the force of gravity. Some is held as a film on the surface of the grains. This is described by the following (self-explanatory) terms and equation:

$$\text{porosity} = \text{specific yield} + \text{specific retention}$$

For example, Triassic sandstone and Cretaceous chalk in southern England may have similar values of porosity but the specific retention of the chalk is high compared with that of the sandstone, because the chalk possesses small voids and has less void space available per unit volume than the sandstone. The sandstone will have a higher permeability than the chalk (see below).

PERMEABILITY

Permeability, k (dimensions L^2), is a measure of the ease of flow through a rock or soil, independent of the properties of the fluid (cf. hydraulic conductivity, K). It is related to hydraulic conductivity by the equation $K = (k\rho g)/\mu$ where ρ is the density, μ the dynamic viscosity of the fluid, and g is the acceleration of gravity.

The unit of permeability is the *darcy*, which is approximately 10^{-8} cm^2 (1 darcy gives a specific discharge of 1 cm s^{-1} of a fluid of viscosity 1 cP under a hydraulic gradient that makes the term $\rho g \mathrm{d}h/\mathrm{d}l$ equal to 1 atm cm^{-1}). The permeabilities of many common rocks are one or two orders of magnitude less than a darcy and are usually expressed in *millidarcies.*

The principal factor controlling permeability is the size of the voids, since the smaller they are the greater is the surface area of contact of water with solid mineral and the greater are the capillary forces restraining flow. In loose soils permeability increases with the (diameter)^2 of the grains. Flow also takes place through secondary voids such as joints, and rocks of this character are usually referred to as pervious rather than permeable.

A rough estimate of k may be obtained from grain size, and a better approximation by using a permeameter for a laboratory test. A reliable value of bulk permeability of a rock body is best obtained by well tests in the field (Section 5.1.2).

HYDRAULIC CONDUCTIVITY

The hydraulic conductivity, K, which is called the coefficient of permeability in some older publications, is a measure of the ease of flow of *water* through a soil or rock under a given hydraulic gradient. It is defined by Darcy's Law, by which it is determined experimentally (Section 5.1.1). Hydraulic conductivity (dimensions $L\ S^{-1}$) is linked to the rock index property permeability (k), and also to the viscosity and density of the fluid (water). It is *not* a rock index property. It has the dimensions of a velocity and is often expressed as metres per day (m day^{-1}). Typical values for soils are: clays 0–1 m day^{-1}; sand 10–260 m day^{-1}; gravel up to 300 m day^{-1}.

SWELLING COEFFICIENT (E_s)

This is a measure of the change of length of a sample which was initially oven dried then soaked in water till it became fully saturated, expressed as a ratio of the original (dry) and final (saturated) lengths. This swelling behaviour is related to the amount of expansive clay minerals contained in a rock. The behaviour of rocks in water can be expressed in terms of their ability to resist slaking when exposed to water, using the **slake durability test** (Franklin *et al.* 1971).

SEISMIC VELOCITIES (V_p and V_s)

These are discussed fully elsewhere (see Section 6.3.2).

REBOUND NUMBER

Rebound number (R) is measured using a **Schmidt Concrete Test Hammer**, type N, and is used to assess *in situ* strength of rocks. The rebound height of the hammer is expressed as a percentage of the forward travel distance of the hammer mass.

UNIAXIAL (OR UNCONFINED) COMPRESSIVE STRENGTH

Before measuring the strength of a specimen of rock it is important to ensure that the rock specimen has been correctly prepared. The procedures are outlined below.

(a) *Specimen size and shape*. Cylindrical test specimens are cored either from a rock mass in the field using EX (22 mm dia.), AX (28 mm) or BX (41 mm) hollow core bits, or from a hand specimen in a laboratory using 25, 38 or 63 mm dia. hollow core bits. The **aspect ratio** of the core (length : diameter) is important and must be >2. The top and bottom surfaces of the core should be smooth, parallel to each other and at right angles to the core length.

(b) *Method of testing*. The core is placed with the flat ends between two platens. Conventionally a rigid platen is used, which will create uneven stress within the specimen during crushing, but flexible or hydrostatic platens can be used. The implications of the use of *stiff* or *soft* testing systems are discussed by Attewell and Farmer (1976). The rate of compression is important and is defined in ASTM specification C170-50 (see Appendix G) as 0.7 N mm^{-2} s^{-1} (stress controlled) and 1 mm min^{-1} (deformation controlled). Uniaxial or unconfined compressive strength (q_u or S_c) is measured in

Table 7.4 Unconfined compressive strengths of the main rock types.

Descriptive term	Compressive strength ($MN\,m^{-2}$)	Indicative rock types
very weak	<1.25	some weakly compacted sedimentary rocks, some very highly weathered igneous or metamorphic rocks boulder clays
weak	1.25–5	
moderately weak	5–12.5	
moderately strong	12.5–50	some sedimentary rocks, some foliated metamorphic rocks, highly weathered igneous and metamorphic rocks
strong	50–100	some low-grade metamorphic rocks, marbles, some strongly cemented sedimentary rocks, some weathered and metamorphic igneous rocks
very strong	100–200	mainly plutonic, hypabyssal and extrusive igneous rocks (medium to coarse grained), sedimentary quartzites, strong slates, gneisses
extremely strong	>200	fine-grained igneous rocks; metamorphic quartzites, some hornfelses

meganewtons per square metre ($MN\,m^{-2}$) or newtons per square millimetre ($N\,mm^{-2}$). Definitions of terms used to describe q_u are given in Table 7.4. (Note that meganewtons per square metre can be converted to pounds force per square inch by multiplying by ~145.) Several specimens should be tested from each rock unit since variations in rock strength (within a single rock unit) may occur owing to:

(i) the properties of the constituent minerals, especially their hardnesses, the presence of cleavages, and the degree of their alteration;
(ii) the presence and shape of any voids within the rocks, and whether these voids are filled with water;
(iii) the nature of the bonding between mineral grains.

A field test called a **point load test** (described by Franklin *et al.* 1971) can be employed to get a good indication both of q_u and of the tensile strength from the point load strength index I_s, which is measured in the field. When testing rock cores or examining rock properties, the weathering grade of each sample should also be recorded, since it will affect the physical properties and change them from the values possessed by the fresh rock. In particular, the saturation moisture content i_s and porosity n increase with weathering, whereas the apparent specific gravity G_b decreases.

The susceptibility of rock to future weathering, either as a rock mass or crushed to form aggregate, must also be considered in engineering projects. Crystalline rocks are less susceptible than cemented and compacted sedimentary rocks, especially if constituents like clay or pyrite are present in the latter. Weathering conditions can be simulated in the laboratory, and the reaction of suspect rocks to factors such as long-term loading and changes of temperature and moisture content can be assessed. Any deterioration in the values of index properties is a sure sign of deterioration in other mechanical properties of the rock. If frost attack is likely, the **frost heave test** should be carried out (p. 232). To assess weathering resistance and durability of the rock, the **sodium sulphate soundness test** is useful. This test is described fully in ASTM C88-76.

The relationships of q_u, R and V_p are shown in Figures 7.3 and 7.4, together with an indication of an empirical engineering property of rocks, '**rippability**'. This is used in deciding whether rocks from a near-horizontal surface can be excavated mechanically by a ripper attached to a tractor (see Fig. 7.5), rather than by explosives. Weak, easily ripped rocks tend to have low values of V_p. (These can be measured, even under a cover of soil, by the seismic-refraction method, Section 6.3.3). An empirical upper limit of $V_p = 2.0$ km s^{-1} arbitrarily defines rocks that can be ripped without difficulty. Rocks with this value of V_p correspond

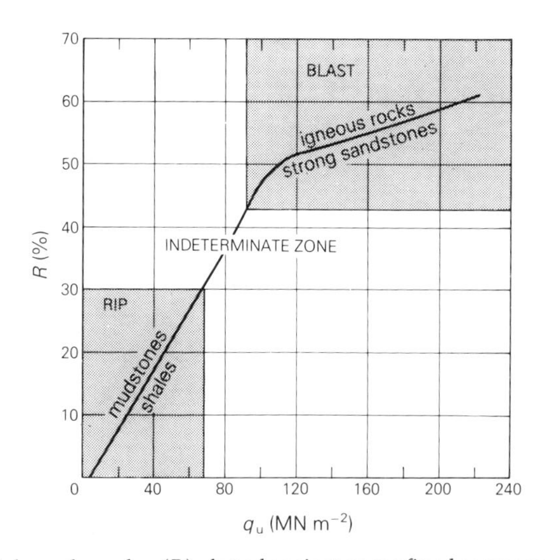

Figure 7.3 Rebound number (R) plotted against unconfined compressive strength (q_u) for various rock types.

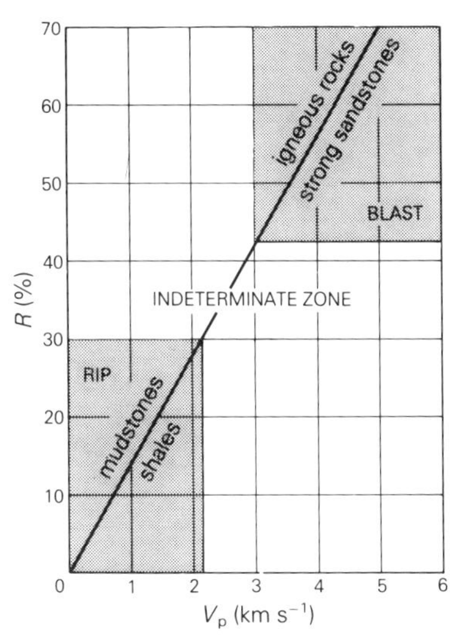

Figure 7.4 Laboratory-determined seismic velocity (V_p) plotted against *in situ* rebound number (R) for various rock types.

Figure 7.5 Excavation of rocks using a mechanical ripper. (Photograph by courtesy of the Caterpillar Tractor Company, Illinois.)

roughly to those with a q_u value of 70 MN m^{-2}. Low velocities may be an indication of poor compaction of rocks, or of extensive jointing (see fracture index, described in Section 6.3.3). Ripping is likely to be successful if the block volume (see Appendix H, Table H.2) is less than 1.0 m^3; that is, the rock mass possesses joint sets such that the rock is fractured into cubes with edges 1.0 m long. Low velocities are found in some sedimentary rocks, in a few metamorphic rocks and in rock bodies of any type which are highly weathered or badly fractured. The ripper performance of various rock types is shown in Figure 7.6, which is based on one in the *Handbook of ripping* (Caterpillar Tractor Co. 1972).

7.2.3 *Rocks as aggregates*

Rocks can be crushed and graded to make aggregate, which can be added to a bonding material such as cement to form concrete, or bitumen to serve as a roadstone. Different civil engineering jobs require different grades of aggregate. The particle size distribution within an aggregate will be specified for a particular situation. This is usually presented as a **particle size distribution curve** in which the cumulative weight percentage passing a particular mesh size can be plotted on a graph. In hard rock quarrying a company is able to provide crushed rock aggregate to a grading curve required by the client. It is important to note that both sand and gravel deposits, and mixtures of the two, will possess a particle size

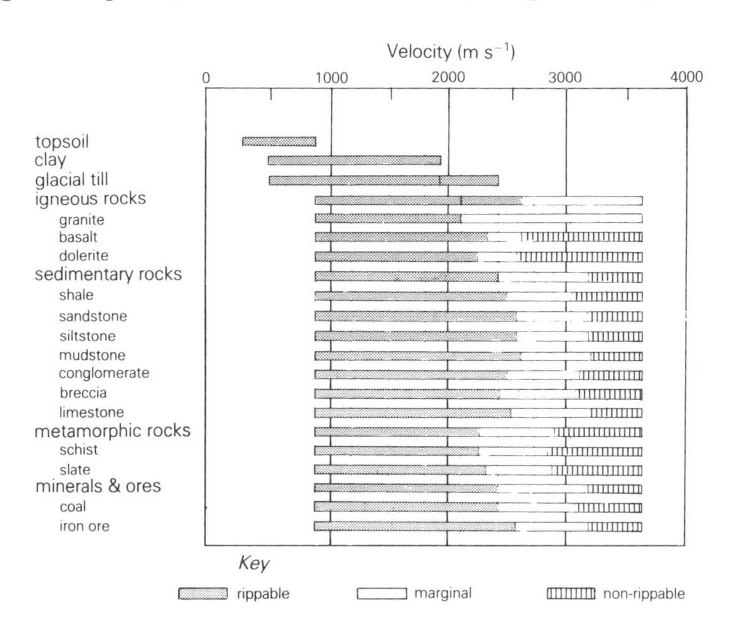

Figure 7.6 Ripper performance in different rock types based on seismic velocities (V_p).

distribution unique to the particular deposit. Thus, the specific grade of material required for an engineering project can always be supplied from a crushed rock quarry, but a sand and gravel opencast pit may not be able to meet the grading specifications required. The suitability of aggregates as components of roadstone or concrete depends upon particular properties, as follows.

ROADSTONE

The standard tests carried out on aggregates to determine their suitability as roadstone (roadmetal) are described in Appendix F. This deals with the main British tests, and Appendix G gives the main aggregate tests employed in Europe and the USA and gives their UK equivalents. The properties of aggregates investigated include the following:

(a) *The composition, texture and degree of alteration* of the rock. The best roadstone is a fresh, fine- to medium-grained igneous rock, with intergrowth of the minerals producing strong bonding, and without phenocrysts or rock glass. Most sedimentary rocks are too easily crushed to use as roadstone, but hard Palaeozoic greywackes (gritstones) that occur in parts of England and Wales are frequently employed for this purpose. Many crystalline metamorphic rocks can also be used, although, apart from hornfels and schistose grit, they are too variable to make good roadstone aggregate.

(b) *The resistance to surface wear*. This is assessed by measuring **aggregate abrasion value** and **polished stone value**. Aggregate abrasion value (AAV) measures resistance to surface wear by abrasion. The lower the value, the greater the resistance. Regulations in the UK demand a value less than 8 for general road use. Polished stone value (PSV) measures the extent to which roadstones are polished by traffic. The greater the value, the greater the resistance. High values (greater than 60) are required in motorways, roundabouts and high-density traffic roads. The relationship between AAV, PSV and AIV (see below) is shown in Figure 7.7, for sedimentary rocks.

There may be a discrepancy between the test results of AAV and PSV and actual behaviour in roads since steel rollers are used instead of tyres. Other factors that may influence resistance to surface wear include the adherence of the aggregate to bitumen. Some rocks (basalt, limestone, greywacke) bond well, whereas others (quartzite, schist) bond poorly. The presence of shear planes in the aggregate will allow water to penetrate and speed its disintegration.

(c) *The resistance to impact and crushing*. This may be assessed by tests of **aggregate impact value** (AIV) and **aggregate crushing value** (ACV). The common rock types range in AIV from good (basic volcanics,

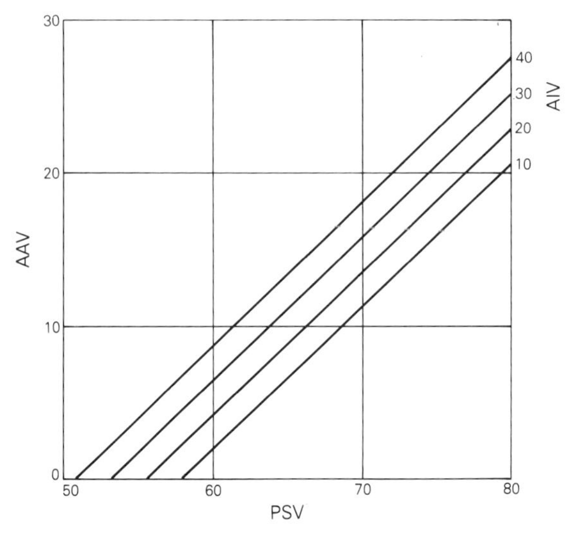

Figure 7.7 Relationships between aggregate abrasion value (AAV), polished stone value (PSV) and aggregate impact value (AIV) for sedimentary rocks.

dolerite, quartzite, quartz gravels), through fair (granite, gravels composed of igneous or metamorphic fragments) to poor (weathered igneous rocks, clay-rich gravels).

For good roadstone aggregate the unconfined crushing strength (q_u, see Section 7.2.2) should be greater than 100 MN m^{-2}.

(d) *Other useful tests.* These include assessment of frost heave. A cylinder of rock aggregate, 150 mm high and 100 mm in diameter, is placed in freezing conditions with its base in flowing water for 250 h. The expansion (or heave) suffered by the specimen must be less than 12 mm.

Tests of 10% fines, specific gravity, water absorption and bulk density are also used and are described in Appendix F. Further, technical details are available in BS 812 (1977). The main US tests are given in Appendix G with their British Standard equivalents.

CONCRETE

The suitability of aggregate for use in concrete can be assessed on the following criteria:

(a) The aggregate should be free from sulphide minerals, especially pyrite. Coal, clay and organic matter should also be absent.
(b) The specific gravity should usually be high, but this criterion depends upon the purpose for which the concrete is needed.

(c) The material should be well graded, with a wide range of particle sizes (see p. 46).
(d) The fragments should have a rough surface, so that a good bond can be achieved between the aggregate and the cement paste.
(e) Chalcedonic silica (flint, chert, agate) and glassy siliceous rocks (rhyolite, pitchstone) are often undesirable in gravel aggregate since they react with highly alkaline cements. (This problem can be overcome by using a low-alkali cement.) An assessment of their relative abundance in the gravel should be made.
(f) The amount of acid-soluble material (sulphate) should be measured.
(g) The **shrinkage** of the concrete as it dries should be measured. This test is made on cubes of concrete prepared from the aggregate and the shrinkage is expressed as a percentage. Low-shrinkage concrete has values less than 0.045%.

The tests on concrete are given in Appendix F.

Special aggregates are required for particular purposes; for example, where a concrete is going to be subjected to marked changes of temperature, the **coefficient of expansion** and the variation of this within an aggregate are important (Table 7.5). Differential expansion on heating produces stresses within the concrete. Fire-resistant concrete often uses artificial aggregate (slag) or limestone because of their consistency and uniformity of expansion. Again, prestressed concrete in high-rise buildings needs low-shrinkage qualities, and limestone and pure quartz gravels provide them. Oil platforms require low-density aggregates (limestone, granite), with low shrinkage. In contrast, atomic reactors require high-density aggregate, and barite has been used in addition, to increase the density.

Table 7.5 Coefficients of expansion of some rock aggregates.

calcareous rocks (limestones)	$1.3 \times 10^{-6}\,°C^{-1}$
crystalline igneous rocks	$2.2 \times 10^{-6}\,°C^{-1}$
compact rocks (shales, marls)	$2.8 \times 10^{-6}\,°C^{-1}$
indurated light-coloured rocks (quartzites)	$3.6 \times 10^{-6}\,°C^{-1}$
cemented light-coloured rocks (sandstones, gritstones)	$3.8 \times 10^{-6}\,°C^{-1}$

7.2.4 Characteristics of some common rock types as aggregates

The following may serve as a practical guide in assessing the suitability of a particular rock mass as a source of aggregate. In all cases, weathered rock and fault zones should be avoided because of loss of rock quality.

AGGREGATES COMPOSED OF FRESH IGNEOUS ROCK

(a) Basic igneous rocks have better PSV and worse AAV than acid igneous rocks.
(b) Fine-grained rocks have poorer PSV and much better AAV and AIV than coarse-grained rocks of the same composition.
(c) A very small amount of weathering of a roadstone aggregate is advantageous as it helps bonding of the aggregate to bitumen. Greater weathering severely reduces the qualities of the rock.
(d) Basic igneous rocks have poorer concrete shrinkage values than acid igneous rocks. The latter are extremely good (less than 0.040%).
(e) Fine-grained, non-vesicular, non-porphyritic rocks have better shrinkage values than their coarse-grained equivalents.

Note that all aggregates required for a project must have current test certificates for the engineering tests specified, and testing should be carried out regularly. Aggregate produced from weathered or highly weathered igneous rock will often be unsuitable for engineering use.

The common sources of aggregate among igneous rocks are varieties of dolerite, basalt and granite. Dolerites have a good AAV where the grain size is very small, and the best PSV where the grain size is very large, such as may occur in a very thick igneous sheet. Basalts are more uniform in grain size, and the PSV and AIV do not usually vary much through a rock mass. The zone between different lava flows should be avoided because of the concentration of vesicles and amygdales. Granites are prone to weathering which has penetrated deep into the rock mass along fractures. There is a consequent drop in crushing strength and impact value. Even in areas of glacial erosion, alteration may affect rock at depths more than 25 m below the surface. Clear of these zones of poor rock, granite is typically uniform in essential properties for an aggregate. Occasionally, mica-rich granites are found, which may be unsuitable for aggregate use, especially if there is more than 10% mica, but tests should be carried out to examine the suitability of the rock as aggregate.

AGGREGATES COMPOSED OF SEDIMENTARY ROCK

The qualities of sedimentary rocks are very variable and most are unsuitable for use as aggregates. The factors influencing PSV, AAV and AIV can be summarised thus:

higher porosity	gives better PSV, worse AAV and worse AIV
lower quartz fines	give better PSV, better AAV and worse AIV
lower quartz fragments	give better PSV, better AAV and better AIV
lower feldspar	gives better PSV, unaltered AAV and better AIV
higher clay	gives better PSV, worse AAV and worse AIV

Greywacke (gritstone) has a good PSV, but the AAV often prevents it from meeting standards in the UK. Conglomerate, sandstone (often from colliery tips) and shale (from oilshale spoil heaps) are occasionally used as rockfill, including road formation beneath the sub-base, where specifications are undemanding. In general, however, their use should be avoided unless extensive tests are carried out. As in other sedimentary rocks, porosity (n) tends to be inversely proportional to unconfined crushing strength; thus aggregates with low n values will be best. Limestones are used as aggregate, particularly the indurated varieties with low porosity. As roadstone aggregate, marbles are more suitable than limestones, but their PSV is low (almost always less than 50). Limestone is, however, an excellent concrete aggregate because of its very low shrinkage quality.

AGGREGATES COMPOSED OF METAMORPHIC ROCK

Among the metamorphic rocks, the fine-grained, non-foliated types, such as hornfelses and quartzites, have the best compressive strengths, low shrinkage values and good AIV and AAV, though their PSV is usually only moderate or poor (less than 57). The foliated metamorphic rocks (slate, schist) have good impact values and crushing strengths perpendicular to the banding, but poor values parallel to the banding. They are only used in concrete or road aggregate as fill material, except in the case of high-grade, coarse-grained gneisses, which have properties similar to those of granite. They may, however, be an ideal **hydraulic stone** for sea defences and dam walls, as they split readily into natural slabs which have high resistance to erosion.

Engineering test data and mean values for several different rock types are given in Appendix F.

GRAVEL

Gravel and sand are widely quarried from natural deposits in the UK. Production of sand or natural gravel aggregate from these sources is inexpensive compared with crushed rock aggregate, since neither drilling and blasting nor (in most cases) crushing are required, only washing and screening. Figures 7.8 and 7.9 show the grading in sands and in gravels at two sites under investigation, and Table 7.6 shows the rock types present in a typical Scottish deposit. However, it must be noted that particle size varies within a single gravel deposit, and rock types comprising the gravel will vary widely, depending upon the type of deposit and its location.

Gravel deposits are widespread in the UK but the deposit usually has limited reserves and often underlies good arable land. The average thickness of gravel deposits is only 5 to 10 m, compared with igneous or other rock quarries (30 m+). In the UK, natural gravels account for about

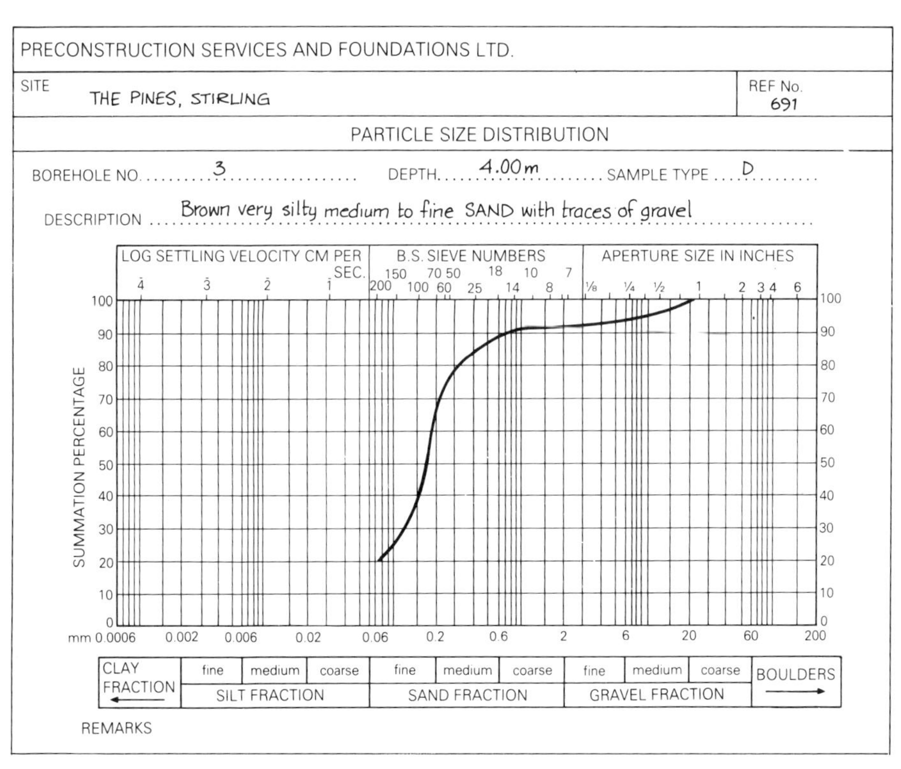

Figure 7.8 Particle size distribution in sands at Stirling, Scotland, at a depth of 4.00 m from an actual site under investigation. Reproduced by kind permission of Bullen and Partners, Consulting Engineers, Glasgow.

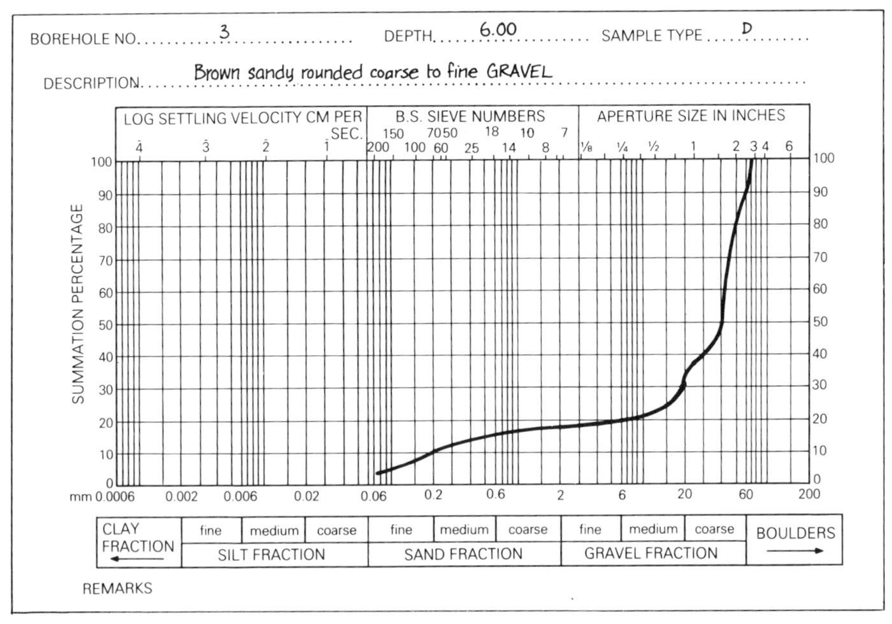

Figure 7.9 Particle size distribution in gravels (same location as Fig. 7.8) at a depth of 6.00 m. Reproduced by kind permission of Bullen and Partners, Consulting Engineers, Glasgow.

half the total aggregate produced (97 million tonnes in 1981 including marine dredged gravels), but the percentage is declining. Gravels have been dredged from offshore deposits during the last 15 years and in 1981, 15 million tonnes were produced from this source. However, considerable information is needed before dredging is carried out, to ensure that removal of these marine gravels does not interfere with local marine conditions (wave and current movement, and so on). Gravel quality is variable even within a single deposit, and the spread of engineering test data on gravels is very large, but flint and quartz gravels should be mentioned as being particularly good for concrete aggregates.

Table 7.6 Rock type percentages in three Scottish Midland Valley gravel pits.

Engineering rock group	Cambusmore Pit	Doune	Cramflat
quartzite	31	14	19
gritstone	23	65	28
schist	16	9	47
porphyry	4	0	0
basalt	26	12	6

7.3 Drilling, blasting and rock excavation

7.3.1 Rock properties related to drilling

The rate and cost of drilling into rock and the procedure adopted are controlled by the so-called hardness and abrasiveness of the rock. Hardness in this usage is *not* as defined for a mineral using Mohs' scale (see Section 2.1.1) but is related to the compressive strength (q_u) of the rock. The compressive strengths of the main rock types, and implicitly the probable ease with which they can be drilled, were given in Table 7.4. The **abrasiveness** of a rock is a measure of how rapidly drill bits are worn down, and it correlates fairly closely with the hardness (in the strict sense) of the minerals present in the rock, and with the grain size. In general, the most abrasive rocks are those with a high content of silica and a fine texture. The life expectancy of a normal 105 mm drill bit is only about 60 m of hole in fine-grained metamorphic quartzite, but may be up to 600 m in fresh quartz dolerite, a comparably strong rock. Table 7.7 gives a guide to abrasiveness of the common rock types.

Table 7.7 Abrasiveness of some rock types.

Descriptive term	Rock types
abrasive	obsidian, rhyolite, aplite, granite, felsite, pegmatite porphyry; sedimentary flints, cherts, quartzite, siliceous sandstones, ganister, quartz or granite-pebbled conglomerate; metamorphic quartzite, hornfelses, quartz-rich gneisses
intermediate	basalt, dolerite, gabbro, andesite, diorite, syenite, pyroclastic rocks; sedimentary sandstones, siliceous limestones, friable sandstones and grits (although friable these rocks are quite abrasive)
least abrasive	all weathered igneous and decomposed rocks, serpentine; metamorphic slates, schists, marbles; sedimentary limestones, clays, shales, mudstones, marls, chalk, coals

7.3.2 *Blasting techniques*

Full accounts of rock blasting methods are given by Langefors and Kihlström (1973) and McGregor (1968). A useful handbook has been produced by Du Pont (1963). There are three main stages in excavating rock by explosives and reducing it to manageable-sized pieces: drilling, blasting, and secondary blasting or drop-balling. The particular technique, including the choice of explosives and how they are placed within the rock, depends on the following factors:

(a) The type of rock: there may be relatively soft sedimentary layers requiring low-density, low-strength explosives, or hard crystalline rocks requiring high-density, high-strength explosives.
(b) The degree of fragmentation required, especially in quarry work where the rock must be suitable for the loading and crushing machinery.
(c) The shot-hole condition, which may be dry or wet. For the latter, a water-resistant explosive, such as gelignite or TNT-based slurry, is required. If the shot holes are partially filled with water, a water-resistant type of explosive is used until the charge column has been brought above the water line.
(d) The size of the shot hole, which can be anything from 75 mm (3 in) to 300 mm (12 in) in diameter (see Table 6.3). In the UK, 105-mm holes are the most common, for both quarrying and other excavation. Very large diameter holes (300 mm or greater) are unsuitable for rock blasting where the compressive strength of the rock is more than 170 MN m^{-2}.

Present methods vary widely in choice and positioning of explosives. A technique which was widespread, but has increasingly been superseded by other methods because of the improved efficiency of modern drilling methods, is **heading blasting**. Small tunnels are driven into the rock mass for about 15 m, then side tunnels are driven parallel to the outside face of the rock mass from the tunnel ends. Chambers are made at intervals along these drives and packed with explosives, and these charges are then fired. This technique originally used black powder (a mixture of charcoal, potassium nitrate and sulphur) but improved safety precautions have stopped the use of this explosive mixture.

Drill holes are commonly of three types:

(a) Small-diameter (75 mm or less) short holes are used where the rock face is being excavated in steps (benches). Vertical holes up to 5 m deep are drilled in a line parallel to the free face. The horizontal distance from the bottom of the holes to the free face must always be less than the depth of the hole. By this method the yield of rock is 8–11 tonnes per kilogram of explosive (the **blasting ratio**).

(b) Small-diameter long holes are used where ground vibration from blasting must be kept to a minimum. Holes are drilled about 2.5 m apart and about 15 m deep parallel to the face. The blasting ratio is 11–18 tonnes kg^{-1}.

(c) Medium-diameter (76–127 mm) holes are most commonly used to blast away rock faces, which may be anything between 6 and 30 m in depth. The length of the hole depends on the depth of the face. Quarrying regulations in the UK result in most holes of this type being about 18 m (60 ft) deep, and spaced about 4 m apart. They are in line parallel to the face and are usually angled to the horizontal at 70 to 80° towards the face as an aid to blasting away the bottom of the face cleanly. Each hole yields up to 660 tonnes of rock, and gives a blasting ratio of about 10 tonnes kg^{-1}.

Blasting is carried out by charging the shot hole with gelignite at its bottom and priming this charge with a detonator. A low-velocity explosive, such as ammonium nitrate mixed with diesel oil (ANFO), is usually poured in on top of this high-velocity explosive to within about 3 m of the mouth of the hole. The remainder is then stemmed (tamped) with sand or other fine material. The ammonium nitrate is set off by the detonation of the more sensitive gelignite. Detonators are made with delays of from 1 to 15 ms so that holes can be exploded sequentially at intervals of 1 ms or more. If more than 15 holes have to be detonated in one blast, then a mechanical or electronic delay setter is used for the same purpose.

Secondary blasting is employed to break up larger blocks produced by the initial blast. In pop-shooting, a small hole is drilled into the block, charged and fired. In blasting-plaster shooting, the charge is attached to the block by clay and, since no hole is required, the cost is less. Most quarries also have a machine capable of dropping a heavy steel ball on to oversize blocks. This technique is called drop-balling.

New developments in blasting include the use of improved detonation cord ('Nonel') and the use of larger drill holes (152 mm). The lower section is loaded separately from the upper section, with inert stemming between them, and the two sections are fired separately. This gives a similar blast ratio to the standard 105 mm hole technique but requires fewer holes.

7.3.3 *Case history of rock excavation: the new Strome Road (A890), Wester Ross, Scotland*

Construction of the new Strome Road involved making cuttings in a variety of metamorphic rocks and ensuring that the rock slopes of the cuttings were stable. The rock consisted of Lewisian feldspathic and hornblende-gneisses (that is, members of the engineering granite group) with intercalated chlorite-schists and hornblende-rich rocks (schist group), and younger Moine quartzite and quartz-rich gneisses (quartzite group). The structure is complex. The Moine rocks have been thrust westwards over the Lewisian rocks along a major plane of faulting, the Moine Thrust, below which there is extensive fracturing (Fig. 7.10). In the southern part of the site, Lewisian rocks are present above, as well as below, the thrust. This part of the new road was examined in detail by one of the authors (C.D.G.) and the discontinuities were logged systematically. They are all joints and comprise five distinct joint sets. The results are shown in the orientation diagram (Fig. 7.11), where the diameters represent the strike (in degrees magnetic) of the sets, and angles of dip are also given. Figure 7.12 shows the relationships of the five

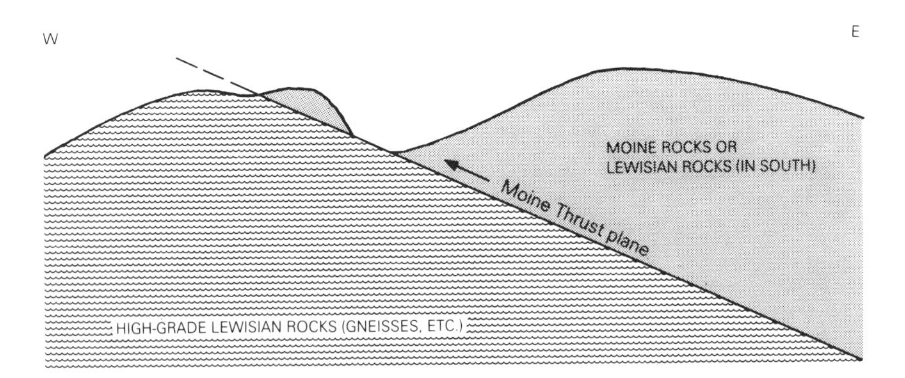

Figure 7.10 Section showing rocks in the area of the new Strome road (the A890). The Moine Thrust plane is shown.

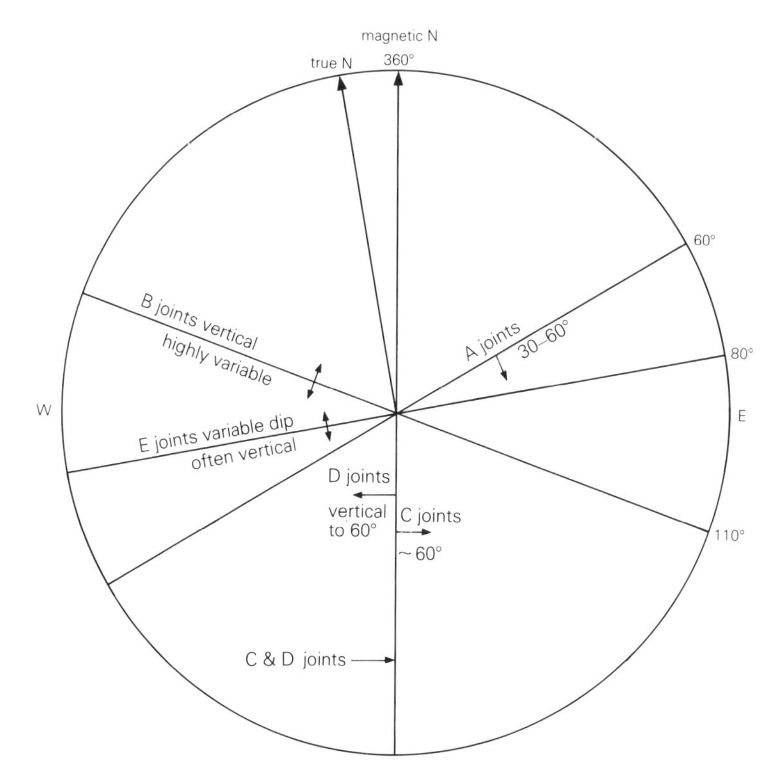

Figure 7.11 An orientation diagram showing the strike directions and dips of the five major discontinuities (joint sets) in the Strome area.

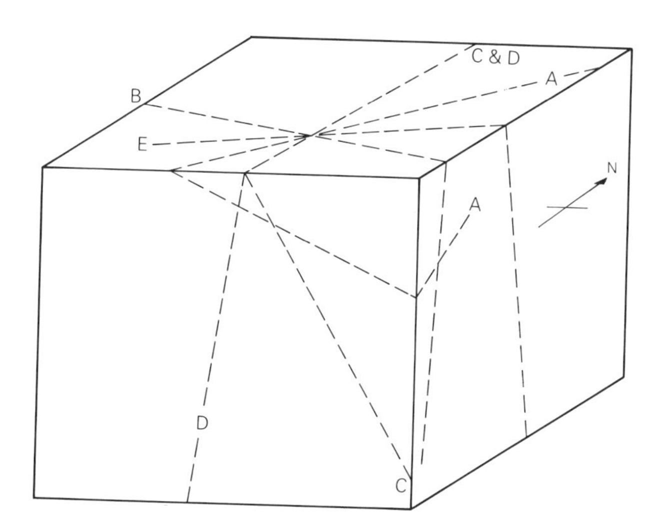

Figure 7.12 A block diagram showing the five main discontinuities in the Strome area in three dimensions and their relationships to each other. Data are from Figure 7.11.

main joint sets in a block diagram. Details of each joint set are listed in Table 7.8. At least three sets were present in each of the nine road cuttings examined, and in one, all five were seen together. The rock mass here possesses a very small block volume, even before blasting. The joints are opened by blasting and a highly unstable slope is produced. At the planning stage, slopes of 70° to the horizontal were considered to be practicable, but in the event, slopes of about 45° were the best that could be achieved in most cuttings. **Pre-splitting** techniques (Langefors & Kihlström 1973, pp. 304–7) were tried as a possible means of producing a safe permanent slope. In this procedure, close-set small-diameter drill holes are drilled along the position of the intended face of a rock cutting. These are then detonated using a weak charge to produce a fracture. Then the rock mass is drilled and blasted in the usual way, but the rock is removed only up the pre-split surface (see Fig. 7.13). A safe angle of slope (see Table 8.1) is probably about 35–40°, so the present slopes may decrease in angle still further before stability is finally reached. To allow for this, the road was constructed as far from the high side of the deeper cuts as possible, and roadside trenches were dug to trap any falling rock. Fortunately, most cuts were only about 10 m deep and the remedial measures appear to have worked. Many of the geological problems encountered could have been predicted at the planning stage if the engineers had fully understood the effect of the Moine Thrust on the rocks that were to be excavated, but this was not brought out in the original site investigation report.

Table 7.8 Discontinuity data (strike measurements are in degrees magnetic).

A discontinuities (joint set)
(a) medium-spaced 600 mm
(b) closed discontinuity (but not tight), smooth surface/no water flow
(c) 060°/58°S

B discontinuities (joint set)
(a) medium-spaced 600 mm
(b) closed discontinuity, smooth surface/no water flow
(c) 110°/vertical

C and D discontinuities (joint sets)
(a) medium- to widely spaced 1 m+
(b) open discontinuities, smooth regular surface/no water flow
(c) 350° to 360°/60°E (joint set C), 80°W (joint set D); both joint sets have similar strike but different dips (see block diagram)

E discontinuities (joint set)
(a) medium-spaced 600 mm
(b) closed discontinuity, smooth surface/no water flow
(c) 080°/vertical but variable – always steep

Figure 7.13 Pre-split rock surface in a road cutting for the M90 motorway near Perth, Scotland.

7.3.4 *Test problem*

The reader can test his or her grasp of this and the previous chapter by analysing the data shown in Table 7.9. These are measurements made on various rock types that crop out in an area where large-scale excavation for a concrete dam is required. The relationships of the rocks are shown on the section (Fig. 7.14).

Table 7.9 Data on rocks seen in Figure 7.14.

Rock types		G_b	i_s(%)	V_{lab} (km s^{-1})	Block volume (m^3)
1		2.0	12.0	0.3	none
2		2.6	0.1	6	2
3	mudstone	2.6	3.0	1.7	2
4	mudstone	2.9	0.35	5.5	3
5	sandstone	2.6	1.5	2.8	1
6	limestone	2.7	1.0	3.6	2

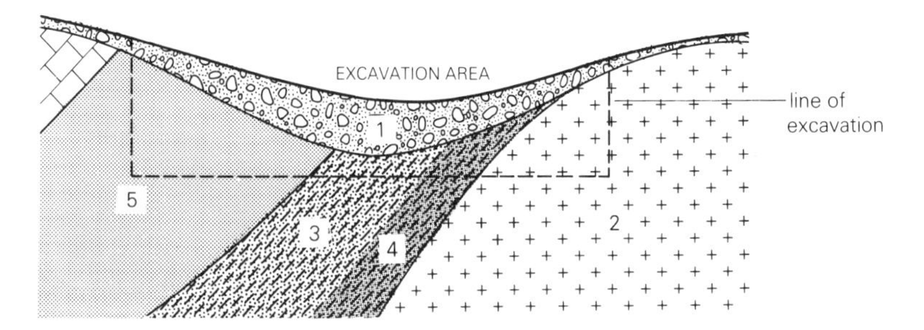

Figure 7.14 A cross section showing rock relationships in an area where large-scale excavation is to be carried out.

Question 1. If rock type 1 is a sedimentary rock and rock type 2 is igneous, what are the *precise* rock types likely to be, and what difficulties will the excavator meet?

Question 2. Calculate the unconfined compressive strengths and the porosities of the various rocks. Are any of them suitable to use as constructional materials for the dam?

Answer 1. Rock type 1 has a solid mineral grain specific gravity of 2.65, which indicates that the rock is composed predominantly of the light-coloured minerals, quartz and feldspar (and clays), rather than of denser dark-coloured minerals. The distribution of the rocks on the cross section (Fig. 7.14) suggests that it is a superficial deposit (soil) such as boulder clay. As such, it does not need to be blasted but can be ripped. Care should, however, be taken when ripping, as large boulders could seriously damage the teeth of the excavator.

Rock type 2 has a very low moisture content and low porosity. Its form on the cross section suggests that it is a granite intrusion which will have to be blasted before excavation commences. The significant difference between the two mudstones 3 and 4 in Table 7.9 is that the latter occurs *within the thermal aureole* of the granite and has been baked. This has altered the engineering properties (lowered porosity, increased specific gravity). As a result, whereas mudstone 3 can be ripped, mudstone 4 will almost certainly have to be blasted. The pattern of blasting of rocks 2 and 4 should be chosen carefully to produce small blocks that are suitable for easy loading. The initial block volume in these rocks is such that huge blocks are defined by the natural fracture pattern and they would be difficult to excavate without further breakage.

Answer 2. The rocks have porosities and compressive strengths as follows:

(1) $n = 24.0\%$, $q_u < 10$ MN m^{-2}. (2) $n = 0.26\%$, $q_u > 250$ MN m^{-2}.
(3) $n = 7.8\%$, $q_u = 50$ MN m^{-2}. (4) $n = 0.96\%$, $q_u = 175$ MN m^{-2}.
(5) $n = 3.9\%$, $q_u = 85$ MN m^{-2}. (6) $n = 2.70\%$, $q_u > 105$ MN m^{-2}.

n values are obtained from the nomogram (Fig. 7.1) by plotting in G_b and i_s values. V_{lab} values are plotted into Fig. 7.4 and R values obtained, which are then plotted into Fig. 7.3 to get q_u.

An outer skin of strong rock is used to form a hydraulic stone facing to an earth-type dam, and large blocks of either the granite (2) or the *metamorphosed* mudstone (4) would be suitable. Types 2, 4, 5 and 6 would provide good coarse aggregate to fill the core of an earth dam. Types 1 and 3 may give rise to problems and should not be used for this purpose. If the structure were a concrete dam, the best aggregate for the concrete would be the granite (2) and the limestone (6), as they have the quality of low shrinkage. Since the granite must be excavated, it makes sense to crush it and use it as aggregate, rather than quarry the limestone (6) only for that purpose.

References and selected reading

Attewell, P. B. and I. W. Farmer 1976. *Principles of engineering geology*. London: Chapman and Hall.

British Standards Institution 1965. *Specification for aggregates from natural sources for concrete (including granolithic)*. . Br. Stand. Inst. Rep. BS 882 and 1201.

British Standards Institution 1967. *Specification for gravel aggregates for surface treatment (including surface dressings on roads)*. Br. Stand. Inst. Rep. BS 1984.

British Standards Institution 1971. *Glossary of mining terms. Section 5: Geology*. Br. Stand. Inst. Rep. BS 3618.

British Standards Institution 1973. *Methods for the sampling and testing of lightweight aggregates for concrete*. Br. Stand. Inst. Rep. BS 3681: Part 2.

British Standards Institution 1977. *Methods for sampling and testing of mineral aggregates , sands and fillers*. Br. Stand. Inst. Rep. BS 812.

Building Research Station 1968. *Shrinkage of natural aggregates in concrete*. Bldg Res. Stn Dig. 35.

Building Research Establishment 1973. *Concrete: materials*. Bldg Res. Stn Dig. 150.

Caterpillar Tractor Co. 1972. *Handbook of ripping*. Illinois: Caterpillar Tractor Co.

Duncan, N. 1969. *Engineering geology and rock mechanics* (2 vols). London: Leonard Hill.

Du Pont, E. J. 1963. *Blaster's handbook*. USA: E. I. Du Pont de Nemours & Co.

Edwards, A. G. 1967. *Properties of concrete made with Scottish crushed rock aggregates*. Bldg Res. Stn Eng. Pap. 42.

Edwards, A. G. 1970. *Scottish aggregates: rock constituents and suitability for concrete*. Bldg Res. Stn 28/70.

Franklin, J. A., E. Broch and G. Walton 1971. Logging the mechanical character of rock. *Trans Inst. Min. Metall.* **80**, A1–9.

Hartley, A. 1974. A review of the geological factors influencing the mechanical properties of road surface aggregates. *Q. J. Engng Geol.* **7**, 69–100.

Hawkes, I. and M. Mellor 1970. Uniaxial testing in rock mechanics laboratories. *Engng Geol.* **4**, 177–286.
Hawkes, J. R. and J. R. Hosking 1972. *British arenaceous rocks for skid-resistant surfaces*. Transp. Road Res. Lab. Rep. LR 488.
Imperial Chemical Industries 1972. *Pamphlets on the use of explosives in different situations*. Stevenston, Scotland: Nobels' Explosive Co.
Jaeger, C. 1972. *Rock mechanics and engineering*. Cambridge: Cambridge University Press.
Langefors, U. and K. Kihlström 1973. *The modern technique of rock blasting*. New York: Wiley.
McGregor, K. 1968. *Drilling of rocks*. London: McLaren.
Matheson, G. D. 1983. *Guidance on pre-split blasting in highway rock excavation*. Working Paper 1982/3. Scottish Branch TRRL.
Orchard, D. F. 1979. *Concrete technology* (Vols 1 and 2), 4th edn. London: Applied Science.
Road Research Laboratory 1959. *Roadstone test data presented in tabular form*. Rd Res. Lab. Rd Note 24.
Williams, A. R. and G. Lees 1970. Topographical and petrological variation of road aggregates and the wet skidding resistance of tyres. *Q. J. Engng Geol.* **2**, 217–35.

8 *Principal geological factors affecting certain engineering projects*

8.1 Stability of slopes and cuttings

8.1.1 Geological factors affecting the stability of a new excavation

The static conditions that control the initial and also the later stability of a steep face cut into soil or rock, and which may determine the need for support or remedial treatment, are as follows.

The properties in bulk, particularly the *shear strength*, of the material forming the cutting: the stability of a cutting in rock is usually dependent on the occurrence of joints and other planes of weakness, and on the amount of cohesion and the friction across these planes.

The *structure* of the rocks and soils, and specifically how any planes of weakness are orientated relative to the newly exposed face: for example, horizontal bedding planes in poorly jointed sandstone often give near-vertical faces which are stable, whereas faults or joints striking parallel to the new face, and dipping steeply towards it, will probably be planes of movement or potential instability.

The *groundwater conditions:* saturation significantly lowers the strength of most soils compared with their values when dry. Certain soils weaken to a stage at which they run like viscous liquids. High pore pressure of ground water in a layer, or in a plane of weakness, lowers frictional resistance to movement. (The mechanism is the same as that which allows a hovercraft to glide over water, or land on a high-pressure cushion of air.) A dramatic and tragic example of instability triggered by high pore pressure within a body of soils was the Aberfan disaster in 1966, when a spoil heap of mine waste slid downwards on to a Welsh village enveloping and destroying a large part of the local school.

Stresses produced by natural loads adjacent to the cutting: steep-sided valleys or mounds affect the state of stress in the ground near the surface, not only below themselves but also for some distance around. This lateral change in stress conditions may be sufficient to cause instability of weak rocks and soils. For example, valley bulges (Section 3.5.8) are produced in this way, and they may be accompanied by instability of an adjacent slope.

8.1.2 Other geological factors causing instability of existing slopes

An initially stable slope may become unstable with the passage of time because of human disturbance. This may consist of adding a fresh load such as a spoil heap, removing support by excavating, or triggering movement by vibrations from nearby heavy machinery. The common geological causes are as follows.

Weathering of the soil or rock of the slope so that it becomes weaker: this may affect the bulk of the material (for example, boulder clay) or may be concentrated along planes in a rock. Chemical alteration of existing minerals is important under certain conditions, as is mechanical breakdown in others (Section 3.2.3), Periglacial weathering in Pleistocene times also affects the stability of some present-day cliffs.

Erosion of the slope by a river or other natural agent, usually at its base but possibly along a weaker layer or plane, may cause undermining to take place.

Change in *water content* of the soil or rock: heavy rain, especially after a drought, saturates the material forming the slope, increasing its mass and the gravitational pull on a given volume, and also reducing the strength of soils and the friction along any discontinuities.

8.1.3 Types of failure of soil slopes

Instability of soils on a slope may take one of the following forms.

Creep occurs on steep slopes and produces a downhill movement at low rates (less than 10 mm per year) of the top few metres of soil (see Fig. 8.1). It is facilitated by the effects of frost, and by heavy rain washing out fines from the soil. Any excavation on a slope affected by creep is likely to increase movement. Creep may be recognisable by displacement of fencing or of a cover of turf, or by drag effects of strata under the soil.

Flow is a rapid movement of waterlogged soil, broken rock and mud down hill, usually after prolonged rain.

Scree or **talus slide** occurs where rock fragments spall off a fractured rock mass which has been subjected to mechanical weathering. They rest at a natural angle of repose at the foot of the slope. Excavation at the foot of a scree slope, or inadequate drainage, which increase pore water pressure in the mass, will lead to instability of the slope.

Translational failure involves movement along a particular plane of weakness in the soils of the slope. For example, in a slope formed of a clay layer in sand layers, deterioration of the shear strength of the clay may lead to movement at the boundary of clay and sand. Failure of this type may occur with slope angles as low as 6°.

Rotational slip affects clays or clay-rich rocks such as mudstones and

Figure 8.1 Minor ridges on a hillside produced by soil creep. (British Geological Survey photograph, C1333, published by permission of the Director; NERC copyright.)

shales. The surface of movement is curved, and is such that shear resistance along it is assumed to be given by the equation

$$S = C_c + (\sigma - u) \tan \phi_c$$

where σ is the stress normal to the surface, u is the pore water pressure, C_c is the cohesion of the clay, ϕ_c is the angle of frictional resistance of the clay in terms of effective stress, and S is the shearing resistance. Figure 8.2 shows rotational slip failure in a slope in California.

Rotational slip is usually preceded by development of vertical tension cracks at the top of the slope. These cracks may reach a maximum depth of $2C/\gamma$, where C is the cohesion of clay and γ is its density. Typical laboratory values for marl (a clay-rich rock) are

$C = 2.5$ MN m^{-2} and $\phi = 42°$ (under dry conditions)
$C = 2.0$ MN m^{-2} and $\phi = 21°$ (under saturated conditions)

Figure 8.2 Slope failure at Sonoma County, California. (Reproduced by kind permission of California Division of Highways.)

8.1.4 *Types of failure of natural rock slopes*

Minor rock falls are produced by weathering acting on unstable rock slopes. The susceptibility of a given rock to weathering processes can be estimated by determining its saturation moisture content (i_s) and swelling coefficient (E_s). Igneous and high-grade metamorphic rocks with i_s values of less than 1% are generally safe from weathering effects. Sedimentary rocks and low-grade metamorphic rocks are considered to be safe on slopes if their i_s values are less than 3%. If exposed rocks on a slope have high saturation moisture contents, then tests should be carried out over a period of time, under both freeze–thaw and wet–dry conditions, and their swelling behaviour noted. Ice action is important if joints are present. If the rock mass has a low block volume, that is, less than 0.5 m^3, minor rock falls may occur, even if the rock has a low saturation moisture content.

Major rock falls usually result from collapse caused by undermining of rocks above a weak layer (see Fig. 8.3). The agent may be weathering, erosion or mining. Common weaknesses in a rock mass which can lead to collapse after weathering and etching out by erosion are layers of clay rock, chlorite in joints, and carbonate rocks, including calcareous sandstones.

Translational failure along a particular plane of weakness may occur in a rock mass cut by faults, master joints or steep bedding planes. Figure 8.4 illustrates spectacular slope failure along San Andreas Fault zone, California. The shear strength of the infill in these discontinuities determines the stability of any slope cut in the rock mass. As in soils, the shearing resistance along any plane is given by the equation

$$S = C_s + (\sigma - u) \tan \phi_s$$

where C is the **cohesion** between the two sides of the discontinuity, σ is the stress acting normal to the plane of discontinuity, u is the water pressure in the discontinuity, and ϕ_s is the **angle of frictional resistance** attributed to the plane of movement. In Table 8.1, two values of ϕ are given: one (ϕ_d) assumes that there is some cohesion, the other (ϕ_{ult}) is for a cohesionless surface. Friction angles for materials commonly found infilling joints are given, as also are some representative values of unconfined compressive strengths of rocks.

In rock masses with discontinuities that are not infilled, a slope will be stable only if the angle of dip of any discontinuities orientated towards the slope is less than the angle of friction, ϕ, of the rock. Water pressure in the discontinuity may reduce the safe slope angle, and must be taken into account. If the angle of dip of a discontinuity is greater than the angle of friction, the slope will only be stable if there is some cohesion, that is, if C has a value greater than zero.

Figure 8.3 A major rock fall in the Scottish Highlands menacing the West Highland Railway. (British Geological Survey photograph, MNS1733(8), published by permission of the Director; NERC copyright.)

Figure 8.4 Housing tract, Daly City, California, located on unstable ground in San Andreas Fault zone. Houses at end of tract (mid-right picture) removed after destruction by slide. (Reproduced by kind permission of US Geological Survey, Menlo Park, California.)

If the discontinuities are infilled with a weak material such as clay, stability of a slope in such a rock mass will depend on its shear strength. Where a rock mass consists of alternating layers of strong and weak rocks, tests *in situ* and in the laboratory are necessary to determine whether failure is likely to occur because of shear failure in the weak bands, or because of sliding along the bedding planes, if they dip steeply.

Rotational failure occurs when the stresses generated by gravity on rocks near the slope exceed the inherent strength of the rock mass. It may be accompanied by translational failure. The extra load of a spoil heap may trigger failure of this type.

Creep failure takes place when certain rocks, such as clays, absorb water, swell, slowly deform and move down hill. If they are overlain by a more stable, stronger layer, translational failure may take place along the bedding plane between them. Cambering (Section 3.5.8) is produced in this way.

Table 8.1 Angles of frictional resistance (ϕ) and unconfined compressive strengths of some common rock types. The data are from Hoek (1970), Hoek and Bray (1974) and Attewell and Farmer (1976).

	ϕ_d (deg)	ϕ_{ult} (deg)	Unconfined compressive strengths ($MN\,m^{-2}$)
Igneous rocks			
basalt	47	<45	150–300
dolerite	similar to basalt		100–350
gabbro	similar to basalt		250–300
andesite	31–35	28–30	similar to basalt
porphyry	40	30–34	similar to basalt
diorite	similar to granite		150–300
granite	>35	31–33	100–250
Sedimentary rocks			
sandstone/greywacke	27–38	25–34	20–170
siltstone	43	43	10–120
shale/mudstone	37	27–32	5–100
limestone	>40	33–37	30–250
Metamorphic rocks			
schist	variable, probably quite low		100–200
gneiss	variable, less than quartzite		50–200
quartzite	44	26–34	150–300
Infill materials	$\phi_{average}$ (deg)		
calcite	20–27		
breccia	22–30		
rock aggregate	40		
shaly material	14–22		
clay	10–20		

8.1.5 *Stabilisation of slopes*

Full accounts of slope stabilisation methods can be found in advanced texts such as that by Attewell and Farmer (1976). In brief, the four common measures taken to prevent slope failure are as follows:

(a) The slope is modified by removing material from the potentially active part of the slope and adding it to the 'toe' of the slope.
(b) The slope is drained to reduce load and increase strength of frictional forces by means of trenches filled with rubble.
(c) The slope is supported by a retaining wall or by embedded piles which are anchored to the rock mass. The soil and rock behind the wall must be drained.

(d) An unstable rock face may be stabilised either by bolting or by using steel mesh. Bolting is used to anchor large unstable blocks, and steel mesh to cover entire sections of an unstable steep rock face (as described in the next section). Where such unstable slopes exist, for example in a new road cutting, sufficient shoulder width should be allowed to 'absorb' debris, and a side ditch excavated with or without a rock fence to protect the new road.

8.1.6 Case history: the Kishorn Dock excavation, Wester Ross, Scotland

Some of the largest excavations of solid rocks made in the UK in the 1970s were in the course of preparation of sites at which oil production platforms could be constructed. Kishorn Dock is a site of this type, excavated entirely in Torridonian sedimentary rocks. A detailed section of these strata at the site is shown in Figure 8.5. The arkose (gritstone group) bands are unaltered, but the thin shaly layers contain some chlorite produced by very slight thermal metamorphism. These strata dip

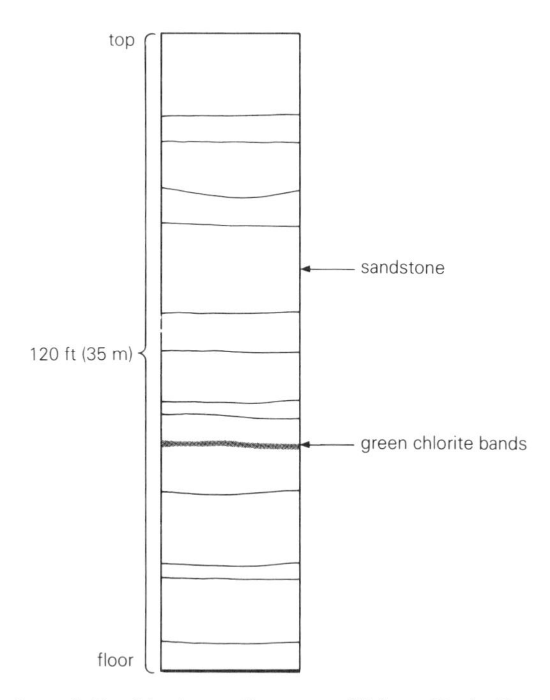

Figure 8.5 Section through Torridonian sediments at Kishorn Dock. Figure 8.6 gives precise location of section. The rocks include red coarse-grained arkoses (gritstone group) with intercalated green bands of chlorite. These bands comprise 5% of the total volume of the rock.

at angles ranging between 13° and 21° to the south-west, and have strikes of between N315 and N330 (Fig. 8.6). The bedding plane discontinuities present little problem of instability in a vertical cut, but there is also a second important set of discontinuities that strike roughly parallel to the highest face of the excavation and dip steeply (at about 65°) in towards it.

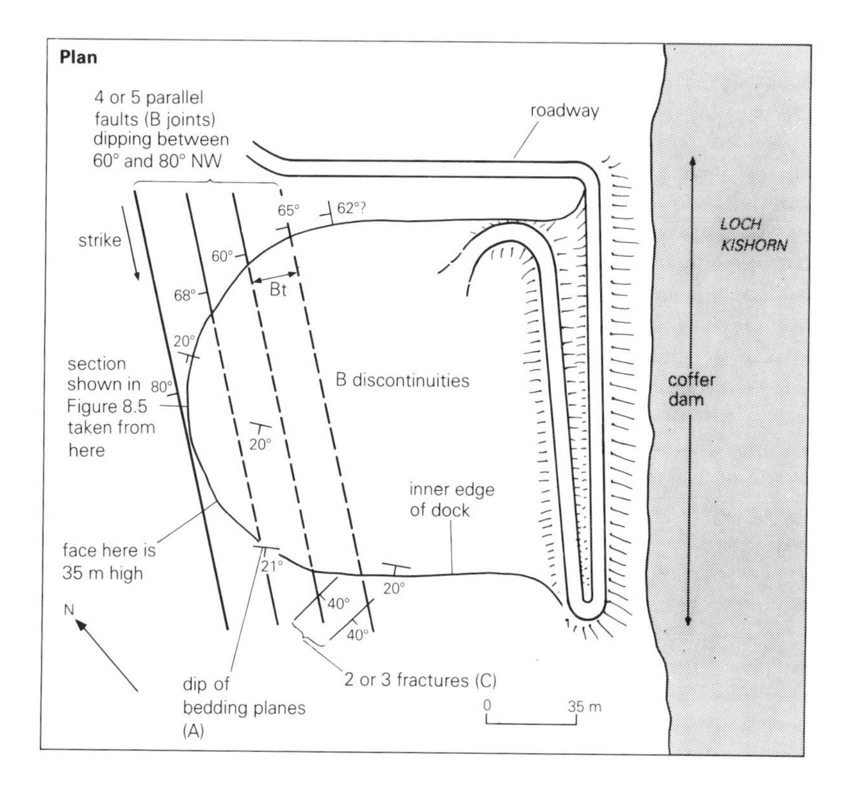

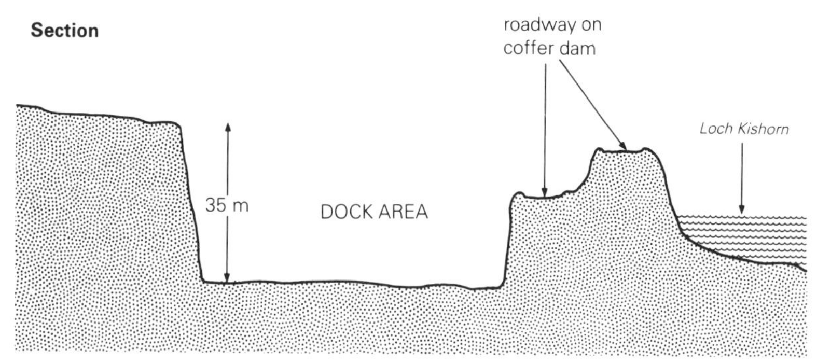

Figure 8.6 Plan and section of the Kishorn Dock excavation showing the main discontinuities, and also the bedding planes of the rocks.

Table 8.2 Discontinuity data (see Appendix H).

A discontinuities (bedding planes)
(a) Spacing variable (see section)
(b) Tight discontinuity, rough irregular surface/no water flow
(c) 330°/20° SW (dip variable between 13° and 21°)

B discontinuities (set of parallel faults – very rare)
(a) Very widely spaced 18 m
(b) Open discontinuity 10 mm, smooth with slickensides/no water flow
(c) 025° to 030°/60° to 80° NW

C discontinuities (joint set) found in south excavation
(a) Very widely spaced 15 m
(b) Open discontinuity, smooth surface/no water flow
(c) 090°/40° S

Table 8.2 gives data on all the discontinuities present. They are small reverse faults, which are secondary features related to the Kishorn Thrust Plane. This is not present at Kishorn Dock but crops out nearby. Before erosion to present land surface level, it lay a few hundred metres above the reverse faults seen in the Dock. The relationship of thrust and minor faults below it is a familiar model in structural geology, and the pattern could have been expected at the stage of preliminary investigation.

These discontinuities and the broken rock adjacent to them make the rock mass at Kishorn Dock inhomogeneous, potentially unstable in vertical cuts, and awkward to excavate. For example, a flat floor is an essential feature for a dock of this type, but was diffficult to achieve as mechanical shovels tended to gouge the fracture zones preferentially and to overexcavate locally. It also proved difficult to blast clean-angled benches on the excavated face, and raised the important question of what angle of slope was needed to make them safe. On the western side of the excavation, the strike of the reverse faults is tangential to the excavated face of the dock, and their dip is away from the face at steep angles (more than 65°). When blasting took place there, the nearest fault plane to the face tended to open and the wedge of rock between the plane and the face would spall off, leaving the Dock face with a dangerous overhang. The most stable dock face there would be one with a low angle of slope (to the horizontal): theoretically a slope at right angles to the angle of dip of the reverse faults (i.e. an angle of about 30°). This angle would fit well with the friction-angle data given in Table 8.1 (see sandstone). In practice, this was not feasible because of cost, and because there was insufficient free ground at the edge of the excavation to allow these shallower-angled benches to be dug. Instead, the slope angle of the main bench was reduced from vertical to 75°, and chain mesh was hung on the face to minimise the risk from any rock fall.

Stability of the face in the western part of the dock was confused and reduced by the presence of another important set of joints (the C joints in Fig. 8.6). The same considerations ruled out a slope of 35° but the angle was reduced from 80° to 45° and the face was hung with chain mesh. These solutions proved to be adequate, and during the first two-year period after construction no rockfall occurred.

8.2 Impounded surface water: geology of reservoir and dam sites

8.2.1 Leakage and other considerations

A reservoir is meant to hold water, hence the principal geological criterion of the suitability of a reservoir site is that the rocks and soils around and below it form an impervious basin naturally, without need of excessive and expensive grouting of potential leakage paths. Other geological factors to consider are the consequence of a change in the position of the water table as the reservoir fills, and the rate at which sediment will accumulate in the reservoir.

The important sources of leakage from a reservoir are (a) through permeable soils, (b) through rock aquifers, and (c) along faults and master joints.

(a) In regions that have escaped glacial erosion, there may be a thick mantle of weathered rock forming a permeable soil. In glaciated regions, however, soils (most notably boulder clay) may form a useful impervious skin below the reservoir, but other deposits (particularly sand and gravel) may offer paths for leakage. Buried channels infilled with sand and gravel are a possible hazard in both types of region. Superficial deposits can present other problems as well. Drift obscures the shape of the rock-head surface. Peat may flavour and colour the water in the reservoir, and may need to be removed before filling. A change in the water table as the reservoir fills may affect the stability of screes and other soils on the hillsides and cause landslides.

(b) Leakage through rock aquifers is controlled by their structure, and its relationship to the hydraulic gradients produced by the head of water in the new reservoir. For example, leakage along the sandstone layer (Fig. 8.7) would take place in structure 'b' but not in 'a'. The hydraulic gradient slopes down from the reservoir end of the aquifer to its outcrop in an adjacent valley. If, however, the water table in an aquifer, which forms one side of the reservoir (Fig. 8.7), were higher than the top water level of the impounded water, then flow would be towards the reservoir, and there would be replenish-

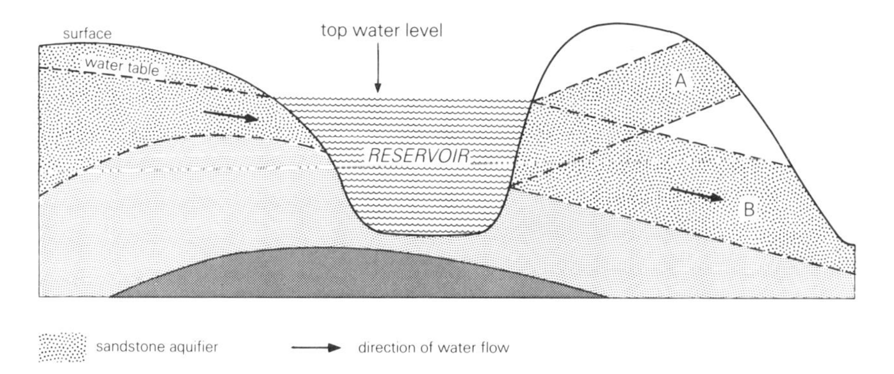

Figure 8.7 A sandstone aquifer (dotted) is a path of leakage from the reservoir if its structure is as represented by B, but not if it dips towards the reservoir as in A. On the left-hand side the water table in the aquifer slopes towards the reservoir and produces flow *towards* it.

ment rather than leakage. In general, dips of the aquifer towards the reservoir inhibit leakage along it. Springs and seepages may be useful in indicating the position of the water table and the movement of ground water around the reservoir. Ideally, a reservoir should be sited on impervious rocks or, if only some layers are impervious, on a syncline plunging upstream. Limestone, and to a lesser extent sandstone with a calcareous cement, may present serious hazards. The calcite dissolves, especially if the water has been in contact with peat and is slightly acidic, and leakage paths progressively widen with time. The rock may become cavernous after several years.

(c) Fault zones with pervious breccia, and open joints, may serve as paths for leakage. If the trend and location of a fracture offer a possible conduit, then its outcrop in the reservoir valley should be sealed with grout. Fractures are particularly hazardous if they affect calcareous rocks, and if a reservoir must be sited on limestone, extra effort and resources must be spent to locate and seal all faults and master joints. In glaciated valleys, incipient plucking of the rock surface by ice has opened fractures to produce **lift joints**. These may be present at depths up to 20 m.

The rocks below and at the sides of a dam should form part of the impervious basin together with the reservoir area, and the same geological considerations about avoiding leakage apply to it, plus some additional criteria. These include the strength of the foundations and the depth to rock head.

Leakage under a dam not only leads to loss of impounded water but may also affect the foundations by the **uplift pressure** of the percolating water, and by erosion where seepage is discharged, on the downstream side of the dam. Progressive erosion of weak rocks or soils backwards along flow paths produces **piping**, which can endanger the dam. The critical hydraulic gradient at which there is a danger of piping is approximately 1 in 1. To lower the gradient to at most 1 in 4 and also reduce leakage, the flow paths from the reservoir should be increased in length by constructing an impervious barrier before, below or behind the dam. This may be an apron of clay on the floor of the reservoir, a curtain of grout injected into the aquifer below the dam, or a weir built on the downstream side. A filter of graded gravel at the outlet for the seepage may be added to prevent washing out of fine material.

Large dams are almost invariably founded on strong solid rock, and the thickness of any cover of soils and the strengths of the rocks at the dam site are important factors in planning and costing the construction and possibly in designing the dam. The additional excavation required at a site over a glacial rock basin, compared with one over a rock barrier which lies a few hundred metres away, may add significantly to the costs. It is important, therefore, to explore the position of rock head within the entire area of choice, and to make a contour map of this. The methods of systematic exploration are described in Chapter 6. Seismic-refraction surveys should be used, combined with borings, to define rock head, and may also be combined with laboratory measurements of specimens to delineate areas of badly fractured or rotten rock. At the later stages of exploration, rocks at the dam site should be inspected *in situ* for weathering and fracture, by digging pits or by boring large-diameter holes. Particular attention should be paid to mapping fractures which might behave as zones of seepage from the reservoir, and as zones of weakness under and at the sides of the dam.

8.2.2 *Case history: leakage from Clubbiedean Dam, Midlothian, Scotland*

An account of a major dam or of a spectacular disaster might be invoked to illustrate these basic geological points, but this can be done more simply by referring to a modestly scaled problem described by Sivasubramaniam and Carter (1969). Clubbiedean Dam is a small earth dam built in 1850 which impounds water for seasonal storage. Leakage had persisted over a long period, with eventually some local collapse and visible free flow into the ground through cracks at the bottom and sides of the depressions. The succession of strata at the dam site (see Fig. 8.8) is as follows.

Carboniferous	Calcareous shales and thin limestones
Old Red Sandstone	Top Sandstone – a soft grey limy sandstone
	Top Marl – calcareous mudstone

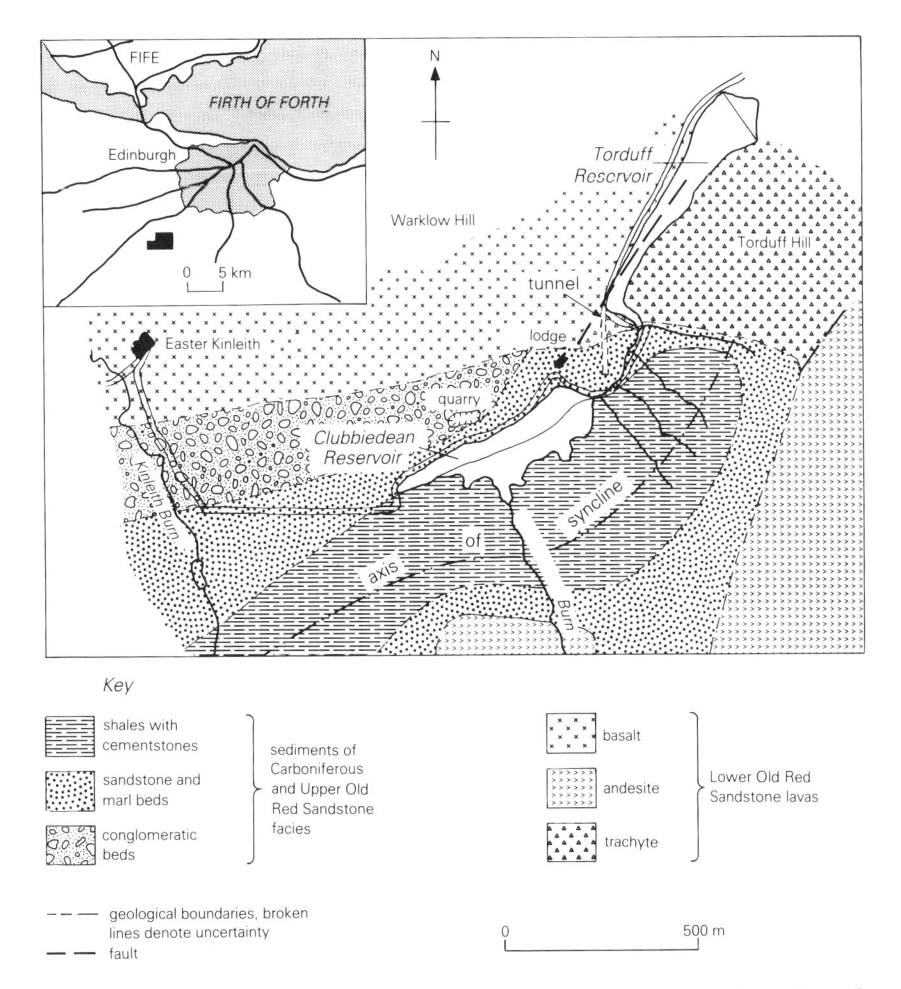

Figure 8.8 Geological map of the area around Clubbiedean Reservoir. (From Sivasubramaniam & Carter 1969, by kind permission of the *Scottish Journal of Geology*.)

Middle Sandstone – soft and permeable near rock head but otherwise well cemented
Middle Marl – calcareous mudstone
Bottom Sandstone – hard pink sandstone
Basal Beds – conglomerate with marl
Lavas – impermeable except where jointed

The impermeable cover of boulder clay on the reservoir site, which might have afforded some protection against leakage, had been mostly stripped away either for use as fill in the dam, or as a result of erosion. The strata are folded into a shallow syncline, and the dam is founded on the northern flank of the syncline.

Leakage and subsidence were due to seepage through the sandstones, which had increased as solution cavities formed and quartz grains were washed away. The Top Sandstone was most affected by cavities, and the Middle and Bottom Sandstones were similarly weakened where they were near the surface. The cavernous zone may have extended to a depth of about 10 m below rock head. During the site investigation by trial pits and bores, there was loss of circulation of drilling water, some sudden drops in drilling rods and poor core recovery.

The treatment used was to inject cement down a line of holes to form a grout curtain under that part of the dam where leakage was taking place. A total of 30 grout holes, some to a depth of 3 m, some to 6 m, and some to 10 m, were used. When the reservoir was refilled, leakage was only 1% of the previous amount.

8.3 Geology of tunnels

8.3.1 Geological considerations in tunnelling

In few engineering projects are the feasibility, the planning, the costing, the design, the techniques used and the risk of serious accidents during construction so dependent on the geology of the site as in tunnelling. The area in which a tunnel is built is determined by its purpose, but the decision to tunnel, rather than, say, build a bridge, is influenced by the relative geological difficulties, and the precise line of tunnel may be determined by a choice of favourable or difficult local geological conditions. The relative ease of extraction of the rocks and the stability of the rock and face are prime factors in rates of progress and setting costs, and also in determining whether a rock-boring machine can be used, whether the ground needs support and whether it is necessary to use compressed air. For example, if a buried channel or deep scour on the sea floor, infilled with saturated sand and gravel, were unexpectedly encountered, the resultant inrush of water at the tunnel face would result in a serious accident.

The geological factors considered in tunnelling projects are as follows:

(a) the ease of extraction of the rocks and soils;
(b) the strength of the rocks and the need to support them;
(c) the amount inadvertently excavated beyond the perimeter of the tunnel outline (that is, **overbreak**), particularly where explosives are used;
(d) the groundwater conditions and the need for drainage; and
(e) the temperature in very deep tunnels and the need for ventilation.

The amount or degree of change in these conditions along the tunnel line can be as important in planning and costs as their average or maximum values. The change is related to structure, which controls not only what type of rock is present in a particular segment of the tunnel but also how its layering and other anisotropic properties are orientated relative to the tunnel face, and how weakened it is by fracture. Ideal geological conditions for excavation of a tunnel are to encounter only one type of rock, devoid of fault zones or intrusions, which is easily excavated but is stiff enough not to need immediate support near the face, which is impermeable and which is not adversely affected on exposure to air. By a fortunate chance, large areas of London are underlain by a layer (the London Clay) which approaches this ideal, and the most extensive system of tunnels in Britain (the London Underground and the London sewerage tunnels) are excavated partly in it. Uniform geological conditions allow a uniform, if moderate, rate of progress without the troublesome and time-consuming need for changes of techniques, imbalances in the effort required from different work sections, and elaborate and vulnerable arrangements.

8.3.2 Methods of excavation

The main problem in constructing a tunnel through non-cohesive soils or weak ('soft') rocks is to support the ground rather than to excavate it. Excavation is usually done by a soft-ground tunnelling machine fitted with a rotary cutting head. This may have a full-face rotary breasting system which remains in contact with the soil face as the cutter head moves. Small slices of soil are fed through slots into the cutter head. The working face is supported by compressed fluid, which may be compressed air either in the tunnel or, where a complex machine is used, restricted to the face area by a pressure bulkhead. The older method of having compressed air in the tunnel itself carries the risk of disabilities to workers, and requires time spent unproductively at the end of each shift on decompression. In recent successful developments, a slurry of mud and water with thixotropic clay added is used at the face instead of air. The clay impedes settlement within the slurry, and tends to form a sealing cake on the face. As the machine advances, supports are installed behind it. The type used depends on the particular geological conditions.

The principal factor controlling rate of progress and costs in constructing a tunnel in strong ('hard') rocks is usually relative ease of excavation. The traditional method is to blast out successive sections of the tunnel by drilling a pattern of holes in the rock, charging them with explosives and firing. The need for any support and the type of support used are dependent on the relative stability of the roof and walls of the tunnel. For example, widely spaced rock bolts and wire mesh might be used as a shield for small loose fragments, and closely spaced ring beams might be employed where there was a danger of a rock fall. This approach is still used for the majority of modest projects of underground excavation, such as short, large-diameter rail or highway tunnels. Since the early 1950s, however, the use of explosives has been increasingly replaced by rock-boring machines for certain types of *major* tunnelling projects. The machines are heavy and expensive and may be designed specially for one major tunnel. Machines equipped with special cutters containing closely spaced tungsten carbide inserts seem to be capable of tackling rocks with compressive strengths in excess of 300 MN m^{-2}. The limiting factor in design is the strength of the cutter edge. The cutters usually wear out but, if they meet a hard block of rock during boring, they may smash on impact. Their rate of wear and replacement is an important economic factor in tunnel costs.

8.3.3 Complications arising from local geological conditions

In soft-rock tunnels, heterogeneous rock or variable conditions at the face can produce serious problems and add to costs. A boulder clay or other soil containing large pebbles creates a near-impossible problem for existing slurry-face machines. Hard-rock rolling cutters can cope with hard boulders but have difficulty in other soft soils. A mixed face of hard rock and wet non-cohesive soil presents an even worse headache to the engineer. The variation in strength of the soils along the tunnel line should ideally be anticipated so that appropriate support can be used while the face is being excavated. Inability to do so can result in overexcavation. Apart from the obvious variations of strength among soil types (for example, between non-cohesive sand and partly consolidated clay), variation related to saturation and porosity may produce significant differences. A small difference in water content can change an otherwise stable soil into running ground. Unstable soils can be consolidated by injecting chemicals or cement into them, or by freezing them.

In hard-rock tunnels, the relative difficulty of excavating particular rocks depends partly on whether explosives or a rock-boring machine is used. Nevertheless, both methods share some important factors. The rate of excavation in both cases is inversely related to the crushing strength of the rocks, and directly related to the amount of fracturing. When explosives are used, the relationship to strength is complicated by the way in which some weak non-brittle rocks, such as mica-schist, react to blast, and do not *pull* well for a given charge; and by the much greater role that fracturing plays. Fractures serve both as paths for expanding gases from the explosion, and as planes of weakness along which the rock will part. In tunnelling, the ease of drilling shot holes is dependent on the hardness and abrasiveness of the rock face (Section 7.3.1), and also on the variation of hardness within it, since, at a sharp boundary between hard and soft, the drill tends to be deflected. The hard mineral most likely to give any trouble is one of the varieties of silica, such as quartz, flint or chert, occurring as veins or nodular concretions. Shales containing ironstone nodules can also be an awkward mixture. Relatively hard minerals and strong rocks are often produced by thermal metamorphism. A weak, soft, calcareous schist may be altered to a strong, hard calc-hornfels. This has proved to be a significant geological factor in certain hydroelectric projects where the reservoir is sited on the high ground corresponding to the outcrop of a large granite intrusion. Tunnelling within the thermal aureole tends to become increasingly difficult as the granite is approached.

Excessive ease of extraction because of overhelpful planes of weakness can lead to overbreak (Fig. 8.9) and to potential rockfalls from the roof. A certain amount of excess excavation between the ideal outer surface of the lining and a pay line is usually covered by contract, but allowance for treating any overbreak beyond that is likely to be borne by the contractor and should be allowed for in his tender. Overbreak depends on the intensity of jointing and the presence of other planes of weakness, such as bedding and schistosity, and also on the orientation of the dominant planes relative to the tunnel face. In general, massive uniform rock, properly blasted, will give a clean section; well bedded, fractured rock will give overbreak. This is greatest in steeply dipping layers where the

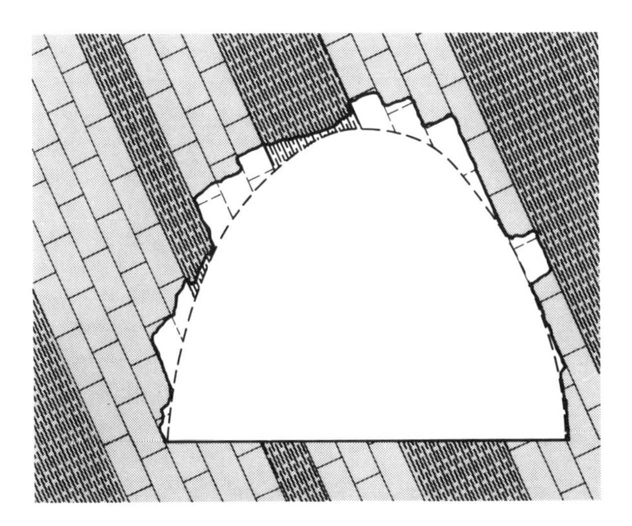

Figure 8.9 The ideal tunnel outline is shown by the heavy broken line. The strata to be excavated are alternating shales and well jointed limestones, dipping steeply and striking parallel to the tunnel. The actual amount pulled out by explosives along the planes of weakness is indicated by the heavy black line. This excess rock, excavated beyond the perfect section, is called overbreak.

tunnel is parallel to their strike. For example, the overbreak in mica-schists was observed to vary in one tunnel from values of 16% across the foliation to 40% along it.

Predicting instability of the roof and the need to support and backfill parts of a tunnel are problems common to excavation both by explosives and by a rock-boring machine – as are seepage, water-filled soils in channels or fissures, swelling and popping rock, and high temperatures.

Excessive overbreak and the risk of rockfalls from the roof are most likely in the following conditions:

(a) at fault zones, especially if they contain loosely cemented breccias;
(b) at dykes, narrower than the tunnel, which have well developed horizontal cooling joints;
(c) at synclinal axes, as the tensional joints developed at the hinge of the fold diverge towards the tunnel, and the blocks are not supported;
(d) at layers of loosely compacted fragmental rocks, such as some volcanic breccias; and
(e) where thin layers of strong and weak rocks (for example, alterations of limestone and shale) are present at the level of the roof, or strike along the tunnel and have a steep dip.

Where a tunnel is likely to follow such a zone of weakness for an appreciable distance, consideration should be give to realigning it. This may have to be done after excavation starts, if exploration has not revealed the structure. For example, the location of every minor dyke in a deep tunnel in a mountainous area cannot always be anticipated, but the trend of the dyke swarm is probably known and may be used to assess the extent to which the dyke and the tunnel line are likely to coincide.

The amount of seepage into a tunnel through pervious rocks and joints may be a significant factor in design. It must be assessed from a knowledge of groundwater conditions, bulk permeabilities of rocks and the geological structure. For example, crystalline rocks, such as granite, gneisses and schists, are typically dry except for flow along joints and faults, and perhaps at the margins of any dykes which cut them. In pervious rocks, the flow of ground water into the tunnel is likely to increase at fault zones and at synclinal axes (which tend to channel flow towards the tunnel). Water-filled fissures, especially in limestones, present a more serious hazard, which must be insured against in certain projects by probing ahead of the working face with small horizontal boreholes. The consequences of driving the tunnel into the wet soils of a buried channel cut into the solid rock are so serious that exploration must be thorough enough to delineate rock head everywhere along most tunnel lines, particularly under the sea, under estuaries and under rivers. Depressions in the rock-head surface may be preglacial channels or trenches scoured by strong tidal currents, or large fissures leached in limestone, which have subsequently been infilled with Pleistocene or recent sediments. In areas which have not been glacially eroded, the bedrock near the surface may be altered by periglacial or other weathering, and consequently have a permeability and strength intermediate between its values when fresh and those of any residual soils formed from it.

The temperatures in a tunnel prior to ventilation are related mainly to distance from the surface, the thermal conductivity of the rocks and the amount of water circulating through them. It is a significant factor only in very deep tunnels. Any sudden drop in temperature at the tunnel face is usually an indication that heat is being drawn from the rocks by circulation of water close to the face, and it may serve as a warning of a water-filled fissure and the threat of a sudden inrush. Inflow of warm water occurred in tunnels through the Alps.

At depths below about 150 m, the pressure of overburden is augmented locally by appreciable secondary stresses in some parts of the excavation. The resultant strain in the rocks around the tunnel may take the form of **swelling** into the tunnel in shales, or some movement along bedding or other planes of weakness in brittle rocks. It may, however, produce explosive scaling from the surface of massive brittle rocks which cannot make such adjustments. This is called a **rock-burst** or **popping rock** and it can be hazardous. It is probably the cause of many of the minor tremors generated as mining subsidence takes place.

The prediction of ground conditions and how they vary along the line of a tunnel follows the general lines of site exploration described in Chapter 6, with appropriate emphases. Zones of weakness and permeability – particularly fault zones, buried channels, lenses of waterlogged running sand, and dykes – must be delineated with especial care. The rocks in many major tunnelling projects are covered by water, and continuous seismic-reflection profiling can be used to advantage. There is always need for a line of borings near the proposed tunnel, to prove the geology and provide samples for physical measurements. As a final check, test drilling may be done ahead of the working face by driving small holes outwards from the face parallel to the tunnel axis to confirm the prognosis and to ensure that no hazards are encountered unexpectedly. These points are illustrated in the case history described in the next section.

Any need to change the method of excavation or support because of an unforeseen change of geological conditions can be a difficult and costly decision for a contractor. The accurate prediction of all such changes before excavation starts is more than can be provided by present exploratory methods. Some geological complications and hazards only become known as tunnelling proceeds. For these reasons, it is very important that the fullest possible appraisal of the geology is made, and that exploration is *not* thought to be finished once excavation has started.

8.3.4 Case history: the proposed Channel Tunnel

The project to build a tunnel under the English Channel and link England to France by rail was shelved as one of the cuts made in British public expenditure in the mid-1970s. This was done, however, at a stage when the geology of the site had been investigated and assessed by members of the Channel Tunnel Study Group as part of a comprehensive study and estimate of costs (well over £1000 m at present). The project seemed reasonably sound in terms of technical feasibility, and the prospects were encouraging and realistic enough to attract some private capital. The following account is drawn mainly from Destombes and Shephard-Thorn (1971), and from an account by Hanlon (1973). The geological investigations and tests were carried out in 1964–65.

Where the Channel narrows, Cretaceous Chalk crops out on both coasts, and its low dips suggested that it might be continuous at the level of the tunnel from one end to the other. The geology below the Channel for some distance on both sides of the shortest route was reconnoitred using shallow seismic-reflection profiling (sparker). A pattern of lines, trending NE–SW at 0.5 km intervals, were linked by 12 tie lines to give several hundreds of kilometres of continuous profile. As a control on the interpretation of these lines, and to provide samples for testing, 73 boreholes were made in the sea floor and, in all, 6 km of cores were recovered. Microfossils from them were used to recognise finer divisions within the Chalk, and so assist in delineating the detailed structure of the layering. Laboratory tests of permeability, water content, density and compressive strength were made, plus some others related to the problems of this particular tunnel; for example, ultrasonic velocity determinations for detailed sparker interpretation, variation in clay and calcite content, static strain modulus and slaking tests.

The investigations showed that the Chalk is folded gently into one major anticline and several lesser folds, all trending east–west. The maximum dips, up to 15°, are found on the French side. One fault with a throw of only 12 m was mapped, and the few others detected were even smaller. Tests in boreholes adjacent to some of them showed no significant increase of bulk permeability, and it is inferred that they are sealed by later mineralisation or by clay. They do not represent a serious hazard. The sea floor is swept by tidal currents and superficial deposits are generally thin. Sand, migrating slowly in giant sand waves, is present near the French coast. There are also several deep hollows cut into the bedrock, which are filled with Pleistocene silt and sandy clay. The largest in mid-Channel extends to 140 m below sea level. Fortunately, the Chalk below them is unaffected by periglacial weathering and there is more than 25 m of sound rock above the tunnel along its entire route. This is considered to be an adequate safety margin. The lack of weathering of the Chalk in the hollows indicates that the infill of drift was already present as a protective cover before the last glaciation (an alternative explanation, which is possible but less likely, is that some of the hollows are scours cut by strong tidal currents, which have been filled with recent sediment).

The lowered sea levels of Pleistocene times left the present sea bed of the Channel exposed to periglacial weathering, which has increased the permeability and lowered the strength of the Chalk by breakage and fissuring. The seismic velocity in the weathered Chalk is up to 30% less than in fresh rock. (This effect must be allowed for in interpretation of the sparker profiles.) The white Upper Chalk is the only layer seriously affected. The penetration of the weathering, particularly along joints, is controlled partly by the minor folds. The badly weathered zones can admit large amounts of water to lower levels to produce serious and unexpected water problems, but the tunnel line is sited safely below all known hazards of this type.

The permeability of chalk is inversely related to its clay content. A Channel Tunnel cut in chalk consisting of more than 80% calcite would probably suffer serious seepage, and this limit was a vital criterion in selecting which layer of the Chalk the tunnel should follow. The Chalk Marl, a layer up to 30 m thick within the Lower Chalk, has a suitable permeability combined with adequate strength to support the forward area until lining can be installed. It consists of a series of units, each 0.3–0.7 m thick, with clay-rich rock at the base grading upwards into purer, harder chalk. The ribs of chalk give strength and the clay acts as a sealant. The permeability of the matrix of the Chalk Marl is nearly zero across the bedding, and the effective bulk permeability *in situ* is entirely due to fissures. Assessments of strength are borne out by the fact that older trial tunnels excavated in the 1880s are still standing although they are unlined. The ease of excavation of the Chalk Marl by rotary cutting machines is shown by results of 30 m per day during trials.

It is possible to route the tunnel through Chalk Marl for most of its length, and the main geological complications are at the coasts. The low dips on the English side give considerable freedom of choice horizontally in siting terminal facilities, but the steep dips and narrower outcrops on the French side are more restrictive, unless a geological penalty is accepted. For example, the tunnel should preferably avoid the water-bearing sandstones that occur below the Chalk.

Despite the careful preliminary investigation and its favourable results, there is still a possibility that some water-filled fissures (joints or bedding-planes) have not been delineated by the survey, and indeed cannot be recognised by present techniques with certainty. They present a potential hazard which must be insured against by stipulating that small-diameter holes should be bored ahead of the tunnel face to probe the rocks ahead. These holes would be fitted with valves so that they could be sealed and, if necessary, grout could be pumped into any pervious zone they penetrated, so as to seal it. The cost of this procedure would be determined largely by time delays when drilling of the main tunnels was halted.

The proposed design includes two main traffic tunnels and a central service tunnel. The main tunnels were costed on the basis of a diameter of 6.85 m, which would accommodate lorries at least 4 m high loaded on rail transporters. The service tunnel would be advanced first, and borings could be made sideways from it to probe conditions along the sites of the two main tunnels. The net progress, including lining, was projected to 0.5 km per month for the service tunnel, and 0.75 km per month for the main tunnels. These estimates are based on an aimed excavation rate of 6 m per hour by the rotary boring machines.

The Study Group also reported on the geological factors relevant to an alternative 'immersed tube' scheme, in which an open trench would be excavated in the Channel floor, and special caissons, which would serve as segments of the tunnel, would be sunk into it. The topography of the sea bed is the most important factor in the choice of route and the feasibility of this scheme. The principal geological consideration is ease of excavation of the trench. It was concluded that no particular difficulties would be likely where Chalk cropped out, but that the silt in the buried channels would be liable to collapse into the trench during construction. Mobile sand waves would also be a problem locally. Otherwise there appeared to be no serious problem of silting up of the trench by the movement of recent sediments.

References and selected reading

Anderson, J. G. C. 1970. Geological factors in the design and construction of the Ffestiniog pumped storage scheme, Merioneth, Wales. *Q. J. Engng Geol.* **2**, 183–94.

Attewell, P. B. and I. W. Farmer 1976. *Principles of engineering geology*. London: Chapman and Hall.

Bishop, A. W. 1973. The stability of tips and spoil heaps. *Q. J. Engng Geol.* **6**, 335–76.

Carter, P. and D. Mills 1976. Engineering geological investigations for the Kielder tunnels. *Q. J. Engng Geol.* **9**, 125–42.

Destombes, J. P. and E. R. Shephard-Thorn. 1971. *Geological results of the Channel tunnel site investigation 1964–65*. Rep. Inst. Geol. Sci. 71/11.

Fookes, P. G. and M. Sweeney 1976. Stabilization and control of local rock falls and degrading rock slopes. *Q. J. Engng Geol.* **9**, 37–56.

Geological Society Engineering Group Working Party 1977. The description of rock masses for engineering purposes. *Q. J. Engng Geol.* **10**, 335–88.

Hanlon, J. 1973. Bore of the century? The Channel tunnel. *New Scientist* **60**, 92–110.
Hoek, E. 1970. Estimating the stability of excavated slopes in opencast mines. *Trans. Inst. Min. Metall.* **79A**, 109–32.
Hoek, E. 1973. Method for the rapid assessment of the stability of three-dimensional rock slopes. *Q. J. Engng Geol.* **6**, 243–56.
Hoek, E. and J. W. Bray 1974. *Rock slope engineering*. London: Inst. Min. Metall.
Hutchinson, J. N. and D. Brunsden 1974. Mudflows: a review and classification. *Q. J. Engng Geol.* **7**, 327–8.
Little, A. L. 1977. Investigating old dams. *Q. J. Engng Geol.* **10**, 271–80.
Londe, P. 1973. Analysis of the stability of rock slopes. *Q. J. Engng Geol.* **6**, 93–124.
McGregor, K. 1968. *Drilling of rocks*. London: McLaren.
Robbins, R. J. 1976. Mechanised tunnelling – progress and expectations. *Trans Inst. Min. Metall.* **85A**, 41–50.
Sivasubramaniam, A. and A. V. F. Carter 1969. The investigation and treatment of leakage through Carboniferous rocks at Clubbiedean Dam, Midlothian. *Scott. J. Geol.* **5**, 207–23.
Stevenson, P. C. and W. R. Moore 1976. A logical loop for the geological investigation of dam sites. *Q. J. Engng Geol.* **9**, 65–72.
Wahlstrom, E. E. 1973. *Tunnelling in rock*. Amsterdam: Elsevier.
Walhstrom, E. E. 1974. *Dams, dam foundations, and reservoir sites*. Amsterdam: Elsevier.
Walters, R. C. S. 1962. *Dam geology*. London: Butterworths.
Watkins, M. D. 1970. Terminology for describing the spacing of discontinuities of rock masses. *Q. J. Engng Geol.* **3**, 193–5.
Weeks, A. G. 1969. The stability of natural slopes in south-east England as affected by periglacial activity. *Q. J. Engng Geol.* **2**, 49–62.
Wood, A. M. M. 1972. Tunnels for roads and motorways. *Q. J. Engng Geol.* **5**, 111–26.
Zaruba, Q. and V. Mencl 1969. *Landslides and their control*. Amsterdam: Elsevier.
Zaruba, Q. and V. Mencl 1976. *Engineering geology*. Prague: Academia.

Appendix A *Descriptions of some important soil groups*

Group	Equivalents	Climate	Weathering	Vegetation	Thickness	Description
(a) Organic rich soils (found in wet, acidic environments)						
tundra soils	tundra, arctic brown soils	freezing, humid	organic decomposition slowed by extreme cold; organic and mineral matter mixed by frost action	lichens, moss, herbs, shrubs	0.5 m	dark-brown peaty layer over grey layers or brown subsoil which are permanently frozen; tundra soils on poorly drained ground and arctic brown soils on well drained higher ground
alpine meadow soils		cold, humid	similar to above but frost action less severe	similar to above	0.5 m	similar to tundra soils, but ground not frozen permanently and not so wet
peat and bog soils	histosols, sols tourbeaux	any (not arid)	organic decomposition slowed by excessive moisture	swamp forest, variable grassland	>1 m	brown, dark-brown or black peaty material over peat or mottled soils of mineral matter; a clayey gley layer may develop

(b) Forest soils (found in acidic environments with alkalis removed)

(light) podzol soils	spodosols	cold, humid, 500–1000 mm rainfall per annum	Fe and Al removed by acid leaching from A horizon and deposited in B horizon	evergreen (spruce, fir) forest	0.5 m	tree remains over a humus-rich O layer containing mineral matter; A leached layer is white–grey sandy; B (deposition) is brown-stained and clayey
(dark) podzol soils	sol brun acide, brown earths	temperate, humid, 750–1000 mm rain-fall per annum	free drainage, rapid runoff	deciduous with some conifers	0.75 m	leaf layers over humus-rich O layer with some mineral matter; A leached layer is brown or greyish brown containing clay; B layer is stained darker brown and clayey, and can be an acidic hardpan
brown forest soils		cool, temperate, 750 mm rain-fall per annum	leaching retarded by high calcium carbonate content	mixed deciduous and conifers		layering obscured; organic layer grades downwards through lighter layer into carbonate-rich parent material
red and yellow podzol soils		warm temperate to tropical, humid	deeply weathered clayey parental material	mixed deciduous and pine evergreens	solum (O, A and B layers) about 1 m; C layer may be many tens of metres thick	thin dark O layer; A leached layer is yellow, grey or brown; B layer is dark clayey; grading down into deeply weathered; C layers brightly coloured, mottled red, yellow, purple, brown, white and grey

Appendix A continued

Group	Equivalents	Climate	Weathering	Vegetation	Thickness	Description
groundwater podzol soils		cool to tropical, humid	strong acidic leaching with organic material transferred to water table	pine forest	1 m	thin O layer; A leached layer is light coloured and sandy; B dark-brown wet layer with organic material overlying water table
(c) Grassland soils (mollisols) (transitional – nearly neutral environments)						
brunizem soils		cool to warm temperate, humid, 600–750 mm rain-fall per annum	weak acid leaching	tall grass (prairie type)	1 m	dark-brown surface soils overlying well oxidised subsoils; grades into light-coloured parent material with no carbonate accumulation
planosols		cool to warm temperate, humid, 600–750 mm rain-fall per annum	acid leaching with development of gleys	grass, some forest (flat ground)		leached surface soils overlying clay-bearing parent material
rendzina soils (calcimorphic soils)	calcareous rendolls	variable	acid leaching slowed by carbonate in soil	grass	variable	very dark-coloured O layer overlying soft, light-coloured calcareous material from parental material such as limestone, marl or chalk

Groups (d) to (i) occur in alkaline environments with alkalis retained						
(d) Chernozems						
	black earths, mollisols, chestnut-brown soils	temperate to cool, sub-humid, 500–600 mm rainfall per annum; chestnut-brown soils require <500 mm rainfall per annum	carbonate accumulation in lower layers	grassland or cultivated	variable	thick, friable dark soils grading downwards into lighter layer and finally into layer of calcium carbonate accumulation
(e) Sierozems						
	desert soils, red desert soils	arid, <250 mm rainfall per annum		sparse shrubs	<0.5 m	red to brown soil with very thin organic layer; carbonate layer very close to surface
(f) Halomorphic soils						
	solonchak, saline, white alkali, sodic soil, solonetz, solods, soloths, soloti	arid to semi-arid but may occur in temperate regions with saline ground	restricted with poor drainage	salt-tolerant species		accumulated salt near ground surface with more than 0.2% salts present; solonetz (black alkali) with less soluble salts – sodium carbonate and sulphate; solonchak (white alkali) with soluble chloride salts (a clay-rich B horizon may be present, especially in solonetz types)

Group	Equivalents	Climate	Weathering	Vegetation	Thickness	Description
(g) Immature soils						
	lithosols, skeletal soils, regosols, entisols	variable	poor	none		these soils depend entirely on parental material
(h) Black tropical soils						
	black cotton soil, vertisols, margalitic soils	tropical monsoon or equatorial	restricted	grassland, savanna	variable, often very thick	occur in soil catenas with black iron-rich soils in depressions; A horizon contains clays that crack and swell with shrinkage and have poor foundational properties
(i) Gleys						
	aquents, aquepts, some meadow soils	dependent only on groundwater conditions	restricted	moisture-tolerant forest, grassland, semi-marsh	variable	brown or grey–brown top layer on grey mottled gley horizon; parent material often impermeable clay-rich rocks
(j) Laterites						
	krasnozems, ferralites, ferruginous soils, latosols, oxisols, ultisols	equatorial to sub-tropical, high rainfall	free drainage, intense surface leaching with permanent waterlogging at depth	forest, thorn, savanna soils are very poor	very thick	red layers with high clay content; matrix often aggregated into concretions (murram) by high alumina and iron content

Appendix B *Hydraulic properties and pumping tests of an aquifer*

The important hydraulic properties of a soil or rock are:

(a) the amount of water it can hold in its voids;
(b) the amount that can be drained from it; and
(c) the ease with which the water can flow through it.

These may be expressed as its porosity, specific yield and coefficient of permeability. These parameters are defined elsewhere (Section 5.1.1). In hydrogeological work, including water supply, it has been conventional, however, to describe the hydraulic properties of an aquifer by other, equivalent parameters. For example, permeability is expressed in practical terms as discharge per unit of time, rather than as velocity of flow. This hydraulic conductivity is usually designated K rather than k. It is expressed as the discharge ($m^3\ day^{-1}$) through a cross-sectional area of $1\ m^2$ under a hydraulic gradient of 1 in 1.

The ability to permit flow through voids may also be expressed by the aquifer's transmissibility, which is the rate of flow of ground water ($m^3\ day^{-1}\ m^{-1}$) through a vertical strip of aquifer 1 m wide extending the full length of the saturated aquifer under a hydraulic gradient of 1 in 1.

The storage capacity of the aquifer is defined by its coefficient of storage (S). It is the ratio of the volume of water *derived from* storage in a vertical column extending through the full thickness of the aquifer to the corresponding volume of the aquifer, expressed as a decimal. The water is assumed to be yielded during a reduction in head equivalent to a fall in water level of 1 m, and the column rests on a base of $1\ m^2$.

While a well is discharging, water is being drawn from the rocks around it to produce a **cone of depression** (**exhaustion**) in the water table, or a **cone of pressure relief** in confined water, with a matching cone-shaped depression in the piezometric surface. Hydraulic gradients are produced radially inwards around the well and induce flow into it. The dimensions of the cone of depression (Fig. B.1) formed by a well being pumped from a depth H below the original water table, in an aquifer of permeability k, are related to the quantity Q yielded under steady-state flow conditions by the equation

$$Q = \pi k[(H^2 - h_0^2)/\ln(R/r)]$$

where H, h and R are dimensions of the cone of depression, shown in Figure B.1a, and r is the radius of the well (ln is the conventional symbol for $\log_e$). The equation is a useful approximation if the aquifer is homogeneous, isotropic and of much larger extent than the cone, and if the slope of the cone is less than 30°. It assumes that the well penetrates the full section of the aquifer. This equation describing what happens in an unconfined aquifer when a single well is pumped can be used to predict the extent of de-watering of ground prior to excavation or to laying foundations. The corresponding equation for a confined aquifer (see Fig. B.1b) is

$$Q2\pi kd[(H - h_0)/\ln(R/r)]$$

Conversely, if the gradients of the cone of depression are studied using observation wells (in Fig. B.2), the permeability k of the aquifer around the

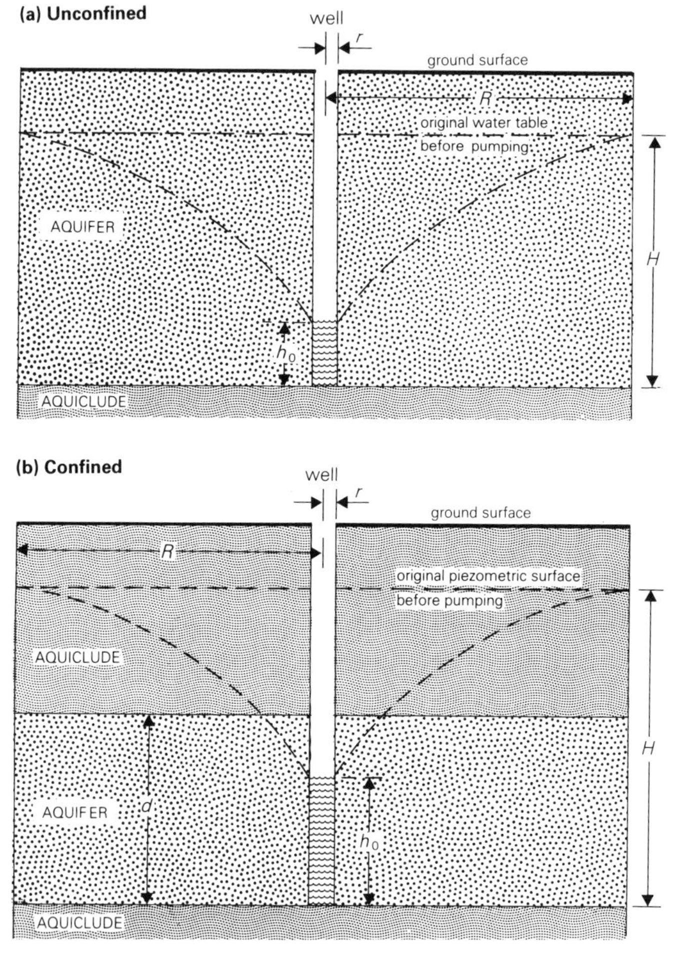

Figure B.1 (a) A cone of depression in the water table has been produced by pumping from the well until steady-state conditions are attained. The saturated rock below the cone is shown by dots. The other symbols are defined in the text. (b) The aquifer of thickness d is confined by an aquiclude above it. Its original piezometric surface has been depressed around the well, and the diagram shows its form when steady-state conditions have been attained. (The head at the well has dropped to a level where part of the aquifer is unsaturated.) The other symbols are defined in the text.

pumped well can be determined. Figure B.2 shows the form of the water table during a pumping test that has lasted long enough (more than 24 h) for steady-state conditions to be reached, that is, where the volume of water Q being discharged per day is constant:

$$k = \frac{Q}{\pi(Z_1^2 - Z_2^2)} \ln(r_2/r_1)$$

The symbols are defined in Figure B.2.

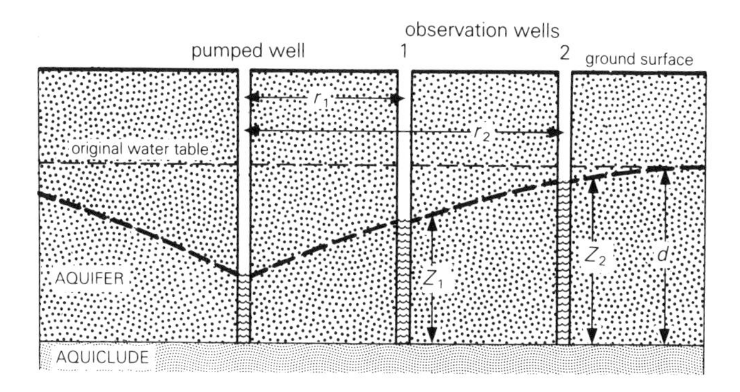

Figure B.2 A pumping test to determine permeability (k) of an unconfined aquifer has lowered the original water table around the pumped well to levels Z_1 and Z_2 in the two observation wells. The relationship between the permeability of the aquifer and these parameters of the cone of depression is given in the text.

Appendix C *The British Geological Survey and other governmental Geological Surveys*

A general account of the British Geological Survey, and the Divisions and Units within it are comprehensively listed in the *Annual Reports of the BGS*. Up to 1983 the British Geological Survey (BGS) was known as the Institute of Geological Sciences (IGS). Field investigations for each region of the UK (for example, Scotland North Lowlands District) are carried out by a team of geologists under the supervision of a District Geologist. In addition, special investigations are the responsibility of a number of specialist units.

The national headquarters are at:

British Geological Survey,
Nicker Hill,
Keyworth,
Nottingham NG12 5GQ

Field staff at this office cover southern and central England. Local offices at Exeter, Newcastle and Aberystwyth deal with south-west England, northern England and Wales.

There is a major regional office at:

Murchison House,
West Mains Road,
Edinburgh EH9 3LA

Field staff here cover Scotland, Northern Ireland (which also has a local office in Belfast) and the northern part of the UK Continental Shelf.

Information about subsidiary offices (for example, in Belfast and Exeter) may be obtained from these two centres. Local data, including unpublished 1/10560 field maps, are deposited in local offices. Some specialist Units have their offices at Keyworth (including the Engineering Geology Unit).

The objects of the BGS include the following:

(a) Preparation and keeping under revision the geological map of Britain and of its surrounding seas. The standard scale of survey on land is 1:10560 or 1:10000. Maps are published and sold of those areas where mining, special academic interest or other activity produces a demand. The price (in 1984) of a 5×5 km^2 map on a scale of 1:10560 was £8.45 from HMSO or its agents. A complete set of maps, including those not available from HMSO, may be consulted at the appropriate BGS office by arrangement. Photocopies of unpublished or out-of-print 6 in sheets may be obtained for £8.05 (in 1984) for National Grid sheets and £5.75 (in 1984) for County sheets (that is, the older 2×3 mile^2 sheets indexed by county boundaries). A range of geological maps, some on smaller scales, others giving detail of certain information such as the nature of superficial deposits is also published.

(b) Investigation by the methods of structural interpretation, geophysics and boring of the subsurface geology of the UK and the British continental shelf. The activities of the specialist Units include geophysical surveys on land and sea and the commissioning of deep borings at localities of special interest.

(c) Conservation, correlation and interpretation of geological information made available under statutory and other arrangements, derived from mining, quarrying, other excavations, boring and prospecting work in Britain and its continental shelf. By law, the logs of all mineral borings penetrating more than 100 ft (30 m) and water borings more than 50 ft (15 m) below surface in the UK must be deposited with the BGS, where they are held for five years on a confidential basis if so requested. This information, excluding that which is confidential, is available to enquirers on an *ad hoc* basis by arrangement. Because of the inclusion of material with different degrees of confidentiality, it is not normally possible for complete files of bore records to be placed at the enquirer's disposal. A separate statute covers data from surveys carried out by the petroleum industry in British waters. Geophysical and other data are deposited with the two Continental Shelf units and the Marine Geophysics unit, and are not immediately available; but under the Petroleum and Submarine Pipeline Act (1975), the Department of Energy is empowered to release certain categories of data, including well logs, six years after the granting of an Exploration Licence. Enquiries should be made to the Petroleum Directorate, Department of Energy, Thames House South, Millbank, London SW1, or to the Continental Shelf Division, British Geological Survey, Keyworth.

(d) Application of the results to practical objectives. The BGS provides information and advice to private commercial parties by publication and on a personal basis. The type and quality of geotechnical information that might be made available from file, for certain urban areas, is indicated by the Special Engineering Geology Sheet, *Geology of Belfast and District*, 1971, on a scale of 1:21120 (3 in to 1 mile). A similar map of Newcastle was published in 1976, and more recent ones include the *Firth of Forth*. The Belfast map contains a contoured map of rock head, contoured thicknesses of the estuarine clays (the most troublesome 'soil' in the area for an engineer), and a table of the geotechnical properties of the rocks. An account of the aims in making the map, and of its use *for preliminary investigations only*, may be found in Wilson (1972).

Advice given and work done may be on a payment basis if there is a significant use of staff's time or facilities. Charges may be made if work exceeds half a day of staff time, but this would be quoted and agreed in advance.

The BGS is, however, unable to undertake a consultant role which might result in legal controversy, that is, where it represented a private client in opposition to, or in competition with, another. Its units undertake investigations and other consultative work for Government departments, National Boards and local authorities, in some cases automatically under statute, in others by special contract. The Ministry of Transport refers all road projects within Britain costing over £1m to the BGS for geological appraisal.

Activities of the BGS related to engineering projects have included:

(a) studies of a prospective site for a third London airport, and of a reservoir storage scheme in the Wash, for the Department of the Environment;
(b) interpretation of marine boreholes for the Channel Tunnel project;
(c) preparation of maps of drift lithology and thicknesses at Irvine New Town;
(d) study of a zone of instability in the Severn Gorge for the Telford Development Corporation;
(e) work on the stability of pit tips in Kent for the NCB; and
(f) assessment of potential sites for toxic and nuclear waste disposal.

A fuller picture of the practical contribution of BGS is given in Robbie (1972).

The organisation and statutory responsibilities of government Geological Surveys in other countries are broadly similar to those of the BGS. Obvious major differences exist for historical, geographical and political reasons. For example, in some of the English-speaking countries with federal systems of government, including the United States, Canada and Australia, Geological Surveys funded by, and accountable to, individual States exist side by side with a Federal Geological Survey. The relationship and division of function between State and Federal Surveys vary from country to country. In general, the State Surveys tend to devote their resources to more practical tasks related to economic and other interests of the State. A Federal Survey's responsibilities characteristically include the systematic surveying of the entire country and the publication of a uniform series of geological maps with a standard stratigraphic nomenclature. Since the three countries instanced have areas several times the size of Britain, the basic scale of mapping is smaller to permit completion of the reconnaissance survey in a reasonable time. Detailed surveys are also undertaken for special, usually economic, reasons. Government Geological Surveys in Canada and Australia have traditionally had a stronger economic bias, particularly as a service to their mining interests, than the BGS, but this difference has narrowed in the last decade.

An account of government Geological Surveys in Australia is given by Johns (1976). There are Geological Surveys of Victoria, New South Wales, Queensland, Tasmania, South Australia, Western Australia and the Northern Territory. Nearly all of them pre-date the foundation of the Federal Survey in 1911. This is one of the five branches of the Bureau of Geology, Geophysics and Mineral Resources. Its activities include systematic mapping of the continent on a basic scale of 1:250000, and the exploration of the continental shelf. Its headquarters are in Canberra. State and Provincial Geological Surveys also exist in the United States and Canada, but they are relatively small compared with the Federal Survey, the US Geological Survey and the Geological Survey of Canada. The headquarters of the former are in Washington, and some of the important activities are based at Denver and at San Francisco. Descriptions of the Geological Survey of Canada and its responsibilities are available in Blackadar (1976), Lang (1970, pp. 270–2) and Zaslow (1975). The headquarters are in Ottawa.

References and selected reading

Bazley, R. A. B. and P. I. Manning 1971. *Belfast*. Sp. Eng. Geol. Sheet 1 : 21 120, Inst. Geol. Sci.

Blackadar, R. 1976. *The Geological Survey of Canada, past achievements and future goals*. Ottawa: Geological Survey of Canada.

Dearman, W. R., M. S. Money, R. J. Coffey, P. Scott, and M. Wheeler 1977. Engineering geological mapping of the Tyne and Wear conurbation, N.E. England. *Q. J. Engng Geol.* **10**, 145–68.

Johns, R. K. (ed.) 1976. *History and role of government Geological Surveys in Australia*. South Australia: A. B. James, Govt Printer.

Lang, A. H. 1970. *Prospecting in Canada*. Ottawa: Geological Survey of Canada.

Robbie, J. A. 1972. The Institute of Geological Sciences and the Scottish environment. *Proc. 24th Int. Geol. Congr.* **13**, 55–63.

Wilson, H. E. 1972. The geological map and the civil engineers. *Proc. 24th Int. Geol. Congr.* **13**, 83–6.

Zaslow, M. 1975. *Reading the Rocks: the story of the Geological Survey of Canada 1842–1972*. Ottawa: Macmillan.

Appendix D *Exploring for old coal workings in the United Kingdom*

If preliminary investigation indicates that exploitable coal is present at a site, then the possibility that old room-and-pillar workings are present should be anticipated, particularly near the outcrop of a well documented seam. The presence and location of abandoned mineshafts are even less predictable at this stage of investigation. Shafts were used for the purpose of ventilation and pumping duties as well as for winding the coal. Since the pumps were always bucket or ram pumps worked by wooden rods from the surface, the shafts containing this type of pumping system had always to be placed at the *deepest* part of the workings.

Where there is a suspicion of mineral workings, allowance should be made for alternative designs and layout, and the results of the site investigation should be relayed quickly to all concerned. In a large investigation, it is sensible to carry out exploration in stages, so that methods are adapted to the specific problems as the overall picture emerges.

An appreciation of the known geology of any mine plans, and of any surface traces (spoil heaps, old building foundations, crown holes or other evidence of subsidence) is essential to make full use of any further survey results. The National Coal Board offers an excellent service of reports on mineral properties containing coal or coal and ironstone. The service is relatively fast and its cost is low.

A proton magnetometer survey is the geophysical method that has been used most often to search for old shafts. An attempt is first made to determine the outline of the colliery yard. Then, as yards often conform to a common pattern, the likely position of the two shafts may be inferred. In many cases no conclusive trace of a yard remains, and a detailed magnetic survey over the entire site, rather than concentrated on a suspect area, may be the only approach. A grid of 1 m × 1 m stations is required. Anomalies additional to those associated with the shafts may be produced by other concealed brickwork and ironwork, and these spurious sources may well outnumber shafts. Each anomaly of approximately the right dimensions must be investigated further by drilling or excavating, bearing in mind that the ground above the shaft may be *dangerously* unstable. The magnetic method has the more serious disadvantage that a shaft will only give a distinctive anomaly if the backfill is ash or another material of higher susceptibility than its surroundings. Difficulty in spotting the presence of a shaft may also arise if partial failure of the soils around its top results in a wider surface area of ash with a tapered margin instead of a vertical boundary between the fill and the surround. Any use of magnetic results must be made with these qualifications in mind, and with *caution*. Magnetic surveys are of little practical value in sites littered with the debris of long industrial or urban usage. A site that has been landscaped to a level surface using industrial waste is lost to the magnetic survey application. It may, however, serve as a rapid and relatively cheap reconnaissance of a less complex site, where a thorough and productive preliminary investigation has been carried out.

The void of a standard mineshaft produces a maximum gravity anomaly of less than 0.1 mgal. This is too small to be detected with certainty by a standard gravity survey, even with stations on a 1 m × 1 m grid. Gravity meters, which are ten times more sensitive than a normal prospecting meter, are being developed, and a few high-resolution surveys have been described in geophysical literature. They are, at least theoretically, capable of detecting the void of a shaft, and high-resolution gravity surveys are likely to develop into a standard exploration service offered by specialist contractors. In comparison with proton magnetometer surveys, they will be informative, but neither rapid nor cheap. To benefit from the meter's sensitivity, and to allow isolation of the small gravity anomalies of the shafts, the levelling of stations and the determination of surface irregularities must be correspondingly refined.

Neither of the above geophysical methods is suitable as a technique to locate room-and-pillar workings, and indeed no geophysical method has yet proved satisfactory as a standard tool for this purpose. Excavation is essential after the preliminary investigations as the principal, or only, method of exploration. Remember that a regular grid of drill holes may match a pattern of pillars.

Rotary drilling with some cored holes *may* provide useful information. A minimum diameter of 76 mm is required to give core recovery of good enough quality.

Trenches excavated by small mechanical excavators are very useful to check shallow workings. The ideal method, especially to locate shafts if the site is restricted or merits intensive exploration, is to use a mechanical back-acting shovel.

Where large underground galleries are preserved, closed-circuit television and photography may be used as additional aids. Remember that *in situ* inspection of underground workings is dangerous and is a job for a mining engineer. In addition to possible instability of the roof, there may be a hazard from coal gas.

Remedial treatment of old workings lies beyond the scope of this book, but it should be mentioned that the most economical method of dealing with *certain* room-and-pillar workings, where overburden is thin, may be to excavate the remaining coal.

For fuller discussions, including case histories, reference should be made to Maxwell (1976), Price *et al.* (1969), Dearman *et al.* (1977) and Taylor (1968). The Department of the Environment has published a useful report by Ove Arup and Partners (1976) on locating shafts.

References and selected reading

Dearman, W. R., F. J. Baynes and R. Pearson 1977. Geophysical detection of disused mineshafts in the Newcastle upon Tyne area. *Q. J. Engng Geol.* **10**, 257–69.

Maxwell, G. 1976. Old mine shafts and their location by geophysical surveying. *Q. J. Engng Geol.* **9**, 283–90.

Ove Arup and Partners 1976. *Reclamation of derelict land: procedure for locating abandoned mine shafts*. Report prepared for the DoE.

Price, D. G., A. B. Malkin and J. L. Knill 1969. Foundations of multistorey blocks on the Coal Measures with special reference to old mine workings. *Q. J. Engng Geol.* **1**, 271-322.

Taylor, R. K. 1968. Site investigations in coalfields: the problem of shallow mine workings. *Q. J. Engng Geol.* **1**, 115–33.

Appendix E *The time–distance graph of first arrivals from a velocity model with two layers separated by a horizontal interface, and where V_2 is greater than V_1*

Derivation of the velocities, V_1 and V_2, from the gradients θ_1 and θ_2 of the corresponding segments of the T–X graph (see Fig. E.1)

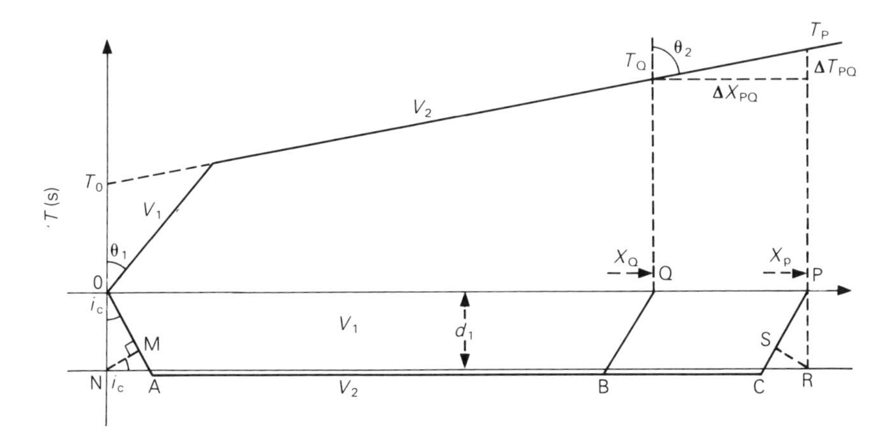

Figure E.1

For the direct ray we have:

$$V_1 = X/T$$

for any pair of values on the V_1 segment, that is:

$$\tan \theta_1 = V_1$$

For the refracted ray we have: the paths to Q and P are similar as far as B, and BQ = CP. Hence:

$$T_{QP} = BC/V_2 = X_{QP}/V_2$$

and

$$V_2 = X_{QP}/T_{PQ}$$

That is, we have:

$$\tan \theta_2 = V_2$$

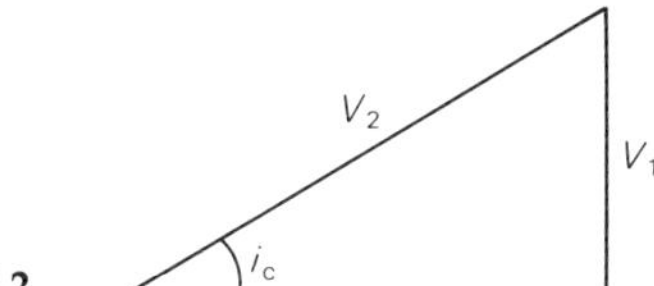

Figure E.2

Derivation of d_1, the thickness of V_1, from the T–X graph

For any point P on that part of the *T–X* graph, where the refracted ray is the first arrival, we have:

$$\begin{aligned} T_P &= OA/V_1 + AC/V_2 + CP/V_1 \\ &= OM/V_1 + MA/V_1 + AC/V_2 + CS/V_1 + SP/V_1 \\ &= (ON \cos i_c)/V_1 + (AN \sin i_c)/V_1 + AC/V_2 + (CR \sin i_c)/V_1 \\ &= (PR \cos i_c)/V_1 \end{aligned}$$

But from Snell's Law, we have:

$$V_1/V_2 = \sin i_1/\sin i_2 = \sin i_1$$

since $i_2 = 90°$. Thus, using Snell's Law:

$$\begin{aligned} T_P &= (d_0 \cos i_c)/V_1 + AN/V_2 + AC/V_2 + CR/V_2 + (d_1 \cos i_c)/V_1 \\ &= (2d_1 \cos i_c)/V_1 + X_P/V_2 \end{aligned}$$

for $X = 0$, $T_0 = (2d_1 \cos i_c)/V_1$. Again, using Snell's Law, $\sin i_0 = V_1/V_2$, we have:

$$\cos i_c = (V_2^2 - V_1^2)^{1/2}/V_2.$$

Therefore, we obtain:

$$T_0 = 2d_1(V_2^2 - V_1^2)^{1/2}/V_2V_1$$

and hence:

$$d_1 = T_0V_2V_1/2(V_2^2 - V_1^2)^{1/2}$$

Appendix F *Quality of aggregates*

Tests on rock aggregates

Classification of an aggregate

The rock used should be given a group classification as described in Section 7.1.2. The main groups used as aggregates are basalt, gabbro, granite, porphyry, hornfels, schist, quartzite, limestone, gritstone, flint and artificial types.

Particle shape, texture and size

Table F.1 gives the main characteristics of particle shape. Generally, rounded particles are easier to work but angular particles bond better with the matrix. The bonding of the aggregate is also related to surface texture. Surface texture is described by one of the six terms listed and described in Table F.2. The category of surface texture is based on the impression gained by a simple visual examination of hand specimens, and is not a precise petrographical classification. Different specimens of the same rock type may fall into adjacent categories.

The particle size is determined by a sieve analysis in which a given weight of aggregate is passed through a series of sieves of standard sizes and each fraction resulting is then weighed and related to the original sample weight, as a percentage passing a certain aperture.

The amount of clay, silt or dust in fine or coarse aggregate

Three methods can be employed in this determination. These are:

(a) the sedimentation method, which is a gravimetric method used for particles up to 20 μm in size (1 μm = 0.001 mm);

Table F.1 Particle shape description.

Descriptive term	Characteristics
rounded	fully water-worn or completely shaped by attrition
irregular	naturally irregular, or partly shaped by attrition and having rounded edges
angular	possessing well defined edges formed at the intersection of roughly planar faces
flaky	material of which the thickness is small relative to the other two dimensions
elongated	material, usually angular, in which the length is considerably larger than the other two dimensions
flaky and elongated	material having the length considerably larger than the width, and the width considerably larger than the thickness

Table F.2 Surface texture description.

Descriptive term		Characteristics
1	glassy	conchoidal fracture
2	smooth	water-worn, or smooth as a result of fracture of laminated or fine-grained rock
3	granular	fracture showing more or less uniform rounded grains
4	rough	rough fracture of fine- or medium-grained rock containing no easily visible crystalline constituents
5	crystalline	containing easily visible crystalline constituents
6	honeycombed	with visible pores and cavities

(b) the decantation method, which is used for material less than 75 μm in size;
(c) the field settling test, which is an approximate volumetric method used for sands and gravels, but is not used for crushed rock aggregate. It gives a rough guide to the percentage of silt, clay or dust in an aggregate.

Flakiness index
The flakiness index of an aggregate is the percentage of weight of particles in it whose least dimension (thickness) is less than 0.6 of their mean dimension. A sieve with elongated slots is used.

Elongation index
Elongation index is the percentage by weight of particles whose greatest dimension (length) is greater than 1.8 times their mean dimension. Neither this test nor the previous one is applicable to fine material passing a 6.35 mm ($\frac{1}{4}$ in) BS sieve.

Angularity number
Angularity (absence of rounding of aggregate particles) affects the ease of handling of a mixture of aggregate and binder, whether concrete or bitumen. The angularity number is a measure of the relative angularity based on the percentage of voids in the aggregate after compaction. The least angular (most rounded) aggregates have 33% voids, and the angularity number of an aggregate is defined as the amount by which the percentage of voids exceeds 33. The angularity number usually ranges from 0 to 12.

Ten per cent fines value
This gives a measure of the resistance of the aggregate to crushing. The higher the value the greater the resistance. The load required to produce 10% fines from the aggregate can be measured using aggregate impact testing equipment.

Aggregate impact value (AIV)
This gives a relative measure of the resistance of the aggregate to sudden impact, which may differ from its resistance to a compressive load. The amount of fine

material produced is expressed as a percentage of the initial weight of sample, each aggregate sample being treated in the same way.

Aggregate crushing value (ACV)
This gives a relative measure of the resistance of an aggregate to crushing under a gradually applied compressive load. The test is similar to the one for AIV except that a load is gradually applied. If the results of the AIV test and this one are 30 or more for an aggregate, the result is unclear and determination of the 10% fines value should be carried out.

Crushing strength (q_u)
Determination of the unconfined crushing (or compressive) strength is usually carried out on the solid rock material.

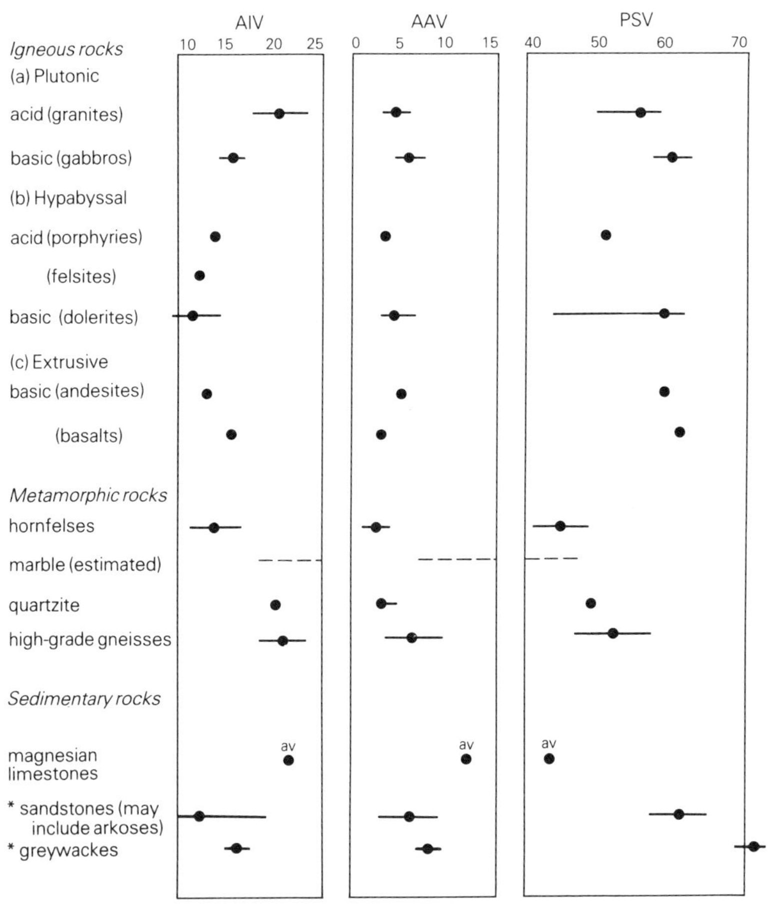

Figure F.1 Selected test values of some rock types: AIV (aggregate impact value), AAV (aggregate abrasion value) and PSV (polished stone value).

Aggregate abrasion value (AAV)
This is a measure of the resistance of aggregates to surface wear by abrasion; the lower the value the greater the resistance.

Polished stone value (PSV)
This measures the extent to which aggregate will polish from traffic; the greater the value the greater the resistance. Figure F.1 gives test values for some rock types.

Tests on concrete
Tests carried out on the rock aggregate include the aggregate crushing value, aggregate impact value, elongation and flakiness indices, specific gravity and water absorption. These have already been described earlier in this appendix and in Chapter 7.

Tests carried out on the actual concrete include drying shrinkage, water absorption, 28 days compressive strength, *E* value and density (slump and moisture expansion). Compressive strengths are carried out both on air-cured and water-cured samples. The results of these tests carried out by the Transport and Road Research Laboratories and the Building Research Station are shown in a series of graphs (Figs F.2–5).

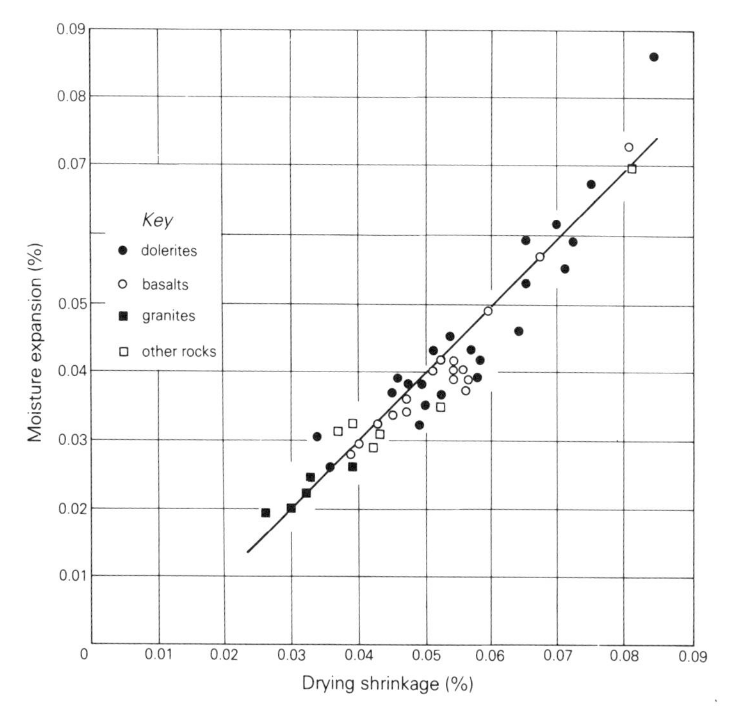

Figure F.2 Relationship between drying shrinkage and moisture expansion; the numbers refer to mixes with fine aggregate (zone 2).

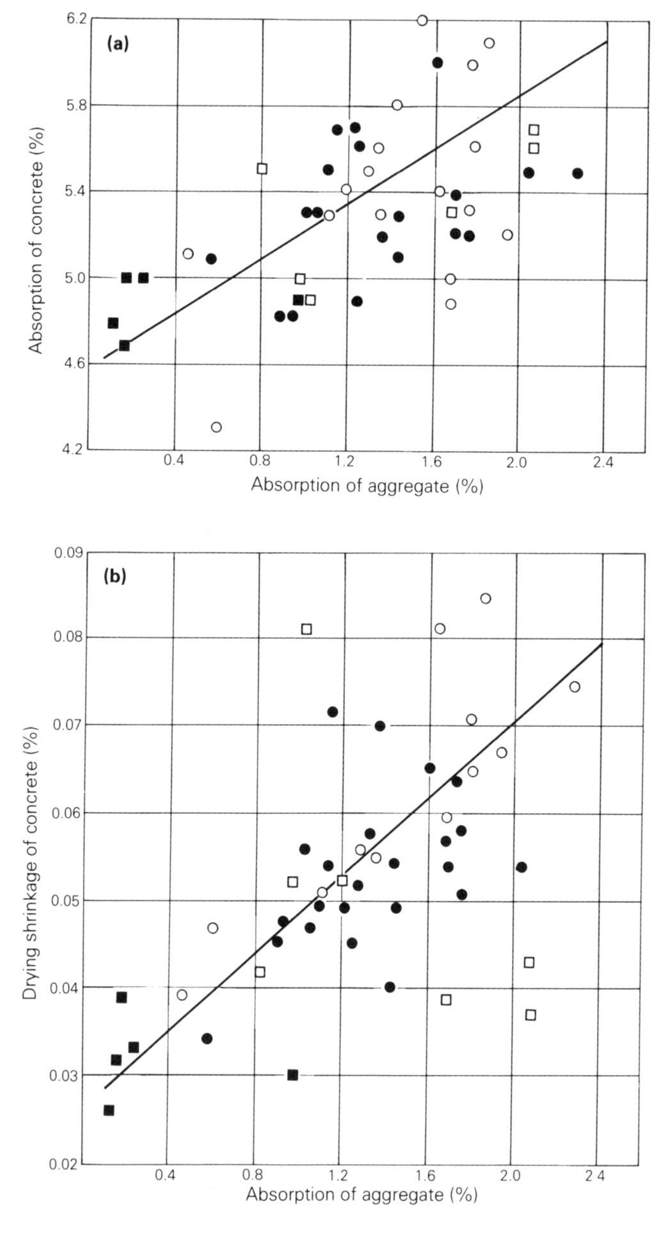

Figure F.3 Relationship between absorption of aggregate and (a) absorption of concrete, and (b) drying shrinkage of concrete. Symbols as in Figure F.2.

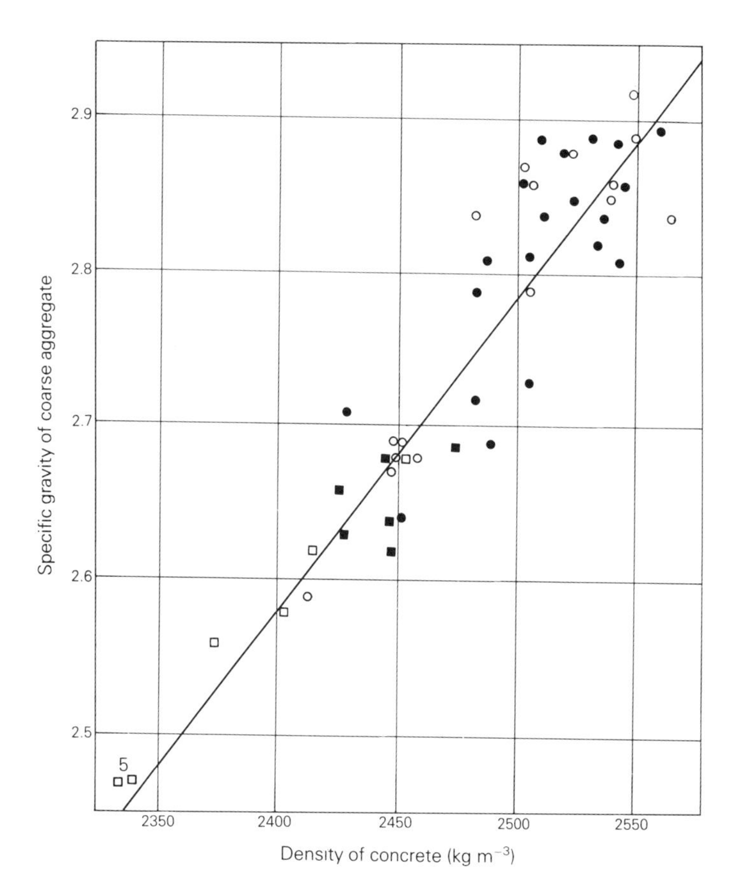

Figure F.4 Relationship between specific gravity of coarse aggregate and concrete density. Symbols as in Figure F.2.

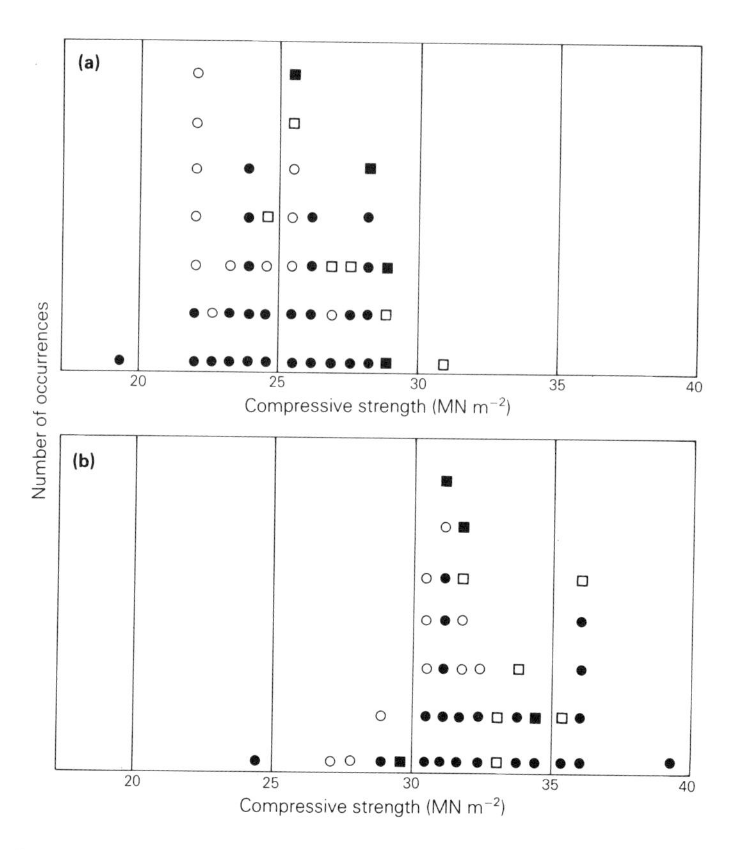

Figure F.5 Compressive strengths of concrete made with different types of crushed rock aggregates. (a) Concrete cured in air at 18°C and 65% relative humidity. (b) Concrete cured in water at 18°C. Symbols as in Figure F.2.

Appendix G *Aggregate quality and tests in different countries*

Aggregate quality is of great importance in determining the uses that can be made of broken or crushed rock. In the UK a series of engineering tests are used to determine the strength and wear of a specific aggregate, in addition to the 'normal' physical tests in which water absorption, density, aggregate shape and other properties are determined. All the UK test procedures are outlined in standards prepared by the British Standards Institution, and other countries have similar institutions with broadly similar functions (e.g. the DIN standards of the Deutsche Normenausschuss of West Germany). In Table G.1 an attempt has been made to compare the aggregate tests employed by the various northern European countries with those of the United Kingdom. Rock and aggregate testing is carefully controlled in the United States, and the American Society for Testing and Materials (ASTM) is the most important body dealing with this. Each year they publish an Annual Book of ASTM Standards in 48 volumes, of which numbers 14, 15 and 19 deal with rock and rock aggregate. The ASTM standard tests are compared with the UK ones in Table G.2.

Table G.1 Aggregate tests: European standards equivalent to UK specifications.

UK BS 812: 1975	Germany DIN	France NF	Belgium NBN	Holland NEN	Denmark DS	Sweden SS
(a) Physical properties of aggregates						
(i) particle size and grading	66100	P18–560	329	Ontw 2871	401	13 21 12
(ii) particle shape elongation flakiness	52114	P18–561	329	?		
(iii) % fines		?				
(iv) density bulk density	52110		B11–255 B11–251			13 21 15
grain density (density grade)	52101			?		13 21 14
(v) porosity	52102			?		
(vi) moisture content			B11–201			13 21 13
(vii) water absorption	52103		B11–255	N1562		
(viii) saturation moisture content (or coeff)	52113			?		

(b) Aggregate strength and hardness:						
(i) crushing strength	52109					
	52105		B11–252	?		13 21 16?
(ii) impact strength	52109					
(iii) 10% fines						
(c) Aggregate wear and skid resistance:						
(i) abrasion value (AAV)	52108	P18–573		Ontw 2873		
(ii) attrition test (see British Rail tests)		P18–572				
(iii) polished stone value (PSV)				Ontw 2872		
(d) Chemical tests and weathering tests:						
(i) weathering assessment (see Engineers Code of Practice BS 5930)	52106					
(ii) frost heave or frost susceptibility PRL LN/765/JCJ	52104	P18–593			401	

Table G.2 Aggregate tests: comparison of US and UK specifications.

Test	British Standard (UK)	ASTM Standard (Vols 15, 19) (USA)
(a) Sampling, size, shape and classification		
(i) particle size and grading	BS 812:1 1975	C136–76
(ii) particle shape, elongation, flakiness	BS 812:1 1975	–
(iii) % fines	BS 812:1 1975	C117–76
(iv) sampling aggregates (for quality control)	–	D 75–71
(b) Physical properties		
(i) density: bulk, grain	BS 812:2 1975	C127–77 (coarse aggregate) C128–73 (fine aggregate)
(ii) porosity	BS 812:2 1975	–
(iii) moisture content	BS 812:2 1975	C127–77 (coarse aggregate)
(iv) water absorption	BS 812:2 1975	C128–73 (fine aggregate)
(v) saturation moisture content	BS 812:2 1975	–
(vi) unit weight of aggregate	–	C29–76
(vii) lightweight pieces in aggregate	–	C123–69 (1975)

Test	British Standard	ASTM
(c) Mechanical properties		
(i) crushing strength (ACV)	BS 812:3 1975	C170–50 (rock) C2938–19 (using drillhole rock cores)
(ii) impact strength (AIV)	BS 812:3 1975	–
(iii) 10% fines	BS 812:3 1975	–
(iv) abrasion value (AAV)	BS 812:3 1975	C535–69 (large size coarse aggregate – withdrawn 1979) C131–76 (small size coarse aggregate)
(v) attrition test	British Rail tests	–
(vi) polished stone value (PSV)	BS 812:3 1975	Numerous tests are involved with skid resistance. Tyres for skid resistance testing: E524–76; E501–76. Methods of testing: E303–74 (UK method); E274–77 Skid resistance on paved surfaces using a full scale tyre; E445–76 Skid resistance (using an automobile). Recommended practices: E559–75T (under trial assessment); E451–72 (1976) Pavement polishing in the laboratory (full scale wheel method); E510–75 Pavement surface frictional and polishing characteristics using a small torque device.
(vii) Scratch hardness of coarse aggregate particles	–	C235–68 (withdrawn 1979)
(d) Chemical and weathering tests		
(i) Weathering assessment	BS 5930	C217–58 (slate only)
(ii) Frost heave or frost susceptibility	Road Research Lab LN/765/JCJ	?
(iii) Sodium sulphate soundness test	–	C88–76

Appendix H *Systematic description of rocks and rock discontinuities*

The **rock mass** should be described first as a rock material, and this description includes:

(a) colour;
(b) texture, including grain size;
(c) structure (fabric);
(d) weathered state;
(e) engineering rock group name and specific geological name (if known);
(f) engineering properties; including strength.

Discontinuities are fractures and planes of weakness in the rock mass and include joints, fissures, faults, cleavages and bedding. Joints frequently appear as parallel sets of joint planes (called a joint set), and there may be more than one joint set in the rock mass. A single fault may be encountered in a rock mass, but seldom more than one. The following data should be recorded for each discontinuity (whether a joint set or a single fault plane):

(a) whether continuous or not;
(b) discontinuity spacing in one dimension (see Table H.1);
(c) nature of discontinuities, particularly whether open or tight (if open, the nature of infill should be described), and type of surface (slickensided; smooth or rough; straight, curved or irregular; whether water flow occurs);
(d) attitude of discontinuity (horizontal; vertical; inclined)

Table H.1 Descriptive scheme for discontinuity spacing in one direction.

very widely spaced	>2 m
widely spaced	600 mm–2 m
medium spaced	200 mm–600 mm
closely spaced	60–200 mm
very closely spaced	20–60 mm
extremely closely spaced	<20 mm

If the discontinuity is inclined, geologists and geological maps record the direction of strike (a line at right angles to maximum inclination or dip) and then the angle of maximum dip, measured to the horizontal.

In Figure H.1 the strike is north–south or 360° and the dip is 50° to the east. This can be written 360°/50°E. Other workers write the dip in this way, but instead of the direction of strike, the direction of maximum dip is given; and the above information can be written 090°/50°. The engineer should be aware of the practice employed in the reports examined, but it is emphasised that nearly all reports written by geologists used the strike/angle of dip method.

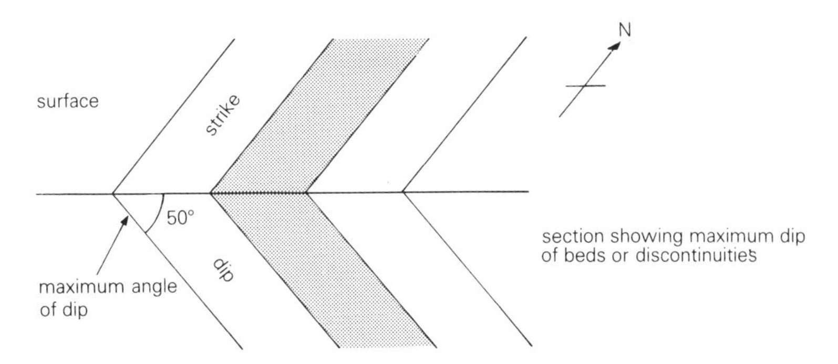

Figure H.1 Block diagram showing a set of discontinuities (bedding planes, joints) striking north–south and dipping at 50° to the east.

To avoid misunderstanding, the information can be written in full thus: strike of discontinuity 360°/dip of discontinuity 50°, or direction of dip of discontinuity 090°/dip of discontinuity 50°.

Discontinuities can be designated by letters or numbers. All the data must be recorded on data sheets with any other relevant information as mentioned at the beginning of this section. For example, in an excavation, the recorded data on the rock mass might be as follows from Figure H.2:

(a) red colour;
(b) medium-grained (1 mm), cemented with graded bedding;
(c) see below;
(d) slightly weathered on surface, some discoloration (Grade II);
(e) gritstone group (arenaceous – red sandstone);
(f) to be examined and engineering properties ascertained.

From (c) above, in the diagram of this rock mass the discontinuities are shown, the data on these being:

A discontinuity (bedding planes; continuous)
(a) thick spacing, 2m
(b) discontinuity – tight; rough and slightly irregular surface; no water flow
(c) horizontal (no strike or dip)
B discontinuity (joint set; continuous)
(a) widely spaced, 1 m
(b) discontinuity – open; smooth, infilled with very weak clay; no water flow
(c) 090°/70°S (i.e. east–west strike with dip of 70° to the south)
C discontinuity (joint set; continuous)
(a) very widely spaced, 3 m
(b) discontinuity – open 15 mm wide; rough, irregular surface; water flow, measured at 15 litres min^{-1}
(c) 360° /vertical (north–south strike/vertical dip)
D discontinuity (joint set; discontinuous)
(a) medium spaced, 300 mm
(b) discontinuity – tight, rough, irregular surface; no water flow
(c) 330°/65°W (strike is along 330° and dip is 65°, approximately to the west)

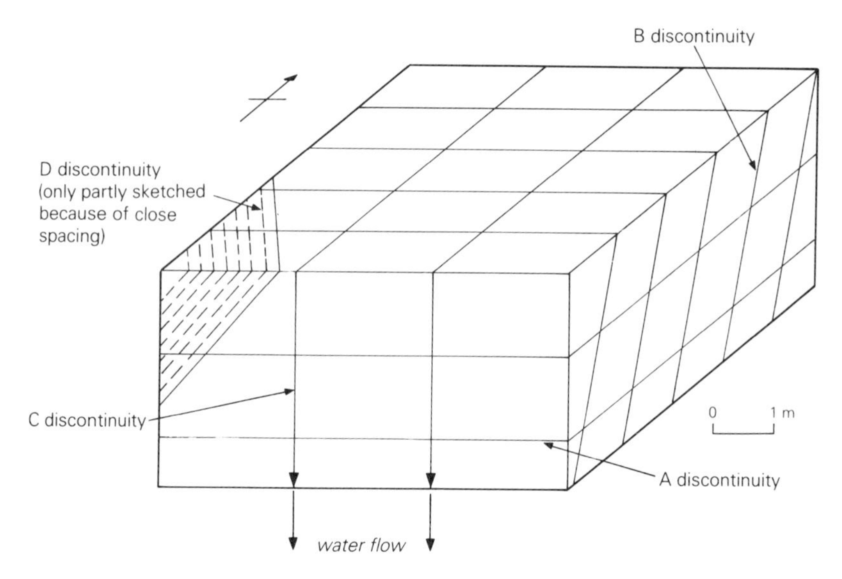

Figure H.2 Sketch of rock mass in block diagram form, showing the four major discontinuities and their relationships to each other.

From the above data, the discontinuity spacing in three dimensions can be estimated. This will give some indication of the size and shape of rock blocks bounded by discontinuities (the **block volume** of the rock mass). Descriptive terms can be used in accordance with Table H.2.

Table H.2 Descriptive terms for block volume.

very large	>2 m
large	600 mm–2 m
medium	200–600 mm
small	60–200 mm
very small	<60 mm

In addition, rock blocks may be described as **blocky** (equidimensional), **tabular** (thickness much less than height or width) or **columnar** (height much greater than cross section).

Thus, the rock mass which has already been described in detail can be summarised as follows:

reddish, medium-grained, slightly weathered sandstone (gritstone group) showing graded bedding; bedding planes (A discontinuity) thick, 2 mm, tight, rough and irregular, horizontal; three joint sets occur:

(a) widely spaced, 1 m, open, smooth infilled with very weak clay, 090°/70°S;
(b) very widely spaced, 3 m, open 15 mm, rough, irregular with strong water flow 15 litres min^{-1}, 360°/vertical;

(c) medium-spaced, 300 mm, tight, rough, irregular, 330°/65°W, producing a large tabular block volume.

Such a description would apply to a reasonably large rock exposure on a site or to part of an exposed quarry face. Several such descriptions are needed to give full data on a site or excavation. All data should be recorded on a map, wherever possible, as well as on data sheets. Engineering properties including strength of material (in MN m^{-2}) can be added to the description later.

Information on joint sets may be given by means of **stereographic representation**. A joint plane is considered to be within an enclosing sphere, with the normal (line at right angles) to the joint plane extended to meet the surface of the enclosing sphere. This point on the surface is then projected towards the north pole of the enclosing sphere, and the stereographic pole of the joint plane is the point where the projection meets the horizontal plane of the sphere. In Figure H.3 the shaded joint plane (A) has its normal (R) meeting the enclosing sphere surface at X. When X is projected to the north pole (P), the stereographic pole of the joint plane is the point Y. The equatorial plane of the enclosing sphere is the stereographic plane. The **stereogram** is shown in Figure H.4, with great circles representing the stereographic traces of a number of planes of different inclination which pass through the centre of the sphere. These planes dip at 2° intervals from the vertical great circle N–S (whose stereographic projection is a straight line) to the horizontal great circle whose stereographic projection is the circumference of the stereogram.

If the joint plane depicted in Figure H.3 has a dip at 70° in a direction of 090° (i.e. exactly due east), then the pole to plane A (Y in Fig. H.3) can be plotted exactly 70° measured from the centre at point Y on Figure H.4.

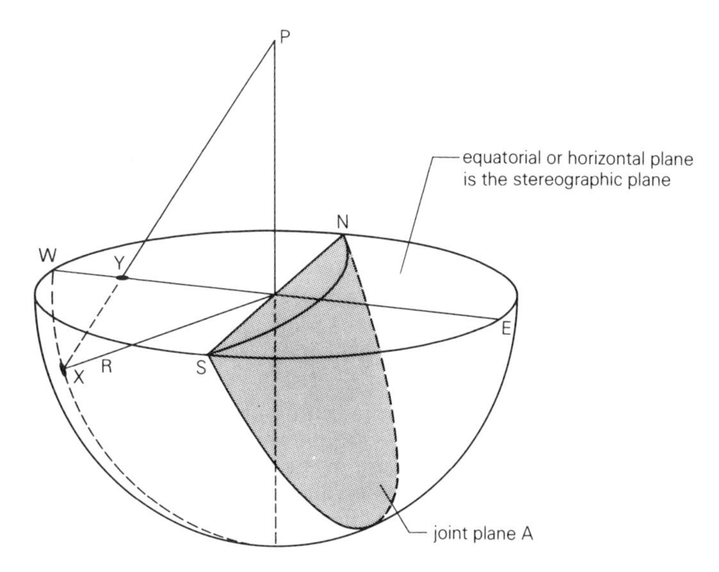

Figure H.3 Polar projection (Y), of pole (R) to joint plane (A) (see exact location of Y on Fig. H.4).

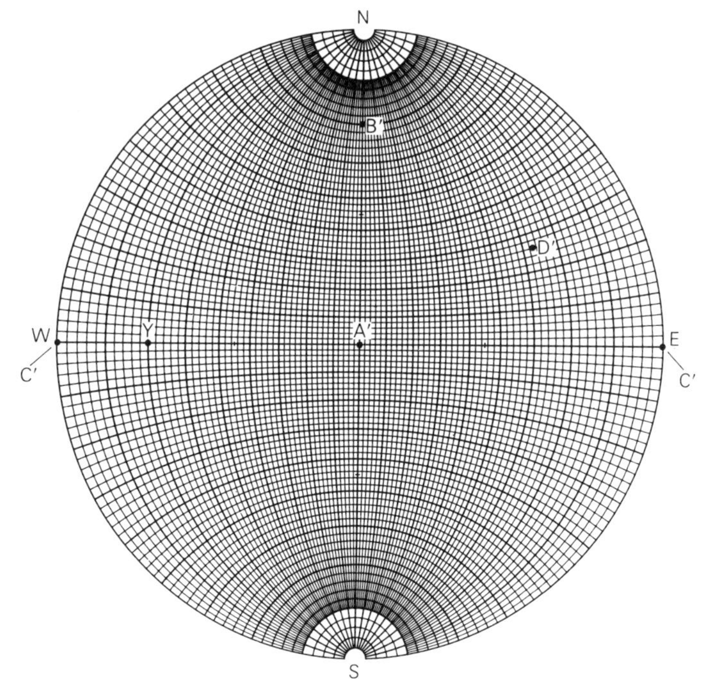

Figure H.4 Stereogram (Wulff Net) with pole Y plotted from Figure H.3 and poles to the discontinuities given in Figure H.2 (A′, B′, C′ and D′ also given.)

The data in Figure H.2 can be summarised as follows:

Discontinuity	Dip	Strike	Direction of dip (the azimuth)
A	none	none	horizontal
B	70°(S)	090°	180°
C	90°	360°	270° or 90°
D	65°(W)	330°	240°

The poles to the discontinuities can be plotted on Figure H.4, at points A′, B′, C′ and D′. A′ is at the centre of the stereogram since its joint plane (A) is horizontal. B and C are plotted in a similar way to the previous pole at Y. The azimuth of D (240°) is first marked on the circumference of the stereogram using an overlying piece of tracing paper, which is rotated until the 240° mark is at the 270° position, and then measuring 65° right from the central point to get the pole (D′). The stereogram is then rotated back to the original position. The use of stereographic plots is becoming much more common in the UK. Several regional authorities present data in this way and the Transport and Road Research Laboratory (TRRL) has described the method in detail (Matheson 1983).

Reference

Matheson, G. D. 1983. *Rock stability assessment in preliminary site investigations – graphical methods.* Working Paper 1982/4, Scottish Branch TRRL.

Index

The pages on which words and phrases of basic importance are most directly defined are shown in **bold** type, so that the index may be used as a glossary of common geological terms and concepts. Numbers in *italic* type refer to text illustrations.